# Multimodality Imaging, Volume 1

Deep learning applications

**Online at: https://doi.org/10.1088/978-0-7503-2244-7**

# Multimodality Imaging, Volume 1

Deep learning applications

**Edited by**

**Mainak Biswas**

*Vignan's Foundation for Science, Technology and Research, India*

**Jasjit S Suri**

*AtheroPoint(TM), Roseville, CA, USA*

**IOP** Publishing, Bristol, UK

ISBN    978-0-7503-2244-7 (ebook)
ISBN    978-0-7503-2242-3 (print)
ISBN    978-0-7503-2245-4 (myPrint)
ISBN    978-0-7503-2243-0 (mobi)

DOI    10.1088/978-0-7503-2244-7

Version: 20221201

IOP ebooks

British Library Cataloguing-in-Publication Data: A catalogue record for this book is available from the British Library.

Published by IOP Publishing, wholly owned by The Institute of Physics, London

IOP Publishing, No.2 The Distillery, Glassfields, Avon Street, Bristol, BS2 0GR, UK

US Office: IOP Publishing, Inc., 190 North Independence Mall West, Suite 601, Philadelphia, PA 19106, USA

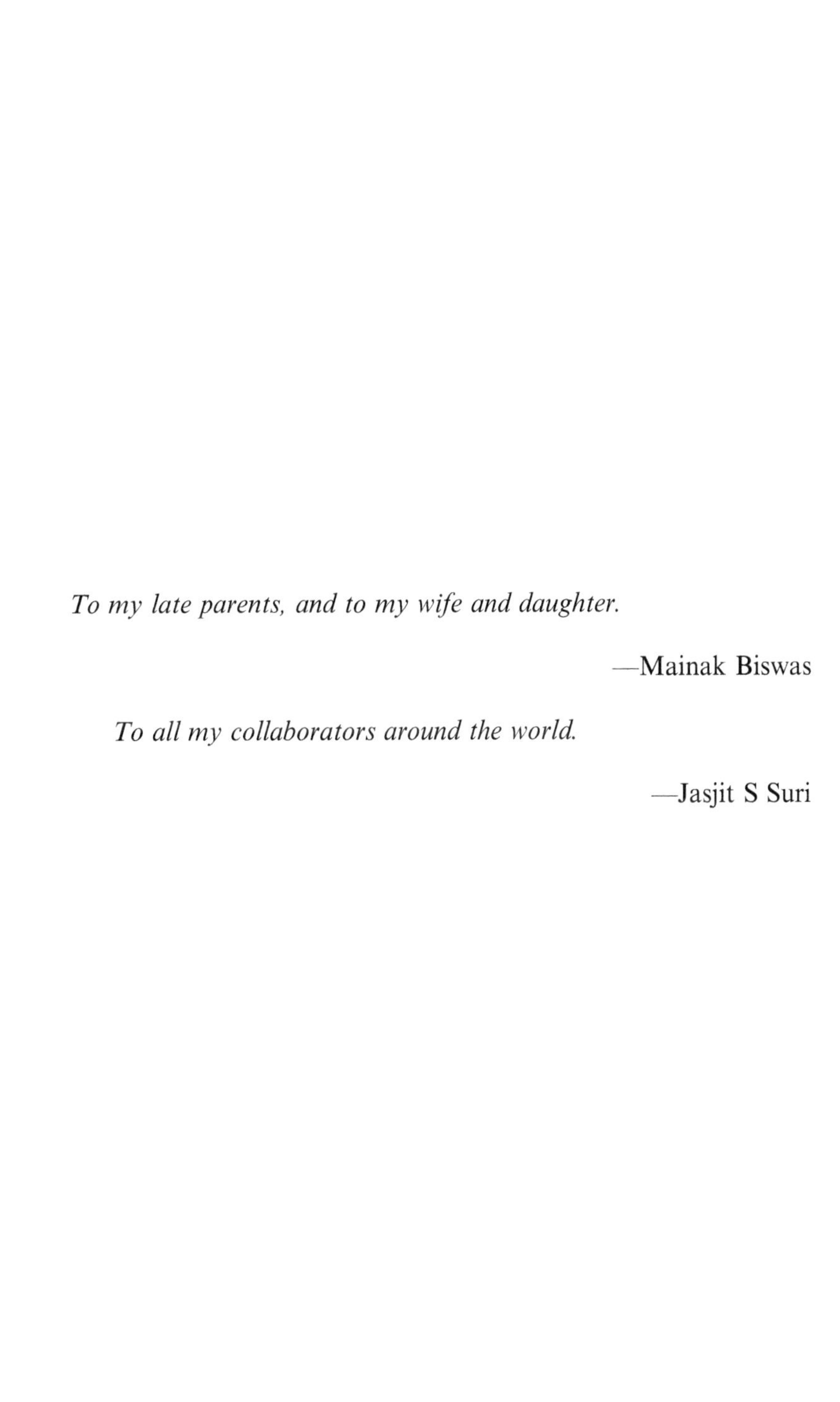

*To my late parents, and to my wife and daughter.*

—Mainak Biswas

*To all my collaborators around the world.*

—Jasjit S Suri

# Contents


Preface xvi

Editor biographies xvii

List of contributors xviii

**Part I    Deep learning and its applications**

**1    Deep learning and augmented radiology** 1-1
*Mainak Biswas and Jasjit S Suri*

1.1    Introduction 1-1

1.2    The present 1-6

    1.2.1    Fatty liver disease risk stratification 1-6

    1.2.2    Carotid intima–media thickness (cIMT) measurement using DL for segmentation 1-7

    1.2.3    Assessment of the treatment effects in acute ischemic stroke (AIS) using DL from MR images 1-8

    1.2.4    Diagnosis of prostate cancer using DL 1-9

    1.2.5    CT-based respiratory disease prognosis using DL 1-10

1.3    Future 1-10

    1.3.1    DL in radiology 1-10

    1.3.2    The future of radiology 1-11

1.4    The potential of deep learning 1-11

    1.4.1    Economy 1-11

    1.4.2    Augmented radiology and DL 1-12

    1.4.3    Further development of DL 1-12

1.5    Challenges and risks in DL 1-12

    1.5.1    Safety 1-12

    1.5.2    Privacy 1-13

    1.5.3    Legality 1-13

1.6    Conclusion 1-13

    References 1-13

**2    Deep learning in biomedical imaging** 2-1
*Mainak Biswas and Jasjit S Suri*

2.1    Introduction 2-1

2.2    Deep learning models                                              2-3
       2.2.1  Deep belief networks                                        2-3
       2.2.2  Autoencoder                                                 2-5
       2.2.3  Convolutional neural networks                               2-6
       2.2.4  Deep residual network                                       2-7
2.3    DL based biomedical imaging systems                               2-8
       2.3.1  Cardiovascular application                                  2-9
       2.3.2  Neurology application                                       2-16
       2.3.3  Mammography application                                     2-20
       2.3.4  Microscopy applications                                     2-22
       2.3.5  Dermatology application                                     2-25
       2.3.6  Gastroenterology applications                               2-28
       2.3.7  Pulmonary application                                       2-29
2.4    Discussion                                                        2-30
       2.4.1  Graphics processing unit and open-source software          2-30
              for deep learning
       References                                                        2-31


## Part II    Deep learning in brain imaging

## 3    A review of artificial intelligence in brain tumor classification and segmentation

*Mainak Biswas and Jasjit S Suri*


3.1    Introduction                                                      3-1
3.2    Brain cancer pathophysiology                                      3-2
       3.2.1  Architecture at the cellular level                         3-2
       3.2.2  Links between brain tumors and genes                       3-6
3.3    Imaging modality                                                  3-6
       3.3.1  Computed tomography imaging                                 3-7
       3.3.2  Magnetic resonance imaging                                  3-7
       3.3.3  Biopsy                                                      3-7
       3.3.4  Hyperstereoscopy imaging                                   3-8
       3.3.5  MR spectroscopy                                            3-9
3.4    Guidelines for tumor grading by the WHO                           3-9
3.5    Brain tumor tests                                                 3-10
       3.5.1  Biomarker test                                             3-10
       3.5.2  Biopsy                                                     3-10
       3.5.3  Imaging test                                               3-11

3.6  Classification methods                                            3-13
   3.6.1  Machine learning                                       3-14
   3.6.2  Deep learning                                          3-17
   3.6.3  Brain image analysis using deep learning               3-18
   3.6.4  A plausible solution for brain cancer classification   3-20
3.7  Brain cancer and other brain disorders                            3-21
   3.7.1  Stroke                                                 3-21
   3.7.2  Alzheimer's disease                                    3-22
   3.7.3  Parkinson's disease                                    3-22
   3.7.4  Leukoaraiosis                                          3-22
   3.7.5  Multiple sclerosis                                     3-22
   3.7.6  Wilson's disease                                       3-22
3.8  Discussion                                                        3-23
   3.8.1  A note on cancer detection biomarkers                  3-24
3.9  Conclusion                                                        3-26
   References                                                     3-26


## 4    MRI based brain tumor classification and its validation: a transfer learning paradigm

4-1


*Luca Saba, Gopal S Tandel, Mainak Biswas, Michele Porcu, N N Khanna, Monica Turk, Christopher K Asare, Annabel A Ankrah and Jasjit S Suri*



4.1  Introduction                                                      4-2
4.2  Background literature survey                                      4-5
4.3  Demographics and data preparation                                4-7
   4.3.1  Patient demographics                                   4-7
   4.3.2  Data preparation                                       4-8
4.4  Methodology                                                       4-10
   4.4.1  CNN model                                              4-10
   4.4.2  The architecture of AlexNet                            4-12
   4.4.3  Transfer learning and workflow                         4-13
   4.4.4  Weight optimization                                    4-14
   4.4.5  A generalization system for tumor classification       4-15
4.5  Experimental protocol, results, and performance evaluation       4-16
   4.5.1  Parameter selection and simulation                     4-16
   4.5.2  Results                                                 4-18
   4.5.3  Performance evaluation                                  4-21

4.6    Model validation and verification                                  4-21

    4.6.1  Hypothesis validation                                         4-21

    4.6.2  Software verification                                         4-24

4.7    Discussion                                                         4-24

4.8    Conclusion                                                         4-27

    Appendix A                                                         4-30

    References                                                          4-32


## 5   Magnetic resonance based Wilson's disease tissue characterization in an artificial intelligence framework using transfer learning   5-1

*Siva Skandha, Luca Saba, Suneet K Gupta, Vijaya K Kumar, Amer M Johri, Narendra N Khanna, Sophie Mavrogeni, John R Laird, Gyan Pareek, Petros P Sfikakis, Athanasios Protogerou, Monica Turk, Aditya M Sharma, Andrew Nicolaides, George D Kitas, Klaudija Viskovic, Tomaz Omerzu and Jasjit S Suri*


5.1    Introduction                                                       5-2

5.2    Background literature                                              5-3

5.3    Methodology                                                        5-4

    5.3.1  Patient demographics                                        5-4

    5.3.2  Data augmentation                                           5-4

    5.3.3  Pre-processing: skull and background removal                5-5

5.4    Global architecture: transfer learning                            5-5

    5.4.1  AlexNet                                                     5-6

    5.4.2  ResNet50                                                    5-8

    5.4.3  DenseNet161                                                 5-9

    5.4.4  XceptionNet                                                 5-10

    5.4.5  InceptionV3                                                 5-10

    5.4.6  SivaSuriNet                                                 5-10

5.5    Results                                                            5-13

5.6    Characterization                                                   5-14

    5.6.1  Mean feature strength                                       5-14

    5.6.2  Higher-order spectrum                                       5-14

5.7    Discussion                                                         5-15

5.8    Conclusion                                                         5-17

    References                                                          5-18

## Part III  Deep learning in cardiovascular imaging


**6    Artificial intelligence based carotid plaque tissue                  6-1
       characterisation and classification from ultrasound
       images using a deep learning paradigm**


*Luca Saba, Sanagala S Skandha, Suneet K Gupta, Anudeep Puvvula,
Vijaya K Koppula, Amer M Johri, Narendra N Khanna, Sophie Mavrogeni,
John R Laird, Gyan Pareek, Martin Miner, Petros P Sfikakis,
Athanasios Protogerou, Durga P Misra, Vikas Agarwal, Aditya M Sharma,
Vijay Viswanathan, Vijay S Rathore, Monika Turk, Raghu Kolluri,
Klaudija Viskovic, Elisa Cuadrado-Godia, George D Kitas, Vijay Nambi,
Deepak L Bhatt, Andrew Nicolaides and Jasjit S Suri*


| | | |
|---|---|---|
| 6.1 | Introduction | 6-2 |
| 6.2 | Methodology | 6-3 |
| | 6.2.1  Patient demographics | 6-3 |
| | 6.2.2  Exclusion criteria | 6-3 |
| | 6.2.3  Ultrasound data acquisition and preprocessing | 6-4 |
| | 6.2.4  Plaque delineation | 6-4 |
| | 6.2.5  Ultrasound plaque data augmentation | 6-5 |
| | 6.2.6  Supercomputer specifications | 6-5 |
| | 6.2.7  Deep learning architecture | 6-6 |
| | 6.2.8  Experimental protocol | 6-7 |
| | 6.2.9  Machine learning for benchmarking deep learning | 6-8 |
| | 6.2.10  Performance parameters using the DL and ML methods | 6-8 |
| 6.3 | Results | 6-9 |
| | 6.3.1  Deep learning data analysis and benchmarking against machine learning | 6-9 |
| | 6.3.2  Plaque characterisation in a deep learning framework | 6-11 |
| 6.4 | Discussion | 6-15 |
| | 6.4.1  A note on the unbalanced datasets for symptomatic and asymptomatic plaques | 6-16 |
| | 6.4.2  Benchmarking against techniques available in the literature | 6-16 |
| | 6.4.3  A special note on the comparison of supercomputer hardware to a local machine | 6-16 |
| | 6.4.4  Strengths, weaknesses, and extensions | 6-18 |
| 6.5 | Conclusion | 6-18 |
| | Disclosure/Conflict of interest | 6-18 |
| | Appendix | 6-19 |
| | References | 6-21 |

**7    Quantification of plaque volume using a two-stage deep**      **7-1**
**learning paradigm**
*Mainak Biswas and Jasjit S Suri*

7.1    Introduction                                                                          7-1
7.2    Background                                                                          7-4
7.3    Data acquisition                                                                    7-6
7.4    Methodology                                                                        7-6
7.5    Experimental protocol and results                                          7-12
7.6    Statistical tests                                                                    7-15
7.7    Discussion                                                                          7-17
7.8    Conclusion                                                                          7-21
         References                                                                         7-21


**8    Stenosis measurement from ultrasound carotid artery images**      **8-1**
**in the deep learning paradigm**
*Mainak Biswas and Jasjit S Suri*

8.1    Introduction                                                                          8-1
8.2    Patient demographics and image acquisition                           8-4
8.3    Methodology                                                                        8-4
8.4    Experimental protocol, performance parameters, and results      8-8
8.5    Statistical tests, variability and error bias analysis, and risk        8-14
         characterization
8.6    Discussion                                                                          8-20
8.7    Conclusion                                                                          8-22
         References                                                                         8-23


**9    A systematic review of conventional and deep learning**      **9-1**
**models for the measurement of plaque burden**
*Mainak Biswas and Jasjit S Suri*

9.1    Introduction                                                                          9-1
9.2    Chronological generation of cIMT regional segmentation and cIMT      9-3
         measurement
9.3    Ml application for cIMT and PA measurement                            9-6
         9.3.1   ANN model for cIMT region detection                          9-8
         9.3.2   Extreme learning machine radial basis neural network model      9-10
                     for cIMT region detection
         9.3.3   Fuzzy $K$-means classifier for cIMT region extraction      9-11

9.4   Deep learning application for cIMT and PA extraction    9-12

    9.4.1  ANN autoencoder based cIMT region segmentation    9-12

    9.4.2  Fully convolutional network for cIMT region estimation    9-14

    9.4.3  FCN for PA measurement    9-16

    9.4.4  Two-stage patching based AI model for cIMT and PA measurement    9-16

9.5   Discussion    9-18

    9.5.1  Benchmarking    9-19

    9.5.2  A short note on cardiovascular risk assessment    9-22

    9.5.3  A note on the clinical impact of AI methods on cIMT/PA techniques    9-23

    9.5.4  A note on inter- and intra-observer variability analysis on the evaluation of AI models    9-23

    9.5.5  A short note on 10 year risk estimation using cIMT and PA    9-24

    9.5.6  Statistical power analysis and diagnostic-odds ratio    9-24

9.6   Conclusions    9-25

    Appendix A    9-26

    References    9-27


## Part IV  Machine and deep learning in liver imaging

## 10  Ultrasound fatty liver disease risk stratification using an extreme learning machine framework    10-1

*Mainak Biswas and Jasjit S Suri*


10.1  Introduction    10-2

10.2  Data demographics, collection, and preparation    10-4

    10.2.1  Sub-sampling of US datasets (S4 and S8)    10-5

10.3  Methodology    10-5

    10.3.1  The three-layered ELM architecture    10-7

    10.3.2  Tissue characterization and risk stratification using ELM and SVM frameworks    10-7

    10.3.3  Feature mining    10-9

10.4  Experimental protocol    10-10

    10.4.1  Experiment 1: the effect of the size of the training data on accuracy using four CV protocols    10-10

    10.4.2  Experiment 2: the effect of the training set size using the sub-sampling strategy    10-10

    10.4.3  Experiment 3: ELM and SVM time complexity    10-11

10.5  Results                                                                 10-11
    10.5.1  Experiment 1: the effect of training data size on accuracy       10-11
        using the four CV protocols
    10.5.2  Experiment 2: the effect of the training data size in parts       10-12
        during the CV protocols
    10.5.3  Experiment 3: time comparison between ELM                         10-12
        and SVM
10.6  Performance evaluations                                                 10-14
    10.6.1  ROC curves                                                       10-14
    10.6.2  Reliability and stability analysis                               10-14
10.7  Discussion                                                             10-17
    10.7.1  Benchmarking                                                     10-17
    10.7.2  ELM and BPNN comparison                                          10-18
    10.7.3  A special note on the ELM and SVM                                 10-18
    10.7.4  Strengths, weaknesses, and future work                           10-20
10.8  Conclusions                                                            10-20
    Appendix A                                                              10-21
    Appendix B                                                              10-22
    References                                                              10-22


## 11  Symtosis: deep learning based liver ultrasound tissue characterisation and risk stratification    11-1

*Mainak Biswas and Jasjit S Suri*


11.1  Introduction                                                           11-1
11.2  Patient demographics and acquisition                                   11-4
11.3  Methodology                                                            11-4
    11.3.1  Risk stratification model                                        11-5
    11.3.2  CNN architecture                                                11-6
11.4  Results                                                                11-7
    11.4.1  Image pre-processing for the DL, ELM, and SVM                     11-7
    11.4.2  The effect of data size on stratification accuracy                11-10
    11.4.3  Stratification analysis using the 'liver segregation index'      11-10
11.5  Performance evaluation: ROC, reliability, and timing analysis          11-12
    11.5.1  ROC analysis                                                     11-12
    11.5.2  Reliability analysis                                             11-13
    11.5.3  Timing analysis                                                  11-13

11.6  Discussion                                                          11-14

11.7  Conclusion                                                          11-16

References                                                               11-16


## Part V    Deep learning in COVID-19


**12    Characterization of COVID-19 severity in infected lungs            12-1
via artificial intelligence transfer learning**



*Sanagala S Skandha, Luca Saba, Suneet K Gupta, Vijaya K Kumar,
Amer M Johri, Narendra N Khanna, Sophie Mavrogeni, John R Laird,
Gyan Pareek, Petros P Sfikakis, Athanasios Protogerou, Monica Turk,
Aditya M Sharma, Andrew Nicolaides, George D Kitas, Klaudija Viskovic,
Tomaz Omerzu and Jasjit S Suri*



12.1  Introduction                                                        12-2

12.2  Methodology                                                         12-3

12.2.1  Patient demographics                                             12-3

12.2.2  Data acquisition                                                 12-3

12.2.3  Baseline characteristics                                        12-3

12.2.4  Segmentation                                                     12-3

12.2.5  Augmentation                                                     12-6

12.2.6  Models                                                           12-6

12.3  Results                                                             12-12

12.4  Characterization                                                    12-16

12.5  Discussion                                                          12-20

References                                                               12-21

# Preface

Deep learning has had a major impact on the healthcare industry and has proven to be a game changer. With the availability of enhanced computational power and the availability of graphical processing units, deep learning has performed better in healthcare than machine learning paradigms. These improvements are further propelled by the open source nature of deep learning and the lower computer hardware prices. The potential in healthcare is great; the processes need to be fully automated and designed to be robust, and error-free paradigms also need to be developed.

## Purpose

This book provides an informative, detailed study of the deep learning paradigms developed to date for the characterization and segmentation of medical images. This book will provide medical professionals with insight into identifying problems early on, and offering patient treatment that is much more tailored and relevant. This book covers different imaging modalities of the brain, carotid artery, heart and vascular system, and also the recent phenomenon of COVID-19.

## Content and organization

The content of this book is spread over five sections consisting of 12 chapters covering deep learning and its applications in different areas of medical imaging. The first section covers the foundation and necessity of deep learning in radiology in chapter 1 and the various types of deep learning models used in medical imaging in chapter 2. The second section includes three chapters on the usage of deep learning in brain imaging. Chapter 3 discusses the pathophysiology of brain tumors and the use of different machine and deep learning models in their diagnosis. Chapter 4 covers the application of deep learning in the detection and grading of brain tumors. In chapter 5 the use of deep learning in the classification of Wilson's disease is discussed.

The third section sheds light on applications of deep learning in the area of cardiovascular imaging. This section includes four chapters. Chapter 6 discusses one of the earliest implementations of segmentation using deep learning in the measurement of carotid intima–media thickness. Chapter 7 discusses a next-generation two-stage deep learning model for measuring plaque volume, including carotid intima–media thickness. In chapter 8 stenosis in carotid arteries is quantified using deep learning. Chapter 9 sheds light on various machine and deep learning models in the quantification of plaque burden and heart and vascular ultrasound.

The fourth and fifth sections discuss machine and deep learning applications in fatty liver disease characterization and COVID-19, respectively. Chapter 10 outlines fatty liver disease detection using a machine learning paradigm, while chapter 11 discusses a deep learning application for the same. Chapter 12 discusses the pedagogy of deep learning in quantifying COVID-19 severity from disease infected lung CT scans.

# Editor biographies

## Mainak Biswas

**Mainak Biswas**, PhD, is a computer scientist with specialization in the application of machine learning and deep learning in the biomedical domain. His research is inspired by providing an effective solution for computer aided diagnosis of diverse diseases. His PhD specialization was in the application of advanced machine learning and deep learning in complex tissue characterization and segmentation from ultrasound images of the liver and carotid arteries. His other interests are the development of advanced machine learning architectures and early warning systems for risk estimation of both symptomatic and asymptomatic patients at high risk of cardiovascular diseases. He has published and presented more than 30 papers on the characterization and segmentation of ultrasound images through machine and deep learning platforms. Dr Mainak Biswas completed a BTech at the Government College of Engineering and Ceramic Technology under the West Bengal University of Technology, Kolkata, an MTech at Jadavpur University, and a PhD at the National Institute of Technology Goa, India. Currently, he is serving as an Associate Professor at Vignan's Foundation for Science, Technology and Research.

## Jasjit S Suri

**Jasjit S Suri**, PhD, MBA, is an innovator, visionary, scientist, and an internationally known world leader in Biomedical Engineering and its Management. Dr Suri received the Director General's Gold medal in 1980 and is a Fellow of (i) the Institute of Electrical and Electronic Engineers (IEEE), (ii) the American Institute of Medical and Biological Engineering (AIMBE), (iii) the American Society of Ultrasound in Medicine (AIUM), (iv) Society of Vascular Medicine (SVM), and (v) Asia Pacific Vascular Society (APVS). He is also the recipient of the Lifetime Achievement Award from Marquis. He is currently Chairman of AtheroPoint, Roseville, CA, USA, dedicated to imaging technologies for cardiovascular and stroke. He has won numerous awards, board member of several industries, has ~28,000 citations, has 1000 publications, co-authored 60 books, has over 50 inventions, has 200 + Artificial Intelligence international journal publications, and has an H-index of 80.

# List of contributors

**Vikas Agarwal**
Department of Clinical Immunology and Rheumatology, SGPGIMS, Lucknow, India

**Annabel A Ankrah**
Department of Radiology, Greater Accra Regional Hospital, Ridge, Accra, Ghana

**Christopher K Asare**
Department of Neurosurgery, Greater Accra Regional Hospital, Ridge, Accra, Ghana

**Deepak L Bhatt**
Brigham and Women's Hospital Heart and Vascular Center, Harvard Medical School, Boston, MA, USA

**Elisa Cuadrado-Godia**
IMIM—Hospital Del Mar, Passeig Marítim, Barcelona, Spain

**Suneet K Gupta**
CSE Department, Bennett University, Greater Noida, UP, India

**Amer M Johri**
Department of Medicine, Division of Cardiology, Queen's University, Kingston, Ontario, Canada

**Narendra N Khanna**
Department of Cardiology, Indraprastha APOLLO Hospitals, New Delhi, India

**George D Kitas**
R&D Academic Affairs, Dudley Group NHS Foundation Trust, Dudley, UK

**Raghu Kolluri**
Ohio Health Heart and Vascular, OH, USA

**Vijaya K Koppula**
CSE Department, CMR College of Engineering and Technology, Hyderabad, India

**Vijaya K Kumar**
CSE Department, CMR College of Engineering and Technology, Hyderabad, Telangana, India

**John R Laird**
Heart and Vascular Institute, Adventist Health St Helena, St Helena, CA, USA

**Sophie Mavrogeni**
Cardiology Clinic, Onassis Cardiac Surgery Center, Athens, Greece

**Martin Miner**
Men's Health Center, Miriam Hospital Providence, RI, USA

**Durga P Misra**
Department of Clinical Immunology and Rheumatology, SGPGIMS, Lucknow, India

**Vijay Nambi**
Michael E DeBakey Veterans Affairs Hospital and Baylor College of Medicine, TX, USA

**Andrew Nicolaides**
Vascular Screening and Diagnostic Centre, University of Nicosia, Nicosia, Cyprus

**Tomaz Omerzu**
Department of Neurology, University Medical Centre Maribor, Slovenia

**Gyan Pareek**
Minimally Invasive Urology Institute, Brown University, Providence, RI, USA

**Michele Porcu**
Department of Radiology, AOU, Italy

**Athanasios Protogerou**
Department of Cardiovascular Prevention, National and Kapodistrian University of Athens, Greece

**Anudeep Puvvula**
Annu's Hospitals for Skin and Diabetes, Nellore, AP, India

**Vijay S Rathore**
Nephrology Department, Kaiser Permanente, Sacramento, CA, USA

**Luca Saba**
Department of Radiology, AOU, Italy

**Petros P Sfikakis**
Rheumatology Unit, National Kapodistrian University of Athens, Greece

**Aditya M Sharma**
Division of Cardiovascular Medicine, University of Virginia, VA, USA

**Sanagala S Skandha**
CSE Department, CMR College of Engineering and Technology, Hyderabad, India

**Siva Skandha**
CSE Department, CMR College of Engineering & Technology, Hyderabad, Telangana, India

**Gopal S Tandel**
Department of Computer Science and Engineering, VNIT, Nagpur, India

**Monica Turk**
Department of Neurology, University Medical Centre Maribor, Maribor, Slovenia
The Hanse-Wissenschaftskolleg Institute for Advanced Study, Delmenhorst, Germany

**Klaudija Viskovic**
University Hospital for Infectious Diseases, Zagreb, Croatia

# Part I

Deep learning and its applications

# Multimodality Imaging, Volume 1
### Deep learning applications
**Mainak Biswas and Jasjit S Suri**

# Chapter 1

# Deep learning and augmented radiology

**Mainak Biswas and Jasjit S Suri**

The effect of deep learning (DL) in today's world is nothing less than dramatic. From self-driving cars, to performing hazardous tasks on inhospitable terrain such as the seabed, to simple chatbots giving directions on a mobile phone, our daily lives have been affected. The cause of this massive development within very few years can be credited to the rapidly decreasing cost of hardware and the availability of open-source software. The healthcare industry is also adapting deep learning technologies to deliver fast and better services to patients. The volume of publications of DL applications in healthcare has exceeded all other domains, in particular in radiology, where one deals with medical images. In this respect, this chapter provides an introduction to DL for radiologists, scientists, academicians, etc. This chapter also provides an understanding of the basic differences between DL and other approaches of AI. In the future, DL will form the foundation of augmented radiology (AR), where patients will be provided real-time, smart, accurate, and cheaper services, reducing mortality rates and improving the quality of life.

A comprehensive study of DL and its applications in radiology is provided in the rest of this chapter. A cumulative survey is also performed on the moral, ethical, and legal aspects of DL in the area of radiology.

## 1.1 Introduction

The concept of machines mimicking the learning functions of the human brain has always interested scientists worldwide. In this respect, a fascinating experiment was carried out on a cat brain by Hubel and Wiesel [1]. The startling revelation of that experiment was the working of the visual cortex. Earlier conceptualizations of the visual cortex assumed it to be holistic, but it was discovered to be hierarchical in nature. This finding of the incremental learning approach of the neuronal network of the brain formed the early foundation of machine and deep learning (ML and DL) [2]. The brain is a complex multi-layered hierarchical neuronal network with lower level neurons passing information to higher level neurons for interpretation and visualization. The perceptron [3] was the first implementation of a single layer artificial neural network (ANN) [4] mimicking the lower level of the visual cortex neural network. The perceptron consists of an input layer, network weights, and an output layer. A training/testing paradigm for training the network weights is followed. The input layer receives the information and passes it to the output layer through the network weights. During training, the output layer computes the sum of the input values and network weights and matches the weighted sum with the desired output. If matched, the network weights remain unchanged; otherwise, they are fine-tuned to as per the bias. This process continues until the bias becomes zero. Geometrically, the perceptron uses the greedy approach of adjusting the weights to draw a hyper-plane between two clusters of instances belonging to different classes. Although the perceptron can accurately characterize if classes are linearly separable, it fails to do so in the case of classes hat are not linearly separable [5]. This problem of non-linearity is resolved by introducing a hidden layer between the input and output layer, i.e. the multilayer perceptron (MLP) [6]. There are multiple variations of MLPs, e.g. extreme-learning machines (ELMs) [7], radial-basis neural networks (RBNNs) [8] etc. The accuracy of each model is determined by the ratio of correctly classified instances to the total classification instances. The other performance parameters are sensitivity, specificity, F-score, area-under-the-curve (AUC), the receiver operating characteristic (ROC) curve etc. The learning laws of MLPs can be characterized into two classes: error correcting codes (ECCs) and gradient descent methods [9]. The ECCs attempt to reduce the bias between the required and computed output at every iteration. The gradient descent algorithms attempt to decrease the error function by manipulating the weight vector. The gradient descent method works by finding the path of steepest descent to find the minimum error. In this respect, Hinton's backpropagation law of propagating error backwards to fine-tune the network weights has become the standard method [10].

In the last two decades AI has been applied to several key areas such as disease diagnosis, disease severity characterization, and the segmentation of regions of interest (ROIs) from radiology images. Some important areas of disease character-ization have been in fatty liver disease diagnosis from ultrasound images [11–13], ovarian cancer diagnosis and risk stratification [14–16], and stroke risk stratification from carotid and coronary ultrasound [17, 18]. Some recent works on image segmentation include carotid artery segmentation from ultrasound images [19–21], plaque classification [22], brain tumour segmentation from MRI images [23], segmentation of the left ventricle [24] etc. The machine learning paradigm is a two-stage process. In the first stage hand-crafted features are extracted from medical images, while in the second stage statistical ML models are used for learning from and testing the features. In some applications another sub-stage of feature selection or feature reduction is applied before the second stage if there is large number of features. This ML paradigm can be interpreted as a low level neural network function of the brain's visual cortex (VC) response to different visual stimuli [25, 26]. In the experiment in [27], Kay *et al* extracted one of the important features, i.e. Gabor filters [28]. This feature can explain the initial primary response of the VC but provide no interpretation. The reason for this is that the perceptron, which is made up of lower level neurons in the visual cortex, responds to geometrical lines at varied angles. The neural network responds to complicated relationships within visual characteristics, such as lines that are perpendicular to one other, as the hidden layer number increases. As seen in figure 1.1, this demonstrates the hierarchical nature of visual cortex learning. The number of hidden layers represents the depth of DL networks, with increasing depth permitting better understanding of patterns, similar to the human brain.

The first ground-breaking paper which led to massive interest in DL models and their applications was presented in 2012 by Krizhevsky and co-authors [29]. In the paper, the proposed AlexNet model achieved 15.2% test error in the top-five characterization of 1.2 million high-resolution images belonging to 1000 labels [30]. AlexNet [30] was the first DL model to be applied to the ImageNet challenge.

Over the years, major corporations entered the ImageNet challenge, bringing the test error rate to 2.25% recently with the DL depth increasing and novel variations in each layer of the DL models [31–35] (shown in table 1.1). One of the major differences between ML and DL models is the nature of feature extraction. As already stated, the ML paradigm is a two-stage process where the first stage assumes the feature extraction part. In general, a pre-processing step is applied before the ML

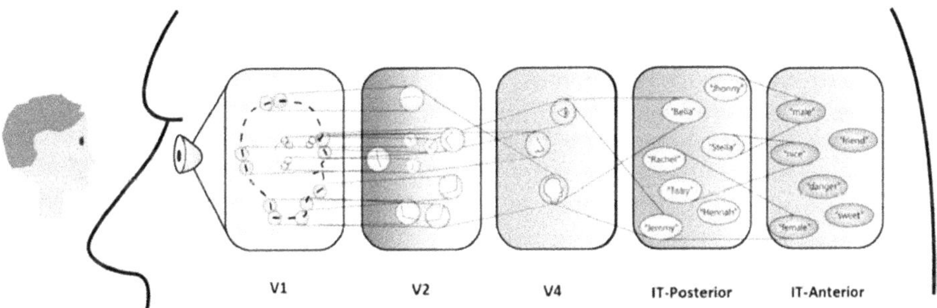


**Figure 1.1.** A neural network illustration of human brain visual cortex. (Image courtesy of AtheroPoint™. Reproduced with permission from [110]. Copyright 2019 Elsevier.)

**Table 1.1.** Top performing DL models in ImageNet from 2012 to 2017. (Reproduced with permission from [110]. Copyright 2019 Elsevier.)

| Model | Year | Top-five error rate |
| --- | --- | --- |
| AlexNet | 2012 | 15.30% |
| VGG16 | 2014 | 7.30% |
| GoogleNet | 2015 | 3.58% |
| ResNet | 2016 | 3.57% |
| Squeeze-and-excitation | 2017 | 2.25% |

paradigm for removing noise before feature extraction. In contrast, DL models are a single-stage system and can be fed raw images for feature extraction which use several rounds of dimensionality reduction, extracting the non-linear relationship within patterns of data for characterization or segmentation. The depth of DL models allows the learning of complex relationships and non-linear dependences within data elements ultimately providing unique high-level features for the classifier. This unique character of extracting features leads to superior learning, thus leading to higher accuracy. Additionally, the automated feature extraction feature allows multi-institutional, cross-country datasets to be used for integrated learning, leading to better generalization when compared to conventional models. The DL models also allow multi-task learning, i.e. allowing classification and segmentation of the same dataset [36]. They also explain how the VC neuronal network responds to visual stimuli [37–39]. This characteristic also allows the DL model trained on a dataset to be applied on a different dataset if the cortical representation of the objects is similar in nature. This particular paradigm of learning is called transfer learning (TL), and is popular amongst the deep learning community [40–45].

The most common DL model is the convolutional neural network (CNN), which is used widely in medical imaging characterization and segmentation [30] (shown in figure 1.2). CNNs use a three layer model repeatedly to mine features from medical images; the layers are the convolutional, rectifier linear unit (ReLu), and pooling layers. The convolutional layer is represented as a group of convolution kernels analogous to the network connections of the ANNs [46]. However, in a CNN these kernels are a matrix of weights multiplied point-wise with each pixel value of the

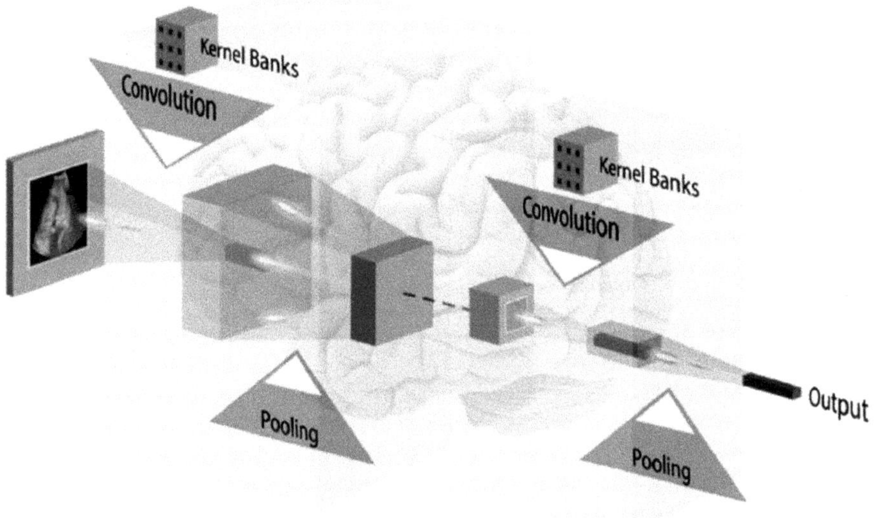


**Figure 1.2.** A convolutional neural network. (Image courtesy of AtheroPoint™. Reproduced with permission from [110]. Copyright 2019 Elsevier.)

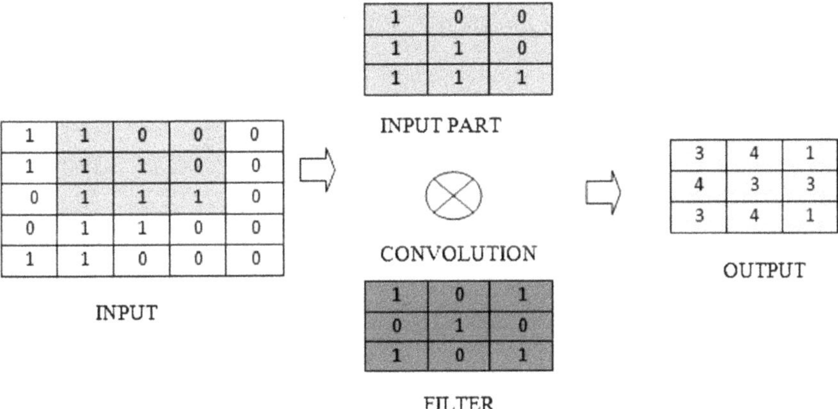



**Figure 1.3.** Convolution operation. (Image courtesy of AtheroPoint™. Reproduced with permission from [110]. Copyright 2019 Elsevier.)

input image to generate feature maps, which are basically a filtered version of the real image. However, in this case each pixel value of the feature map is a function of its neighbouring pixels. This is done to find the contextual relationship present within patterns in the image.

A representative diagram of the convolution operation is shown in figure 1.3. The purpose of the ReLu layer is to filter the feature maps such that they contain positive values only. The reason for this is to overcome the vanishing gradient problem. The pooling layer is used for dimensionality reduction where only the significant aspect of the high-dimensional feature map is reserved in the low-dimensional version of it when applied. These layers are applied multiple times to mine high-level features. These features are used by the fully connected network for characterization. Autoencoders are unsupervised DL models used specifically for image restoration and regeneration [47]. Autoencoders have also been used to remove noise or generate an ROI. The autoencoders are a two-stage DL model: the encoder for image dimension reduction and the decoder for dimensionality restoration. Therefore, the number of input and output neurons is the same in the autoencoder. The encoder performs dimensionality reduction so that the network learns the compact representation of the input image. Residual neural networks (ResNets) [34] are DL networks which are able to avoid the issue of vanishing gradients by skipping learning in some layers, which helps to avoid accuracy saturation. Deep belief networks (DBNs) are unsupervised DL models which use multiple layers of restricted Boltzmann machines [48]. The lower layer connections are directed, whereas the top layers are undirected, forming associative memory. Learning happens layer by layer. DBNs, like autoencoders, are used for image restoration, recognition etc. The classification tree of DL models is given in figure 1.4. In the following sections the current and potential scenarios for DL in radiology are discussed, including the potential for DL in radiology, the dangers of DL are evaluated, and finally the conclusion is presented.

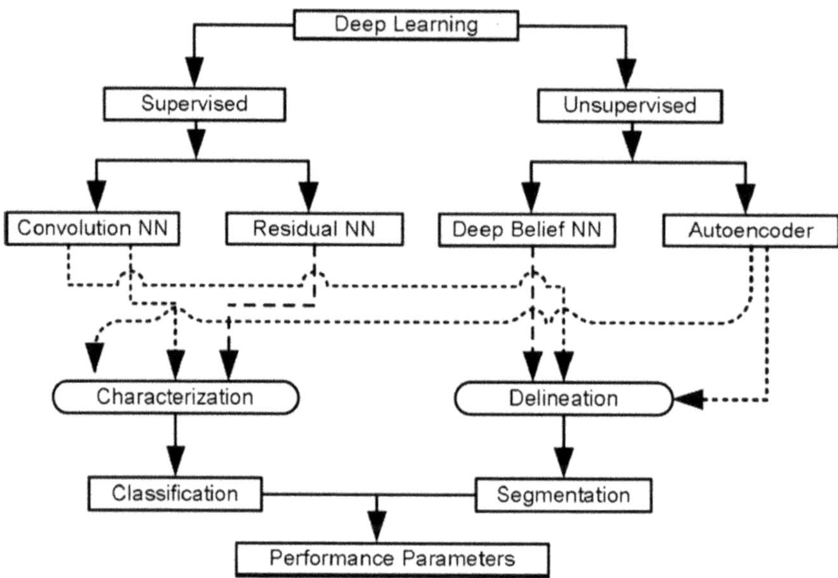


**Figure 1.4.** Deep learning systems classification tree. (Image courtesy of AtheroPoint™. Reproduced with permission from [110]. Copyright 2019 Elsevier.)

## 1.2 The present

DL has affected the medical imaging field largely due to its diagnostic accuracy and robustness. Some recent papers in radiology are presented in the following sections.

### 1.2.1 Fatty liver disease risk stratification

Biomedical imaging related to liver diseases has been drawing the attention of DL researchers. The diagnosis of liver diseases is generally invasive, which is risky and dangerous. Significant R&D has been performed on the automated identification of liver diseases through DL based applications as well as the segmentation of diseased ROIs from medical images. In 2016 Hu *et al* proposed a DL based liver segmentation model from CT images [49]. Li *et al* developed a DL based model for liver tumour segmentation which is considered a significant leap in liver cancer diagnosis [50]. Amongst numerous liver diseases, fatty liver disease (FLD) is one of the major causes of death amongst the USA populace in the age range of 45–54 years [51]. Early diagnosis of FLD can increase the survival rate and speed up treatment plans. Biswas *et al* developed a 22-layered DL network for characterization of FLD from ultrasound images [52]. Figure 1.5 shows diseased and normal liver ultrasounds. The model used unique layers called inception layers [53]. The inception layer combines numerous convolution layers to enable for the application of kernels of various sizes to input image/feature mappings. Allowing many convolution layers within a single layer speeds up convergence and generalization while keeping the DL network's depth and complexity modest. Using ten-fold cross-validation, the model was able to attain 100% accuracy.

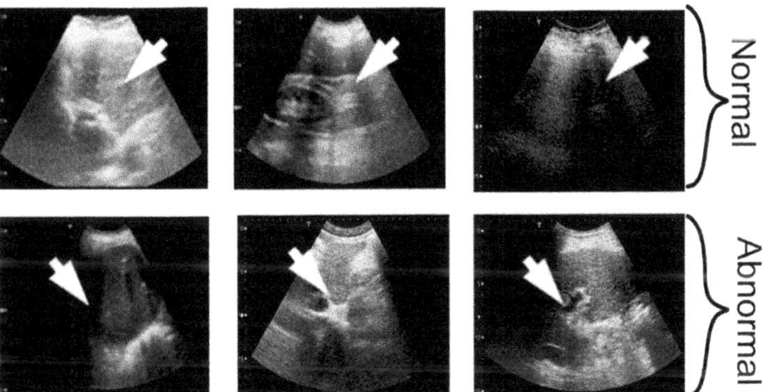


**Figure 1.5.** FLD ultrasound images: normal liver in the top row and abnormal FLD images in the bottom row. (Image courtesy of AtheroPoint$^{TM}$. Reproduced with permission from [110]. Copyright 2019 Elsevier.)

### 1.2.2 Carotid intima–media thickness (cIMT) measurement using DL for segmentation

Cardiovascular diseases kill about 17.9 million people annually worldwide [54]. There are several causes of CVDs. One of them is atherosclerosis, an arterial inflammation disease. Endothelial dysfunction initiates the atherogenesis process, leading to the accumulation of macrophages and fats within the arterial wall. As atherosclerosis progresses, this accumulation leads to the formation of a complex necrotic core with a thin fibrous cap. With time, the fibrous cap ruptures, leading to thrombosis and ultimately stroke. The invasive measures for diagnosis are risky and therefore non-invasive measures such as imaging are preferred, in particular ultrasound. Ultrasound, unlike MRI and CT, is relatively cheap, radiation free, and portable. The huge volume of ultrasound images require DL intervention due to a scarcity of experienced radiologists. In this regard, Biswas *et al* introduced a DL model of a fully convolutional network for extracting the plaque volume from 396 carotid ultrasound images collected from 203 patients [55]. The cross-validation protocol selected was the K10 protocol (nine parts training and one part testing). Two ground truths from two different radiologists were used for the experiment. The DL model consisted of 13 layer convolution and pooling layers for extraction of DL features. These features were up-sampled using three up-sampling layers to generate a segmented image containing the lumen intima (LI) (inner wall) and media adventitia (MA) (outer wall) boundaries [56]. Once generated, the plaque volume was computed in the form of carotid intima–media thickness (cIMT). The experiment was repeated twice for both ground truth images. The pixel-to-pixel characterization was performed using the softmax classifier.

The results showed a performance enhancement of 20% over sonographer readings. Figure 1.6 shows the ground truth boundaries (yellow dashed line), and the DL delineated LI (red dashed line) and MA (green dashed line) boundaries for low, medium, and high-risk patients.

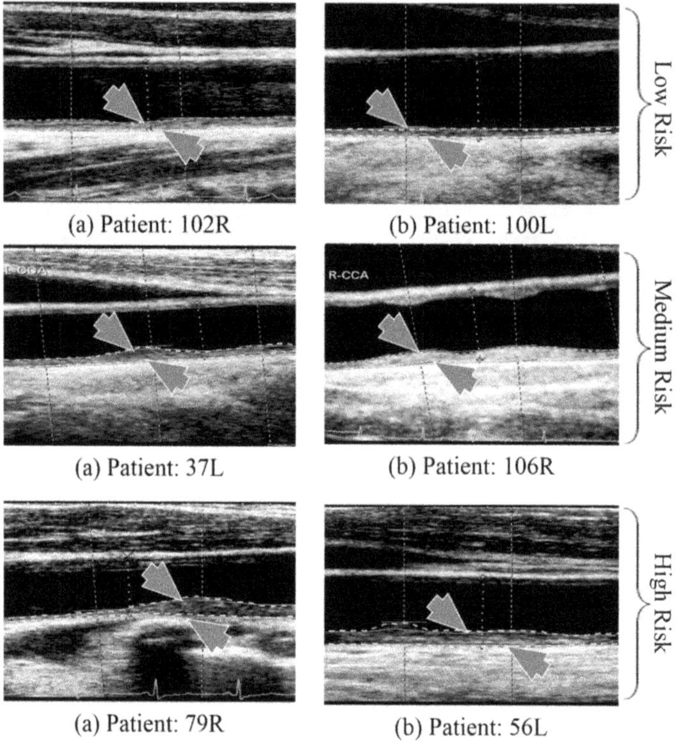

(a) Patient: 102R        (b) Patient: 100L

(a) Patient: 37L        (b) Patient: 106R

(a) Patient: 79R        (b) Patient: 56L

**Figure 1.6.** The LI and MA borders are shown as the red (5 o'clock arrows) and green (11 o'clock arrows) dashed lines using the DL based approach. The yellow dashed tracings represent gold-standard tracings. Low-risk patients are in the top row, medium-risk patients are in the middle row, and high-risk patients are in the bottom row. The cIMT errors were found to be 0.126 ± 0.134 mm and 0.124 ± 0.100 mm, respectively, for the two-patient ground truth information. The LI error was determined to be 0.077 ± 0.057 mm and 0.077 ± 0.049 mm, respectively, when compared to the two-patient ground truth information. The MA error for the two-patient ground truth information was 0.113 ± 0.105 mm and 0.109 ± 0.088 mm, respectively. (Image courtesy of AtheroPoint™. Reproduced with permission from [110]. Copyright 2019 Elsevier.)

### 1.2.3 Assessment of the treatment effects in acute ischemic stroke (AIS) using DL from MR images

The salvageable tissues of AIS patients need to be segmented from radiology images in order to prepare a treatment plan. In this regard, a deep learning CNN model consisting of 37 layers [57] was trained to segment MR images using nine imaging biomarkers. Two separate CNN models were used for comparison of the results. The first was two convolution layers deep and used the nine imaging biomarkers. The other DL network was based on the Tmax biomarker and consisted of two layers [58]. A total of 222 patients were selected for the experiment, 187 of whom were treated with recombinant tissue-type plasminogen activator (rtPA). In a subsequent follow-up, the patients were scanned after symptom onset. It was seen that the high depth CNN model (37 layers) gave a better segmentation accuracy

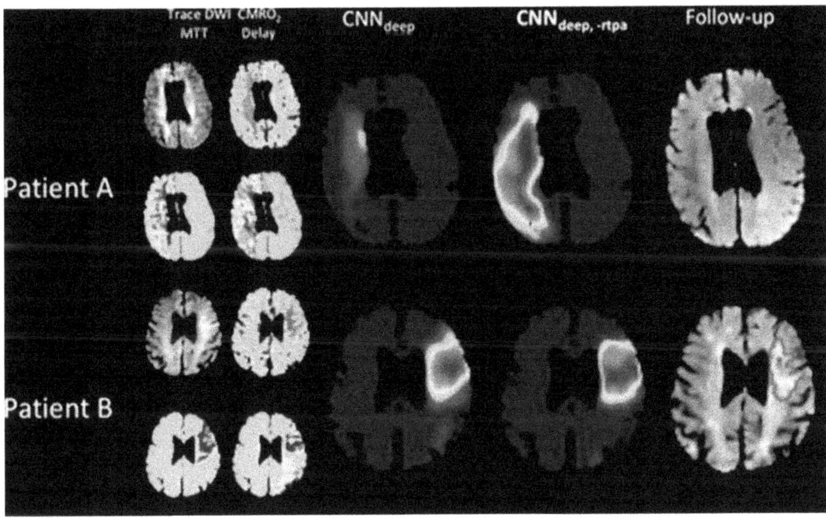


**Figure 1.7.** Deep CNN, shallow CNN, and Tmax CNN (lesions indicated by 7 o'clock arrows) output data from the first follow-up scan for patient A (age: 58, male) and patient B (age: 44, female). Deep CNN properly predicts that patient A has no lesions, however, the shallow CNN and Tmax CNN did not predict accurately. All models properly predicted for patient B. However, the deep CNN's prediction is superior. Reproduced with permission from [57]. Copyright the American Thoracic Society.

than the two other models. The results for two patients, A (age: 58) and B (age: 44), are displayed in figure 1.7. It can be seen that patient A did not have lesions. This was accurately predicted by the high depth CNN model, while the other two shallow models did not achieve higher accuracy. For patient B the deep model again gave better results (AUC: $0.88 \pm 0.12$) than the other two CNNs (AUC: $0.85 \pm 0.11$ and AUC: $0.72 \pm 0.14$). This experiment proved that having deeper (more) layers leads to better accuracy.

### 1.2.4 Diagnosis of prostate cancer using DL

In the USA there are approximately 33 330 deaths due to prostate cancer (PC) per year. The invasive procedure of diagnosing PC is risky, painful, and dangerous. In this respect, a DL network was proposed based on multiparametric MR images [59]. Three blocks of three deep layers were followed by a fully connected layer in the deep CNN. A cohort of 195 patients was used for this study. PC and non-cancer (NC) zones were labelled by an experienced (11 years) radiologist. An ROI was also drawn around the zones. To increase the cohort size for better training, the photos were further enhanced by horizontal flip, random stretching, random shift, and rotation. For each image, the ROI patch was extracted. A total of 159 patients' data (444 ROIs: 215 PC/229 NC) was used for training, 17 patients' data (48 ROIs: 23 PC/25 NC) for validation, and 19 patients' data (55 ROIs: 23 PC/32 NC) for testing. The AUC performance was 0.944 for the experiment. The segmentation results are shown in figure 1.8.

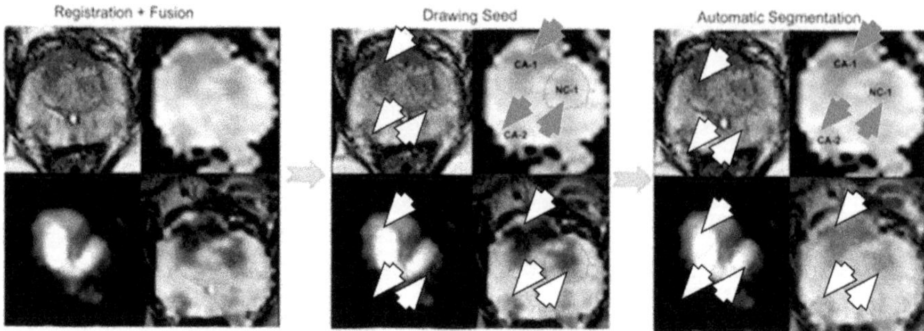


**Figure 1.8.** The registration, pre-processing, and ROI placement outputs. T2-weighted, diffusion-weighted, and apparent diffusion coefficient images were employed. White arrows indicate lesions in the photos. Radiologists would eventually classify the lesions. The 7 o'clock red arrows point to PC ROIs, while the arrows at 1 o'clock point to NC ROIs. (Reproduced with permission from [110]. Copyright 2019 Elsevier.)

### 1.2.5 CT-based respiratory disease prognosis using DL

A three layer DL based model [30] was employed to characterize obstructive pulmonary disease (COPD) from chest CT images. This work also included provision for the detection of acute respiratory disease (ARD) events and mortality [60]. The data were collected from 7983 COPD gene participants for training, and 1672 ECLIPSE (COPD evaluations longitudinally to identify predictive surrogate endpoints) and 1000 COPD gene participants for testing. The performance parameters were given as follows: the c-statistic for detection was 0.856, and the overall c-statistics for the ECLIPSE and COPD gene datasets were 0.55 and 0.64, respectively. The outputs are shown in figure 1.9.

## 1.3 Future

### 1.3.1 DL in radiology

The great success of DL research can be attributed to the computer gaming industry which led to the decreasing price of GPUs [61, 62]. The reduction in overall computation time due to multicore GPUs has led to quicker generalization and convergence and higher accuracy than for conventional models. Further, the easy access to open-source software platforms catapulted DL to broaden its horizons [63–77]. Thereafter, the effect of DL on the information technology (IT) and IT-enabled service (ITES) industries has been so enormous that it is changing the course of other industries, such as automobiles and space exploration. The effects of DL in the medical imaging industry has been so significant that it can be called revolutionary [78]. This is evident from the huge volume of research conducted in this field. DL has seen considerable success in disease characterization [79, 80], brain cancer detection [81–83], the segmentation of medical images [84–90], haemorrhage detection [91, 92], and tumour detection and segmentation [93–95].

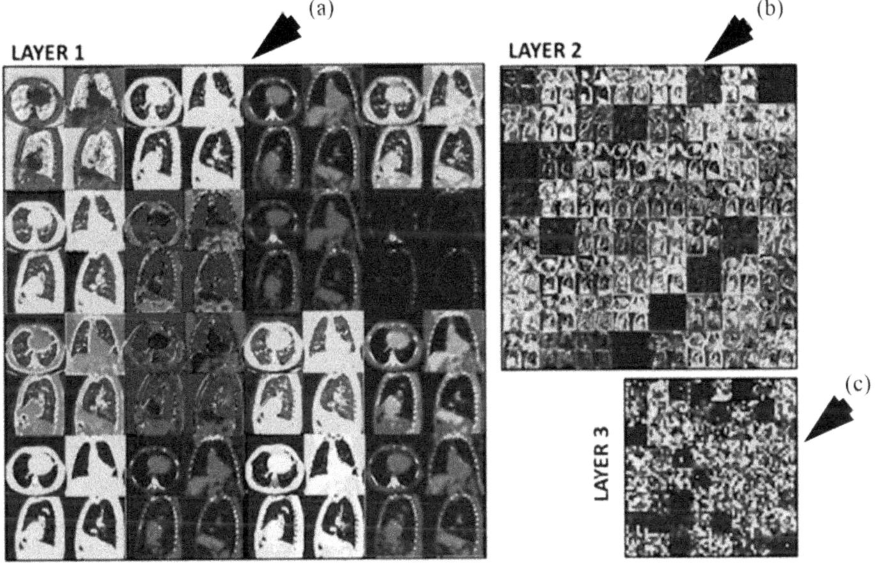



**Figure 1.9.** The results of the three given layers (7 o' clock arrows) of the DL system. Layer 1 output: arrow (a); layer 2 output: arrow (b); layer 3 output: arrow (c). (Reproduced with permission from [110]. Copyright 2019 Elsevier.)

### 1.3.2 The future of radiology

The future advances of medical imaging sciences can be attributed to DL. Hospitals and diagnosis centres should upgrade and update their laboratory infrastructure to accommodate DL. A central data repository needs to be created for each disease, while also protecting the privacy of each patient. This will help in interstate and cross-country accumulation of data which will be of great help for the DL community to train their models better. This will help in building up a robust system with pinpoint diagnostic accuracy and future treatment plans. The real-time availability of results will help to save lives. Overall, the automated diagnosis process will become a trend in the future and will have the capability to improve the well-being of human society in general.

## 1.4 The potential of deep learning

### 1.4.1 Economy

There is huge potential for DL in the biomedical imaging industry [96]. The average rate of image analysis for a radiologist is approximately 200 images per day. DL could help in preliminary findings, such as segmentation of the infected region, which can help the radiologist to make faster decision for patients both near and far (using telemedicine [97]). In the USA, the total number of generated images amounts to 260 million per year. A small investment of 1000 dollars in the computational systems linked to DL based cloud systems can reduce late treatment and medical expenses. The automated DL based diagnosis industry is slated to generate revenue of 1.9 billion dollars by 2022 [98].

### 1.4.2 Augmented radiology and DL

The coming of age of driverless cars is slowly poised to make the person behind the wheel disappear. Henceforth, the discussion on the table is the replacement of radiologists by DL based systems, owing to the multibillion dollar investment by huge corporations such as IBM and GE. However, it has been seen that a robust system can be built with collaboration of both machine and human intelligence. The experimental models developed to date have been built in controlled environments with every aspect of training and testing suited to correct functioning. However, in real time, the environment is highly dynamic and data may become too noisy, and complicated patterns may arise leading to failure of these statistical models. In such cases, human intelligence input might be required to augment machine accuracy. A middle ground is sought where both human and machine intelligence can be combined to generate better diagnosis results. In some cases, the radiologist's work is expedited by including preliminary findings from a DL based system. In such cases, the radiologist can benefit immensely and this timely intervention may save multiple patients' lives [99–102]. Altogether, both are required for faster and accurate diagnosis. The profession of radiology is also poised to benefit from both ML and DL in picture archiving and communication system (PACS), making image availability faster, accurate, and reliable [103–110].

### 1.4.3 Further development of DL

In the future, DL will take over the routine jobs of radiology, leaving the radiologist with the higher decision making, such as final diagnosis. The consultation between the patient and doctor will tend to become more interactive, with DL providing clinical findings as evidence during times of diagnostic uncertainty. DL will tend to become real-time and take into account every detail of a patients' medical history, e.g. histology reports and genetics, in making predictions. This will help in round-the-clock monitoring and better patient care.

## 1.5 Challenges and risks in DL

Several moral and ethical concerns related to DL in radiology are discussed next.

### 1.5.1 Safety

The first concern is safety. The medical profession is guided by the simple code 'first, do no harm'. In this regard, AI based systems will be used in some of the most critical phases of patient health in the most vulnerable state, leaving no room for error. Therefore, before applying such a system it is strictly advised that all rules must be followed to make the system robust and accurate. The highest standards need to be followed in maintaining the reliability and robustness of the system.

### 1.5.2 Privacy

Another chief concern is privacy. Patient health data security must be given utmost priority and data must be stored using modern techniques such as blockchain. The strictest punishment in law must be applied by national bodies for violation or breach of data privacy. Data transmission control protocols should be secured so that no third party can snoop on data and the receiver can only read with proper authority.

### 1.5.3 Legality

Another avenue of discussion is the liability in the case of failure leading to loss of life or property. One cannot blame the machine entirely and if it is to be blamed why use it in the first place? Therefore a shared liability model must be developed to ensure the accountability of designer during installation as well as the physician allowing the DL to take control.

## 1.6 Conclusion

This chapter presented many aspects of DL in healthcare and in particular radiology. Different DL models were discussed and reasons for their success in radiology were described briefly. Different applications of DL were presented for both classification and segmentation in the brain, lung, heart etc. Further, several moral and ethical issues were also discussed, such as safety, privacy, legality etc. Several points were made on the future of DL in radiology. The economic viewpoint was also presented. This chapter also emphasized the foundation of augmented radiology being in combining both machine and human intelligence to save lives in an efficient manner.

## References


[1] Hubel D H and Wiesel T N 1963 Shape and arrangement of columns in cat's striate cortex *J. Physiol.* **165** 559–68

[2] Goodfellow I, Bengio Y, Courville A and Bengio Y 2016 *Deep Learning* vol 1 (Cambridge, MA: MIT Press)

[3] Rosenblatt F 1958 The perceptron: a probabilistic model for information storage and organization in the brain *Psychol. Rev.* **65** 386

[4] Nasrabadi N M 2007 Pattern recognition and machine learning *J. Electron. Imaging* **16** 049901

[5] Haykin S 1994 *Neural Networks, A Comprehensive Foundation* (New York: Macmillan)

[6] Cybenko G 1989 Approximation by superpositions of a sigmoidal function *Math. Control Signals Syst.* **2** 303–14

[7] Huang G-B, Zhu Q-Y and Siew C-K 2006 Extreme learning machine: theory and applications *Neurocomputing* **70** 489–501

[8] Park J and Sandberg I W 1991 Universal approximation using radial-basis-function networks *Neural Comput.* **3** 246–57

[9] Kumar S 2004 *Neural Networks: A Classroom Approach* (Nouda, Uttar Pradesh: Tata)

[10] Hinton G E and Salakhutdinov R R 2006 Reducing the dimensionality of data with neural networks *Science* **313** 504–7

[11] Acharya U R, Sree S V, Ribeiro R, Krishnamurthi G, Marinho R T, Sanches J and Suri J S 2012 Data mining framework for fatty liver disease classification in ultrasound: a hybrid feature extraction paradigm *Med. Phys.* **39** 4255–64

[12] Saba L, Dey N, Ashour A S, Samanta S, Nath S S, Chakraborty S, Sanches J, Kumar D, Marinho R and Suri J S 2016 Automated stratification of liver disease in ultrasound: an online accurate feature classification paradigm *Comput. Methods Programs Biomed.* **130** 118–34

[13] Kuppili V, Biswas M, Sreekumar A, Suri H S, Saba L, Edla D R, Marinhoe R T, Sanches J M and Suri J S 2017 Extreme learning machine framework for risk stratification of fatty liver disease using ultrasound tissue characterization *J. Med. Syst.* **41** 152

[14] Acharya U R, Sree S V, Kulshreshtha S, Molinari F, Koh J E W, Saba L and Suri J S 2014 GyneScan: an improved online paradigm for screening of ovarian cancer via tissue characterization *Technol. Cancer Res. Treat.* **13** 529–39

[15] Acharya U R, Krishnan M M R, Saba L, Molinari F, Guerriero S and Suri J S 2013 Ovarian tumor characterization using 3D ultrasound *Ovarian Neoplasm Imaging* (Boston, MA: Springer) pp 399–412

[16] Acharya U R, Sree S V, Saba L, Molinari F, Guerriero S and Suri J S 2013 Ovarian tumor characterization and classification using ultrasound—a new online paradigm *J. Digital Imag.* **26** 544–53

[17] Acharya R U, Faust O, Alvin A P C, Sree S V, Molinari F, Saba L, Nicolaides A and Suri J S 2012 Symptomatic vs asymptomatic plaque classification in carotid ultrasound *J. Med. Syst.* **36** 1861–71

[18] Saba L et al 2007 Plaque tissue morphology-based stroke risk stratification using carotid ultrasound: a polling-based PCA learning paradigm *J. Med. Syst.* **41** 98

[19] Delsanto S, Molinari F, Giustetto P, Liboni W, Badalamenti S and Suri J S 2007 Characterization of a completely user-independent algorithm for carotid artery segmentation in 2-D ultrasound images *IEEE Trans. Instrum. Meas.* **56** 1265–74

[20] Suri J, Saba L, Dey N, Samanta S, Nath S S, Chakraborty S, Kumar D and Sanches J 2015 2083667 Online system for liver disease classification in ultrasound *Ultrasound Med. Biol.* **41** S18

[21] Molinari F, Zeng G and Suri J S 2010 Greedy technique and its validation for fusion of two segmentation paradigms leads to an accurate intima–media thickness measure in plaque carotid arterial ultrasound *J. Vasc. Ultrasound* **34** 63–73

[22] Saba L, Lippo R S, Tallapally N, Molinari F, Montisci R, Mallarini G and Suri J S 2011 Evaluation of carotid wall thickness by using computed tomography and semiautomated ultrasonographic software *J. Vasc. Ultrasound* **35** 136–42

[23] Suri J S 2011 Two-dimensional fast magnetic resonance brain segmentation *IEEE Eng. Med. Biol. Mag.* **20** 84–95

[24] Suri J S 2000 Computer vision, pattern recognition and image processing in left ventricle segmentation: the last 50 years *Pattern Anal. Appl.* **3** 209–42

[25] Paninski L, Pillow J and Lewi J 2007 Statistical models for neural encoding, decoding, and optimal stimulus design *Prog. Brain Res.* **165** 493–507

[26] Nishimoto S, Vu A T, Naselaris T, Benjamini Y, Yu B and Gallant J L 2011 Reconstructing visual experiences from brain activity evoked by natural movies *Curr. Biol.* **21** 1641–6

[27] Kay K N, Naselaris T, Prenger R J and Gallant J L 2008 Identifying natural images from human brain activity *Nature* **452** 352

[28] Jain A K and Farrokhnia F 1991 Unsupervised texture segmentation using Gabor filters *Pattern Recognit.* **24** 1167–86

[29] Krizhevsky A, Sutskever I and Hinton G E 2012 ImageNet classification with deep convolutional neural networks *Adv. Neural Inf. Process. Syst.* **2012** 1097–105

[30] LeCun Y, Bengio Y and Hinton G 2015 Deep learning *Nature* **521** 436

[31] Zeiler M D and Fergus R 2014 Visualizing and understanding convolutional networks *European Conf. on Computer Vision* (Cham: Springer) pp 818–33

[32] Simonyan K and Zisserman A 2014 Very deep convolutional networks for large-scale image recognition, arXiv:1409.1556

[33] Szegedy C, Liu W, Jia Y, Sermanet P, Reed S, Anguelov D, Erhan D, Vanhoucke V and Rabinovich A 2015 Going deeper with convolutions *IEEE Conference on Computer Vision and Pattern Recognition (CVPR) (Boston, MA, 7–12 June)* (Piscataway, NJ: IEEE)

[34] He K, Zhang X, Ren S and Sun J 2016 Deep residual learning for image recognition *Proc. of the IEEE Conf. on Computer Vision and Pattern Recognition* pp 770–8

[35] Hu J, Shen L and Sun G 2017 Squeeze-and-excitation networks, arXiv:1709.01507

[36] Ching T *et al* 2018 Opportunities and obstacles for deep learning in biology and medicine, bioRxiv:142760

[37] Yamins D L K, Hong H, Cadieu C F, Solomon E A, Seibert D and DiCarlo J J 2014 Performance-optimized hierarchical models predict neural responses in higher visual cortex *Proc. Natl Acad. Sci. USA* **111** 8619–24

[38] Cadieu C F, Hong H, Yamins D L K, Pinto N, Ardila D, Solomon E A, Majaj N J and DiCarlo J J 2014 Deep neural networks rival the representation of primate IT cortex for core visual object recognition *PLoS Comput. Biol.* **10** e1003963

[39] Güçlü U and Marcel A J G 2015 Deep neural networks reveal a gradient in the complexity of neural representations across the ventral stream *J. Neurosci.* **35** 10005–14

[40] Hasson U, Nir Y, Levy I, Fuhrmann G and Malach R 2004 Intersubject synchronization of cortical activity during natural vision *Science* **303** 1634–40

[41] Lu K-H, Hung S-C, Wen H, Marussich L and Liu Z 2016 Influences of high-level features, gaze, and scene transitions on the reliability of BOLD responses to natural movie stimuli *PLoS One* **11** e0161797

[42] Pan S J and Yang Q 2010 A survey on transfer learning *IEEE Trans. Knowl. Data Eng.* **22** 1345–59

[43] Bengio Y 2012 Deep learning of representations for unsupervised and transfer learning *Proc. of ICML Workshop on Unsupervised and Transfer Learning* pp 17–36

[44] Mesnil G *et al* 2011 Unsupervised and transfer learning challenge: a deep learning approach *Proc. of the 2011 Int. Conf. on Unsupervised and Transfer Learning Workshop* vol 27 pp 97–111

[45] Sun Y, Wang X and Tang X 2014 Deep learning face representation from predicting 10,000 classes *Proc. of the IEEE Conf. on Computer Vision and Pattern Recognition* pp 1891–8

[46] O'Neil R 1963 Convolution operators and $L(p,q)$ spaces *Duke Math. J.* **30** 129–42

[47] Liou C-Y, Cheng W-C, Liou J-W and Liou D-R 2014 Autoencoder for words *Neurocomputing* **139** 84–96

[48] Hinton G E 2009 Deep belief networks *Scholarpedia* **4** 5947

[49] Hu P, Wu F, Peng J, Liang P and Kong D 2016 Automatic 3D liver segmentation based on deep learning and globally optimized surface evolution *Phys. Med. Biol.* **61** 8676

[50] Li W, Jia F and Hu Q 2015 Automatic segmentation of liver tumor in CT images with deep convolutional neural networks *J. Comp. Commun.* **3** 146

[51] Browning J D, Szczepaniak L S, Dobbins R, Horton J D, Cohen J C, Grundy S M and Hobbs H H 2004 Prevalence of hepatic steatosis in an urban population in the United States: impact of ethnicity *Hepatology* **40** 1387–95

[52] Biswas M, Kuppili V, Reddy Edla D, Suri H S, Saba L, Tato R M, Miguel Sanches J and Suri J S 2018 Symtosis: a liver ultrasound tissue characterization and risk stratification in optimized deep learning paradigm *Comput. Methods Programs Biomed.* **155** 165–77

[53] Szegedy C, Vanhoucke V, Ioffe S, Shlens J and Wojna Z 2016 Rethinking the inception architecture for computer vision *Proc. of the IEEE Conf. on Computer Vision and Pattern Recognition* pp 2818–26

[54] Lloyd-Jones D *et al* 2010 Heart disease and stroke statistics—2010 update *Circulation* **121** e46–215

[55] Biswas M *et al* 2018 Deep learning strategy for accurate carotid intima–media thickness measurement: an ultrasound study on Japanese diabetic cohort *Comput. Biol. Med.* **98** 100–7

[56] Long J, Shelhamer E and Darrell T 2015 Fully convolutional networks for semantic segmentation *Proc. of the IEEE Conf. on Computer Vision and Pattern Recognition* pp 3431–40

[57] Nielsen A, Hansen M B, Tietze A and Mouridsen K 2018 Prediction of tissue outcome and assessment of treatment effect in acute ischemic stroke using deep learning *Stroke* **49** 1394–401

[58] Straka M, Albers G W and Bammer R 2010 Real-time diffusion-perfusion mismatch analysis in acute stroke *J. Magn. Reson. Imaging* **32** 1024–37

[59] Song Y, Zhang Y D, Yan X, Liu H, Zhou M, Hu B and Yang G 2018 Computer-aided diagnosis of prostate cancer using a deep convolutional neural network from multi-parametric MRI *J. Magn. Reson. Imaging* **98** 1570–7

[60] González G, Samuel Y A, Vegas-Sánchez-Ferrero G, Onieva Onieva J, Rahaghi F N, Ross J C, Díaz A, San José Estépar R and Washko G R 2018 Disease staging and prognosis in smokers using deep learning in chest computed tomography *Am. J. Respir. Crit. Care Med.* **197** 193–203

[61] Luebke D and Humphreys G 2007 How GPUs work *Computer* **40** 96–100

[62] Fialka O and Cadik M 2006 FFT and convolution performance in image filtering on GPU *Information Visualization, 2006. IV 2006, Tenth Int. Conf. on, IEEE* pp 609–14

[63] Bojarski M *et al* 2016 End to end learning for self-driving cars, arXiv:1604.07316

[64] Silberg G, Wallace R, Matuszak G, Plessers J, Brower C and Subramanian D 2012 Self-driving cars: the next revolution *White paper* 36 KPMG LLP and Center of Automotive Research

[65] Araujo L, Manson K and Spring M 2012 Self-driving cars, a case study in making new markets—London, UK *Big Innovation Centre* **9** 1–9

[66] Narla S R 2013 The evolution of connected vehicle technology: from smart drivers to smart cars to self-driving cars *ITE J.* **83** 22

[67] Newton C 2014 Uber will eventually replace all its drivers with self-driving cars *Verge* **5** 2014

[68] Guizzo E 2011 How Google's self-driving car works *IEEE Spectrum Online* **18** 1132–41

[69] Kessler A M 2015 Elon Musk says self-driving Tesla cars will be in the US by summer *New York Times* B1

[70] Hurley J S 2018 Beyond the struggle: artificial intelligence in the Department of Defense (DoD) *ICCWS 2018 13th Int. Conf. on Cyber Warfare and Security* p 297

[71] Graesser A, Chipman P, Leeming F and Biedenbach S 2009 Deep learning and emotion in serious games *Serious Games: Mechanisms and Effects* (London: Routledge) pp 81–100

[72] Min S, Lee B and Yoon S 2017 Deep learning in bioinformatics *Briefings Bioinform.* **18** 851–69

[73] Panda M 2012 Deep learning in bioinformatics *CSI Commun.* **4** 18–20

[74] Park Y and Kellis M 2015 Deep learning for regulatory genomics *Nat. Biotechnol.* **33** 825

[75] Abadi M *et al* 2016 TensorFlow: a system for large-scale machine learning *Proc. 12th USENIX Symp. on Operating Systems Design and Implementation* vol 16 *(Savannah, GA, 2–4 November)* pp 265–83

[76] Bergstra J, Breuleux O, Bastien F, Lamblin P, Pascanu R, Desjardins G, Turian J, Warde-Farley D and Bengio Y 2010 Theano: a CPU and GPU math compiler in Python *Proc. 9th Python in Science Conf.* **vol 1**

[77] Jia Y, Shelhamer E, Donahue J, Karayev S, Long J, Girshick R, Guadarrama S and Darrell T 2014 Caffe: convolutional architecture for fast feature embedding *Proc. of the 22nd ACM Int. Conf. on Multimedia* (New York: ACM) pp 675–8

[78] Ravi D, Wong C, Deligianni F, Berthelot M, Andreu-Perez J, Lo B and Yang G-Z 2017 Deep learning for health informatics *IEEE J. Biomed. Health Inform.* **21** 4–21

[79] Anthimopoulos M, Christodoulidis S, Ebner L, Christe A and Mougiakakou S 2016 Lung pattern classification for interstitial lung diseases using a deep convolutional neural network *IEEE Trans. Med. Imaging* **35** 1207 16

[80] Fakoor R, Ladhak F, Nazi A and Huber M 2013 Using deep learning to enhance cancer diagnosis and classification *Proc. of the Int. Conf. on Machine Learning* vol 28

[81] Li R, Zhang W, Suk H-I, Wang L, Li J, Shen D and Ji S 2014 Deep learning based imaging data completion for improved brain disease diagnosis *Int. Conf. on Medical Image Computing and Computer-Assisted Intervention* (Cham: Springer) pp 305–12

[82] Brosch T and Tam R 2013 Alzheimer's disease neuroimaging initiative. Manifold learning of brain MRIs by deep learning *Int. Conf. on Medical Image Computing and Computer-Assisted Intervention* (Berlin: Springer) pp 633–40

[83] Brosch T, Yoo Y, Li D K B, Traboulsee A and Tam R 2014 Modeling the variability in brain morphology and lesion distribution in multiple sclerosis by deep learning *Int. Conf. on Medical Image Computing and Computer-Assisted Intervention* (Cham: Springer) pp 462–9

[84] Zhen X, Wang Z, Islam A, Bhaduri M, Chan I and Li S 2016 Multi-scale deep networks and regression forests for direct bi-ventricular volume estimation *Med. Image Anal.* **30** 120–9

[85] Geremia E, Menze B H, Clatz O, Konukoglu E, Criminisi A and Ayache N 2010 Spatial decision forests for MS lesion segmentation in multi-channel MR images *Int. Conf. on Medical Image Computing and Computer-Assisted Intervention* (Berlin: Springer) pp 111–8

[86]  Wang L, Gao Y, Shi F, Li G, Gilmore J H, Lin W and Shen D 2015 LINKS: Learning-based multi-source IntegratioN frameworK for Segmentation of infant brain images *NeuroImage* **108** 160–72

[87]  Lynch M, Ghita O and Whelan P F 2006 Automatic segmentation of the left ventricle cavity and myocardium in MRI data *Comput. Biol. Med.* **36** 389–407

[88]  Wachinger C, Reuter M and Klein T 2017 DeepNAT: deep convolutional neural network for segmenting neuroanatomy *NeuroImage* arXiv:1702.08192

[89]  Wolterink J M, Leiner T, Max A and Viergever I I 2016 Dilated convolutional neural networks for cardiovascular MR segmentation in congenital heart disease *Reconstruction, Segmentation, and Analysis of Medical Images* (Cham: Springer) pp 95–102

[90]  Avendi M R, Kheradvar A and Jafarkhani H 2016 A combined deep-learning and deformable-model approach to fully automatic segmentation of the left ventricle in cardiac MRI *Med. Image Anal.* **30** 108–19

[91]  van Grinsven M J J P, van Ginneken B, Hoyng C B, Theelen T and Sánchez C I 2016 Fast convolutional neural network training using selective data sampling: application to hemorrhage detection in color fundus images *IEEE Trans. Med. Imaging* **35** 1273–84

[92]  Dou Q, Chen H, Yu L, Zhao L, Qin J, Wang D, Mok V C T, Shi L and Heng P-A 2016 Automatic detection of cerebral microbleeds from MR images via 3D convolutional neural networks *IEEE Trans. Med. Imaging* **35** 1182–95

[93]  Cruz-Roa A, Basavanhally A, González F, Gilmore H, Feldman M, Ganesan S, Shih N, Tomaszewski J and Madabhushi A 2014 Automatic detection of invasive ductal carcinoma in whole slide images with convolutional neural networks *Proc. SPIE* **9041** 904103

[94]  Roth H R, Lu L, Seff A, Cherry K M, Hoffman J, Wang S, Liu J, Turkbey E and Summers R M 2014 A new 2.5 D representation for lymph node detection using random sets of deep convolutional neural network observations *Int. Conf. on Medical Image Computing and Computer-Assisted Intervention* (Cham: Springer) pp 520–7

[95]  Wang D, Khosla A, Gargeya R, Irshad H and Beck A H 2016 Deep learning for identifying metastatic breast cancer, arXiv:1606.05718

[96]  IBM Research 2013 IBM research accelerating discovery: medical image analytics *YouTube* 10 October https://youtube.com/watch?v=0i11VCNacAE

[97]  Bos L 2005 Economic impact of telemedicine: a survey *Med. Care Compunetics* **2** 140

[98]  2016 Computer aided detection market worth \$1.9 billion by 2022 *Grand View Research* 8/2016 http://grandviewresearch.com/press-release/global-computer-aided-detection-market

[99]  Liew C 2018 The future of radiology augmented with artificial intelligence: a strategy for success *Eur. J. Radiol.* **102** 152–6

[100]  Nagar Y, Malone T W, De Boer P and Garcia A C B 2016 Essays on collective intelligence *PhD Thesis* Massachusetts Institute of Technology, Cambridge, MA

[101]  Kremer M 1993 The O-ring theory of economic development *Q. J. Econ.* **108** 551–75

[102]  https://www.enlitic.com/ai-in-radiology/

[103]  Cooke R E Jr, Gaeta M G, Kaufman D M and Henrici J G 2003 Picture archiving and communication system *US Patent* 6574629

[104]  Vapnik V N 1999 An overview of statistical learning theory *IEEE Trans. Neural Netw.* **10** 988–99

[105]  Vincent P, Larochelle H, Lajoie I, Bengio Y and Manzagol P-A 2010 Stacked denoising autoencoders: learning useful representations in a deep network with a local denoising criterion *J. Mach. Learn. Res.* **11** 3371–408

[106] Cireşan D C, Giusti A, Gambardella L M and Schmidhuber J 2013 Mitosis detection in breast cancer histology images with deep neural networks *Int. Conf. on Medical Image Computing and Computer-assisted Intervention* (Berlin: Springer) pp 411–8

[107] Sirinukunwattana K, Shan E, Raza A, Tsang Y-W, Snead D R J, Cree I A and Rajpoot N M 2016 Locality sensitive deep learning for detection and classification of nuclei in routine colon cancer histology images *IEEE Trans. Med. Imaging* **35** 1196–206

[108] Wang H, Roa A C, Basavanhally A N, Gilmore H L, Shih N, Feldman M, Tomaszewski J, Gonzalez F and Madabhushi A 2014 Mitosis detection in breast cancer pathology images by combining handcrafted and convolutional neural network features *J. Med. Imaging* **1** 034003

[109] Molteni M 2017 If you look at x-rays or moles for a living, AI is coming for your job *Wired* https://wired.com/2017/01/look-x-rays-moles-living-ai-coming-job/

[110] Saba L *et al* 2019 The present and future of deep learning in radiology *Eur. J. Radiol.* **14** 14–24

**IOP** Publishing



# Chapter 2

# Deep learning in biomedical imaging

**Mainak Biswas and Jasjit S Suri**

In the last few decades, deep learning (DL) and allied artificial intelligence (AI) technologies have taken the world by storm with driverless cars, self-recovery and reusable rockets, intelligent chatbots etc. The core of DL is the networked computing neurons, similar to the brain's visual cortex and also functioning like it, i.e. recognizing patterns. The depth of the DL network allows the identification of complex relationships within data. Open-source software and the reduction in the cost of hardware such as GPUs has led to faster implementation of these DL based modules.

This review chapter is dedicated to studying the different DL models available in biomedical imaging and their applications in the cardiovascular, neurological, mammography, microscopy, dermatology, pulmonary, and gastroenterology fields. This chapter also discusses the reasons for this paradigm shift in automated computer-aided diagnosis. Our special focus is on the availability of different open-source software and their uses. Several complications in this paradigm shift and the benefits to users are also discussed in detail.

## 2.1 Introduction

The computational approach to the primary human function of learning and its application has come to full circle through deep learning (DL) [1]. As currently understood, DL has already made inroads into image processing [2], driverless cars [6], thinking and judging robots [3], the gaming industry [4], language processing [5] etc. The DL models are highly popular and are used every day to build real-life technologies never thought possible outside the laboratory environment. This is due to their robust and better performance with respect to conventional algorithms. In fact, the advent of Big Data has led to greater emphasis on DL based technologies as the performance of conventional automated algorithms remained static [7]. DL applications have been further propelled by multicore CPU architectures and the dawn of graphical-processing units (GPUs) allowing a high degree of parallel computations, leading to faster convergence. The identification of patterns by the brain can be ascribed to the functioning of the multi-layered neural network of the visual cortex (VC) [8]. This functioning was adopted in the form of McCulloch Pitt's neural model consisting of a single neuron [9]. This led to the development of multiple neural models based on different parameters, such as learning function, number of hidden layers, activation functions etc. Some of the famous models are the perceptron [10], feed-forward neural networks [11], and feedback neural networks [12]. The earlier neural networks were singular, or included one hidden layer. The DL models, in contrast, use multiple layers (hidden) to increase the depth of the network for robustness and better results.

There are two forms of learning: supervised and unsupervised. In supervised learning, the DL parameters iteratively learn/train using labeled instances, until the learning reaches convergence or the minimum error. In unsupervised learning, the DL model learns complex interrelationships within data without the intervention of any learned samples. One of the famous DL models used both for characterization and segmentation of images is the convolutional neural network (CNN) [20]. The CNN applies repeated layers of convolution, pooling layers (explained later), to mine meaningful information from the images. The meaningful information is in the form of spatially local correlation within the image patterns, also called features or feature maps. There are many famous CNN models which took part in and even won the ImageNet competition, such as LeNet [21], AlexNet [22], GoogleNet [23], etc. However, it has been found that increasing the depth beyond a certain limit stalled the accuracy of the DL model and even decreased it due to the issue of vanishing gradients. DL models such as deep residual networks (ResNets) overcome this issue by skipping layers, and thereby increasing the depth as well as the accuracy [24]. Deep belief networks (DBNs) are unsupervised DL networks and represent the deeper version of restricted Boltzmann machines [13, 14]. DBNs have important applications in the regeneration and identification of images [15, 16], video sequences [17], and mocap data [18]. Autoencoders are another type of unsupervised DL model where the number of input layer neurons equals the number of output layer neurons [19]. There are multiple hidden layers within an autoencoder which train to learn a compressed representation of the input image. Once trained the autoencoders help in denoising images and also help in regenerating the input image.

There are different biomedical imaging modalities to diagnose diseases or scan an affected body part. The most common are computed tomography (CT) [25], magnetic resonance imaging (MRI) [26], and ultrasound (US) [27]. Radiologists detect the abnormalities and trained physicians diagnose and prescribe the treatment plan for the patients. However, the shortage of radiologists and the advancement of technologies has led to the development of a plethora of automated computer-aided diagnosis tools based on machine learning (ML) and computer vision. In this regard, Suri and his team are pioneers in developing automated CAD tools for different critical diseases. In 2008 the team developed a deformable model for cervical tumor delineation [28]. In the year 2011 the team constructed a feature and edge-based mechanism for intima–media thickness from carotid ultrasound images [29]. In the year 2014 the same team developed an ML based tool for classifying ovarian tissue [30] and Hashimoto thyroiditis detection from a Polish dataset of ultrasound images [31]. Also, in 2014 the team developed an automated system to estimate carotid artery thickness from MRI images using a level set [32].

The ML based statistical models for characterization are a two-stage process wherein the first stage handmade features are mined from the images, upon which a statistical ML model is applied for characterization. The handmade features do not cover all comprehensive details and are therefore incomplete and result in poor accuracy. However, in DL, the characterization is performed in the single stage where feature mining is performed internally, bypassing the handmade stage. In segmentation, deformable models have long been used for delineating the boundaries of the region-of-interest (ROI). However the performance of deformable models is severely affected by the presence of noise [33]. Segmentation through ML follows the extraction of the 3D atlas vector from each image voxel. Thereafter, probability maps are computed for each pixel through a training/testing paradigm and the border shape is inferred [34]. However, such measures are task-specific and cannot be generalized for different datasets. DL, on the other hand, applies pixel-to-pixel characterization to infer the shape of the infected/diseased area. All feature extraction is performed internally and, therefore, a single DL model can be trained on a variety of different datasets without worrying about pre-processing. Therefore, the generalization and robustness of DL based models are far better than those of ML models. In the following sections, a study of the four most-used DL models, discussion of the different DL applications in biomedical imaging, a literature review, and a conclusion are provided.

## 2.2 Deep learning models

We briefly discuss all the DL models displayed in figure 2.1 in this section.

### 2.2.1 Deep belief networks

The restricted Boltzmann machine (RBM) serves as the basis of the DBN network for regression, characterization, and feature mining [35]. The RBM is a double layer based system where the first is the input/visible layer and the second is the concealed/hidden layer, as displayed in figure 2.2. The Gibb's energy function is used to define the probability distribution function on the visible/hidden nodes, which is specified by

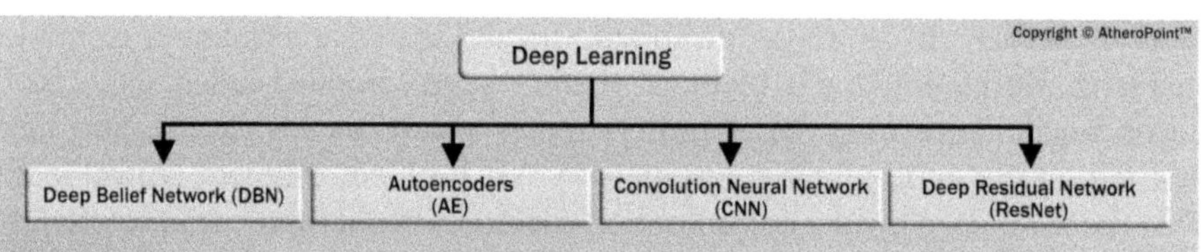


**Figure 2.1.** Deep learning and its various adaptations.

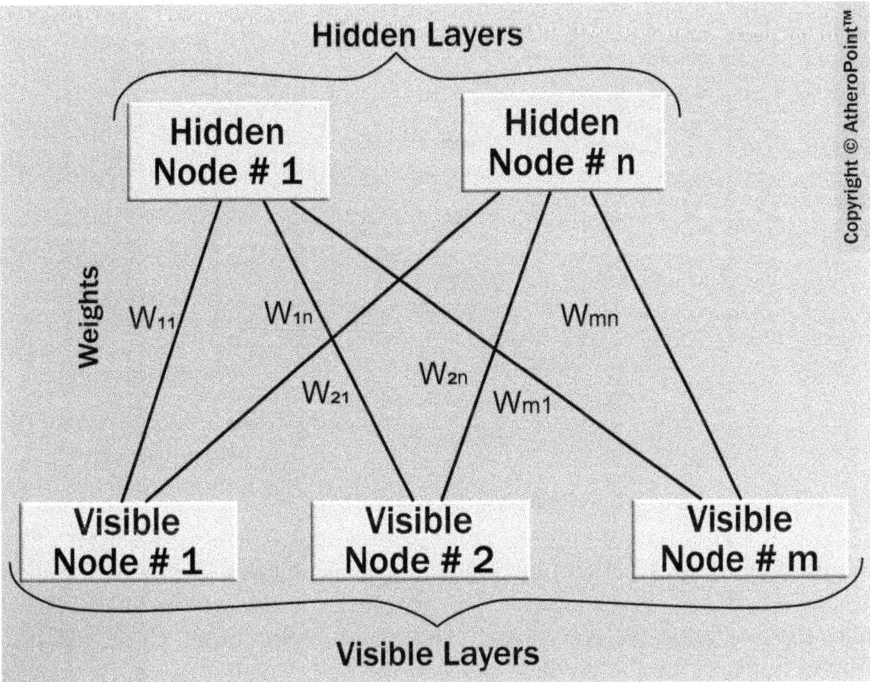


**Figure 2.2.** A two-layer restricted Boltzmann machine.

$$p(v, h) = \frac{1}{Z}e^{-E(v, h)}, \tag{2.1}$$

where the function $E(v,h)$ is given by

$$E(v, h) = \sum_{i=1}^{m}\sum_{j=1}^{n} w_{ij}h_i v_j - \sum_{j=1}^{n} b_j v_j - \sum_{i=1}^{m} c_i h_i, \tag{2.2}$$

where $w_{ij}$ represents the weight value between the visible node $v_j$ and hidden node $h_i$, $Z$ is the normalization constant, and $b_j$ and $c_i$ are bias terms related to the visible and hidden nodes. Since the intra-nodes are not connected the phrase 'restricted' is used. Therefore, the hidden node variables are autonomous if the input node values are provided and vice versa. This allows the states of variables in one layer to be sampled jointly. The Gibb's sampling block defines the sampling of the new states of hidden and visible nodes which are based on conditional probabilities of $p(hv)$ and $p(v|h)$, respectively. This is represented mathematically as

$$p(h|v) = \prod_{i=1}^{m} p[h_i|v] \tag{2.3}$$

and

$$p(v|h) = \prod_{j=1}^{n} p[v_j|h]. \tag{2.4}$$

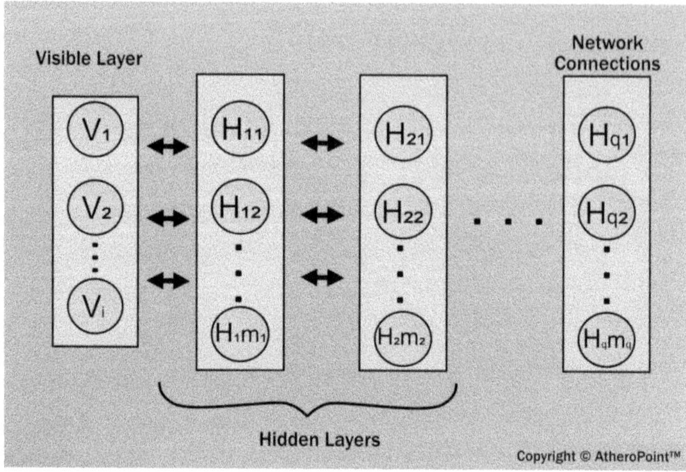



**Figure 2.3.** A multilayer DBN.

The probability distribution function $p(v)$ is measured as a likelihood function for parameter $\theta \in \{W,b,c\}$, given input $v$. This parameter is also written as $p(v, \theta)$, which changes the likelihood for every input. The $p(v, \theta)$ is commonly employed as a log likelihood function for the gradient descend method, which is given as

$$\theta(n + 1) = \theta(n) + b \times \left( -\frac{\partial\ \ln p[v, \theta]}{\partial\theta} \right). \tag{2.5}$$

A DBN is a stack of RBMs. An illustration of the model is displayed in figure 2.3 with a single visible layer with $i$ nodes, $q$ hidden layers, with $m_1$, $m_2$, ..., $m_q$ nodes. Different strategies are proposed for effective training:

    i. The DBM is trained layer-wise.
    ii. Fine-tuning is performed using a backpropagation network.

Gauging the generalization is difficult in a DBN since the probability of the data follows a normalizing constant, also called the partition function. The estimation of this constant helps in controlling the generalization and complexity of the DBN.

### 2.2.2 Autoencoder

An autoencoder is an unsupervised DL network used for image restoration, noise removal, segmentation etc [36]. The input nodes and output nodes are numerically equal in an autoencoder, as shown in figure 2.4. The number of hidden layers differs with the kind of application. In the case that the hidden layer nodes are fewer than the input/output layers, the network learns a compressed form of the input image. In this kind of network the interesting relationships between patterns of picture elements are known. In the case that the hidden layer nodes are greater in number than the input/output layer nodes, then important characteristics can be known through the employment of sparsity constraints [37]. Mathematically, the

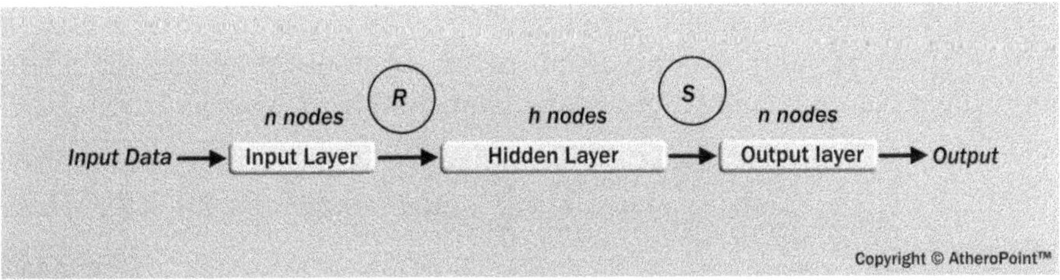



**Figure 2.4.** A generic $n$–$h$–$n$ autoencoder model.

autoencoder with an a–b–a formation can be generalized as a nine element tuple [$a$, $b,m,\mathbf{P},\mathbf{Q},\mathbf{R},\mathbf{S},X,\delta$] where:

1. $\mathbf{R}$ and $\mathbf{S}$ are sets.
2. $a$ and $b$ are positive integers.
3. $\mathbf{R}$ is a class of functions from $\mathbf{P}^a$ to $\mathbf{Q}^b$.
4. $S$ is a class of functions from $\mathbf{Q}^b$ to $\mathbf{P}^a$.
5. $A = \{c_1, c_2, \dots, c_m\}$ is a set of $m$ training vectors in $\mathbf{P}^a$. If external targets are present then $\mathbf{B} = \{d_1, d_2, \dots, d_m\}$ denote the corresponding set of target vectors in $\mathbf{P}^a$.
6. $\delta$ is the dissimilarity function defined over $\mathbf{F}^a$.

Given any input, the autoencoder converts a given input vector $c_n \in \mathbf{P}^a$ to an output vector $\mathbf{R} \cdot S(c_n)\epsilon\mathbf{P}^a$. The optimization involves finding suitable nominee functions, i.e. $r \in \mathbf{R}$ and $s \in \mathbf{S}$ to minimize the dissimilarity function $\delta$ which is specified by

$$\min G(\mathbf{R}, \mathbf{S}) = \min_{R,S} \sum_{i=1}^{m} G[c_i] = \min_{R,S} \sum_{i=1}^{m} \delta[\mathbf{R} \cdot S(c_i), c_i], \tag{2.6}$$

where $G$ is the distortion function. Here the autoencoder approximates the similarity function where the output is similar to the input. The minimization function for both input and output is given as

$$\min E(\mathbf{R}, \mathbf{S}) = \min_{R,S} \sum_{i=1}^{m} E[c_i, d_i] = \min_{R,S} \sum_{i=1}^{m} \delta[\mathbf{R} \cdot S(c_i), d_i]. \tag{2.7}$$

Autoencoders have been helpful in object recognition and image restoration.

### 2.2.3 Convolutional neural networks

The most popular DL based model that is used widely by the biomedical imaging community is the convolutional neural network (CNN) [38]. The design of the CNN is derived from the architecture of the VC, where neurons are arranged hierarchically, and are responsive to minute details. The ability of the VC to exploit local correlation spatially within visual images is utilized by the CNN. The CNN uses three repeating functions of convolution, a rectifier linear unit (ReLu), and pooling to achieve its aim by enforcing local connectivity between patterns in adjacent layers. The convolution layer applies a set of filters over the input images, multiplying itself point-wise with corresponding pixels to create a feature map. Each point in the feature map represents the spatial relationship among the adjoining pixels of the input image. Mathematically, convolution is given as

$$f(x, y) = \text{Img}(x, y) \odot W(x, y) = \sum_{p=-\frac{m}{2}}^{\frac{m}{2}} \sum_{q=-\frac{m}{2}}^{\frac{m}{2}} \text{Img}(x + p, y + q) \times W(x, y), \quad (2.8)$$

where the image 'Img' is convolved with filter $W$, outputting feature value $f$, and $\odot$ symbolize the convolution process. $f(x,y)$ symbolize the co-ordinates of the feature map which are summed values of the product of the convolution filter and neighboring pixels.

The contents of the filters update themselves for every iteration which represents the learning of the CNN. The convolution product output is then passed through the ReLu layer to remove the negative values to prevent the issue of vanishing gradients. The pooling procedure is applied for dimensionality reduction. The pooling process output either keeps the most relevant part of the feature map (max-pooling) or averages them. These three functions, i.e. convolution, ReLu, and pooling, are applied repeatedly to mine the most relevant features. Once mined, a fully connected network (multilayer perceptron) is attached for characterization. There is another adaptation of CNN where instead of a fully connected layer, up-sampling and skipping layers are used. This kind of network is also known as a fully convolutional network (FCN) and is used for the purpose of segmentation. The up-sampling layers represent the inverse of convolution where feature maps are translated into high dimensional representation using the filters. The skipping layers are used to fuse feature maps from the adjoining layers of the CNN and the up-sampled layers. This is exclusively done to retain the spatial information lost during the down-sampling stage. Finally, pixel-to-pixel characterization is applied with a corresponding ground truth to generate a smooth segmentation map. This kind of segmentation is also called semantic segmentation.

### 2.2.4 Deep residual network

The performance of DL models does not improve with an increasing number of deep layers beyond a certain point. In fact, the performance deteriorates rapidly. The reason for this is that the parameters of the shallow layers of the DL network do not update themselves due to the issue of vanishing gradients. One solution is provided by deep residual networks (ResNets) where the training error remains same for the shallow layers. The model works as follows. Supposedly, if $G(i)$ is the desired output from two adjacent layers, displayed in figure 2.5(a), then the residual mapping of the intermediate layer will be $G(i) = F(i) - i$, as shown in figure 2.5(b). Instead a recast of the original mapping can be done, i.e. $G(i) = F(i) + i$, as displayed in figure 2.5(c). This mapping, however, does not decrease the training error, but it does not increase it either, allowing the performance not to degrade with increasing layer numbers.

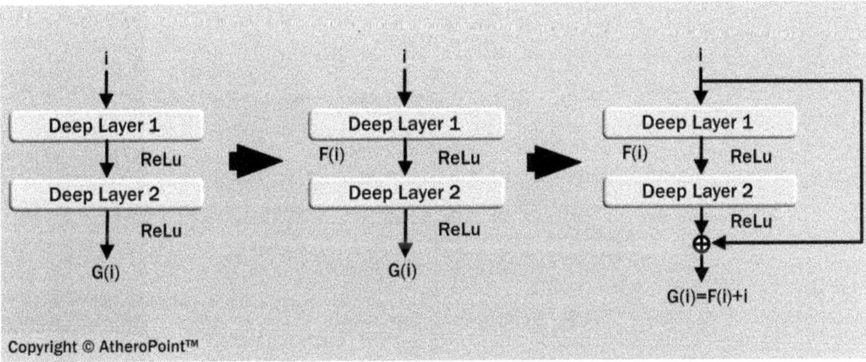



**Figure 2.5.** Deep residual network conceptual model.

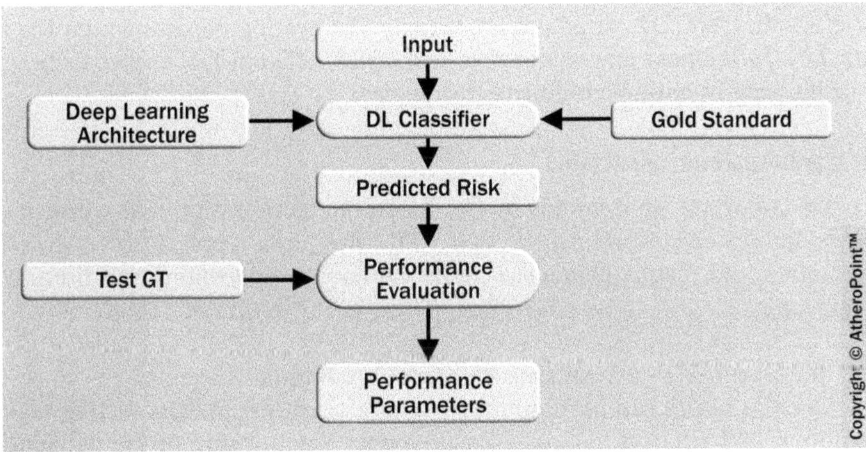



**Figure 2.6.** Generalized classification model in the DL framework.

## 2.3 DL based biomedical imaging systems

The rapid increase in DL publications on biomedical imaging is due to many factors: rapid generalization, better performance, high scalability, robustness etc. The chief advantage of DL tools over ML tools is their plug and play character, without the need for any intermediate stages. The feature mining of the ML algorithms stage is completely integrated within the DL architecture, providing a uniform interface for all kinds of images. DL has been used in biomedical imaging for two purposes: (i) characterization (figure 2.6) and (ii) segmentation. In this section, we will look into the characterization of several diseases using DL models. The segmentation of biomedical images is a complicated operation. DL based segmentation is either a single or double stage operation. The single stage employs either FCN based semantic segmentation or patch based characterization. In the two-stage model, the first stage is usually global ROI detection employing an ML/DL paradigm. The next

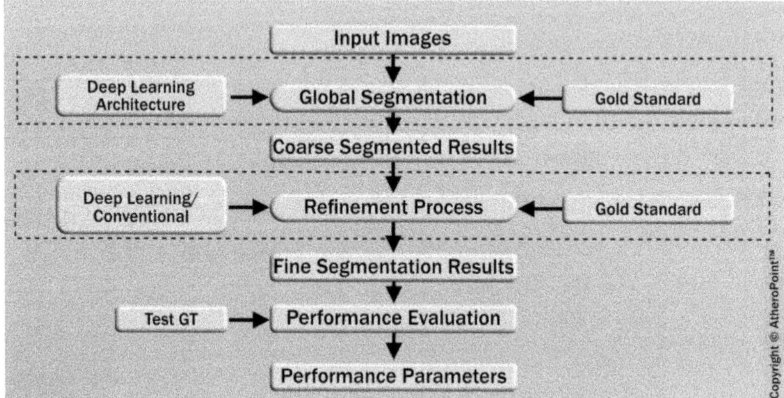


**Figure 2.7.** Generalized two-stage segmentation models in the DL framework.

stage is another ML/DL paradigm for actually delineating borders within the ROI (figure 2.7). In the next few sections we will cover different DL applications in the field of imaging in multiple biomedical domains.

### 2.3.1 Cardiovascular application

There are important applications of DL based characterization and segmentation methods in the cardiovascular domain. DL has been applied to left-ventricle segmentation and plaque characterization. Different applications are discussed in the following.

#### 2.3.1.1 Left-ventricle endocardium tracking in US data

Cardiovascular health can be monitored using different parameters such as regional wall motion and ejection fraction measurement. The tracking of the left-ventricle (LV) endocardium from US scans can provide these details. Some attempts were made by scientists to use this method to determine the size and contour, but each approach had its own limitations [32]. These models depended on Gaussian distribution of the contour and shape from learning samples. These models did not take into account temporal factors such as wall motion leading to variations. Therefore, a DBN based model was applied for the LV tracking to capture all variations of wall motions [39]. This DL network applied segmentation by taking into consideration past segmentation contours of current images. In this regard, the present image is considered as a set of previous states defined by the heart functions of systole and diastole. Therefore, given the cardiac phase and contour, the probabilistic model of shape is given as

$$p\left[\mathrm{Img}_t \mid k_t, c_t\right] \approx p\left[k_t, c_t \mid \mathrm{Img}_t\right] p\left[\mathrm{Img}_t\right], \tag{2.9}$$

where $p[\mathrm{Img}_t]$ is constant and $p[k_t, c_t \mid \mathrm{Img}_t]$ can be described as a combination of affine detection $p(k_t \mid \theta, \mathrm{Img}_t)$, non-rigid probabilistic segmentation $p[c_t \mid \theta, k_t, \mathrm{Img}_t]$, and the prior of affine parameters $p[\theta \mid \mathrm{Img}_t]$:

$$p(k_t, c_t|\mathrm{Img}_t) = \int p(k_t|\theta, \mathrm{Img}_t)\, p\, [c_t \mid 0, k_t, \mathrm{Img}_t]\, p\, [\theta \mid \mathrm{Img}_t] d\theta, \qquad (2.10)$$

where $\theta$ represents the parameters of the affine transformation, $k_t$ symbolizes the cardiac phase, and $c_t$ denotes the LV contour at time stamp $t$. The affine transformation is marginalized using prior information in order to estimate the segmentation and cardiac phase. The segmentation of the heart is performed in two stages by two different sets of DBNs, as shown by equations (2.9) and (2.10). The initial rough contour is determined by the first DBN and then smoothening of the LV shape is carried out. The primary stage is implemented by a cascade of three DBNs trained on three different priors, i.e. systole, diastole, and non-LV. The second stage produces an improved delineation using a principle component analysis based shape model. The LV borders were traced by four cardiologists and one technician. All the DBNs were trained using the same ground truth. The parameters of performance considered for the experiment were average error (AV), mean absolute distance (MAD), Jaccard distance (JD), and average perpendicular error (AVP), which were be 0.83, 0.91, 0.95, and 0.83, respectively. The results were better compared to previous models [40, 41]. The process model and output images are shown in figures 2.8 and 2.9, respectively.

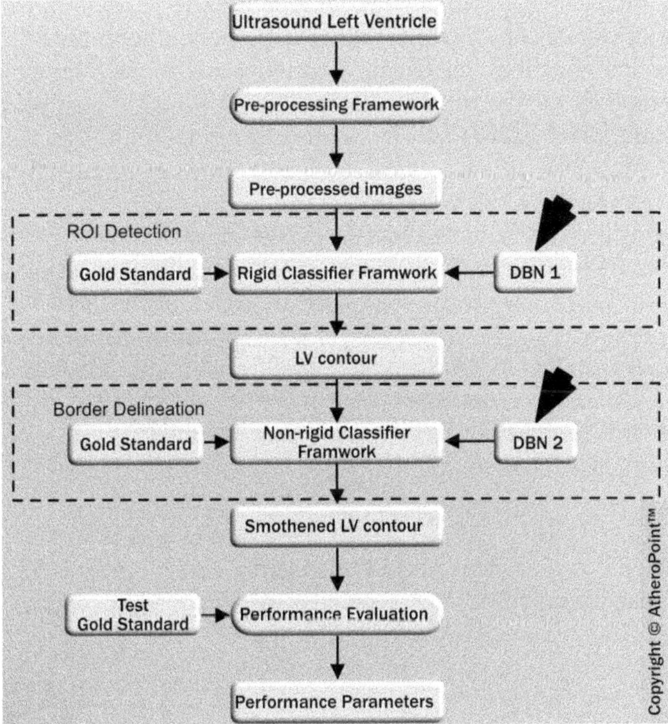


**Figure 2.8.** DBN based LV contour tracking model.

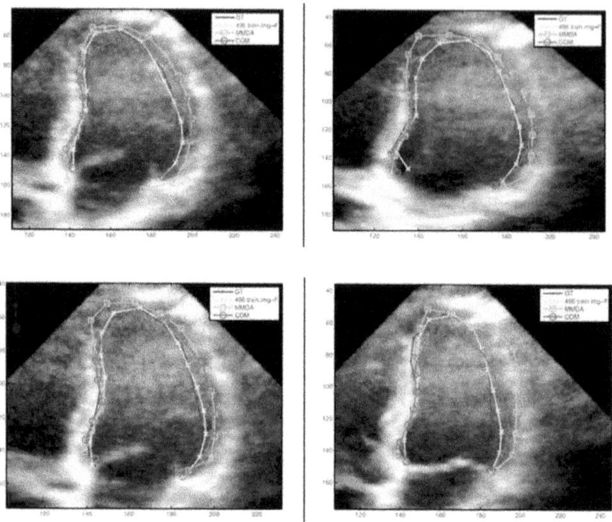

**Figure 2.9.** Output of DBN based segmentation of the LV. The auto-detected outline is shown in yellow, COM [40] is outlined in magenta, multiple model data association [41] is outlined in cyan, and the ground truth outline is displayed in blue. (Reproduced with permission from [39]. Copyright 2013 IEEE.)

### *2.3.1.2 Automatic segmentation of the left ventricle in cardiac MRI*

There are several limitations to conventional LV segmentation models, i.e. leakage, shrinking, and initialization. Therefore, a combination of deep learning along with conventional models, i.e. deformable methods, is proposed to resolve the issues of a lack of data and lower accuracy. The issue of restricted data size is tackled using data augmentation techniques and better design. In this section LV segmentation from an MRI dataset is considered [42].

The dataset is derived from the MICCAI 2009 dataset challenge and consists of 45 MR cardiac images. Initially, a binary mask was obtained using a CNN deep learning model. The binary mask was used to generate the ROI within the MR dataset. The parameters of the CNN model were trained using an autoencoder. The ROI was applied to the MR images to obtain an initial LV shape $\xi^{[0]}$ using an autoencoder architecture. Thereafter, the conventional technique of the deformable model was applied to delineate the final LV shape. Finally, a 3D model of the LV contour was obtained. The equation is

$$\xi^* = \arg \min_\delta E[\xi], \tag{2.11}$$

where $\xi^*$ denotes the contour shape and $E$ represents the energy function. The final contour shape was obtained iteratively using the gradient descend method, which is given as

$$\xi^{[i+1]} = \xi^{[i]} + \eta \frac{d\xi}{dt}, \tag{2.12}$$

where $\eta$ denotes the step size, $i$ represents successive iterations, and $\frac{d\xi}{dt}$ represents the change in $\xi$ with respect to time.

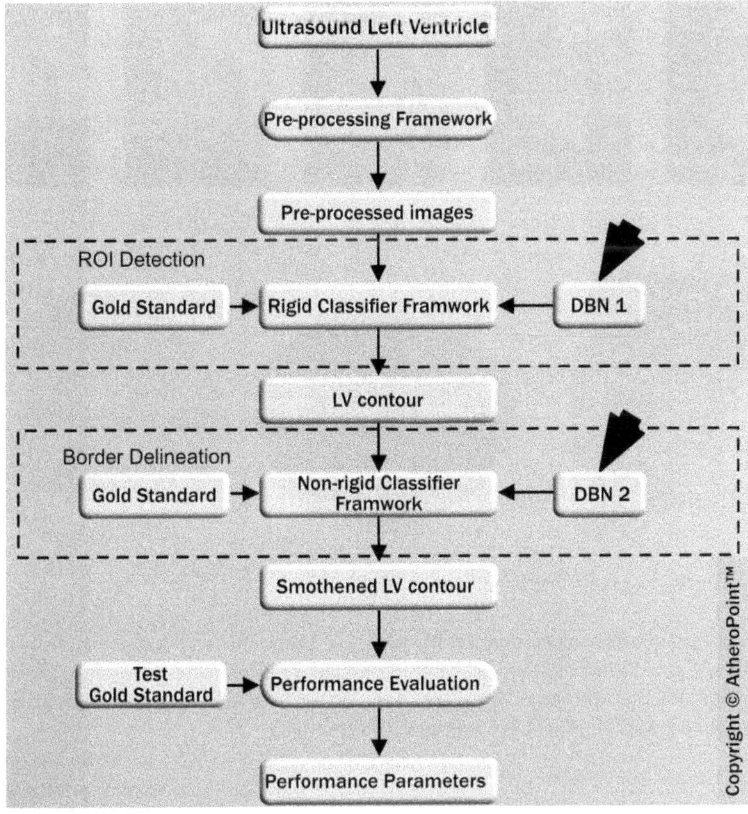


**Figure 2.10.** DL based LV contour tracking model for MRI images.

The process model, and 2D and 3D segmentation outputs are displayed in figures 2.10, 2.11(a) and (b), respectively. The performance parameters of the Dice metric, APD, and conformity are 0.94, 1.81, and 0.86, respectively. Although the proposed model performance was better compared to conventional models, the hardware used was a CPU, and therefore slow. The DL models used were one or two layers deep and hence cannot reach generalization for a larger corpus.

### 2.3.1.3 Volumetric left-ventricle image parsing

Feature mining in ML based algorithms for 3D images is complicated, since the system tries to capture features/parameters in 3D translation, orientation, and anisotropic scale. Therefore, the ultimate search space of feature mining is 9D, which is difficult considering contemporary models. In this respect, a marginal space deep learning model (MSDLM) was proposed [43] that is capable of incrementally learning features and thereby determine the shape of the LV contour. The MSDLM consists of two stages, whereby in the first stage of the model the ROI is detected and in the second stage the LV boundary is delineated. The localization of the ROI is performed using DL based classifiers (DLCs).

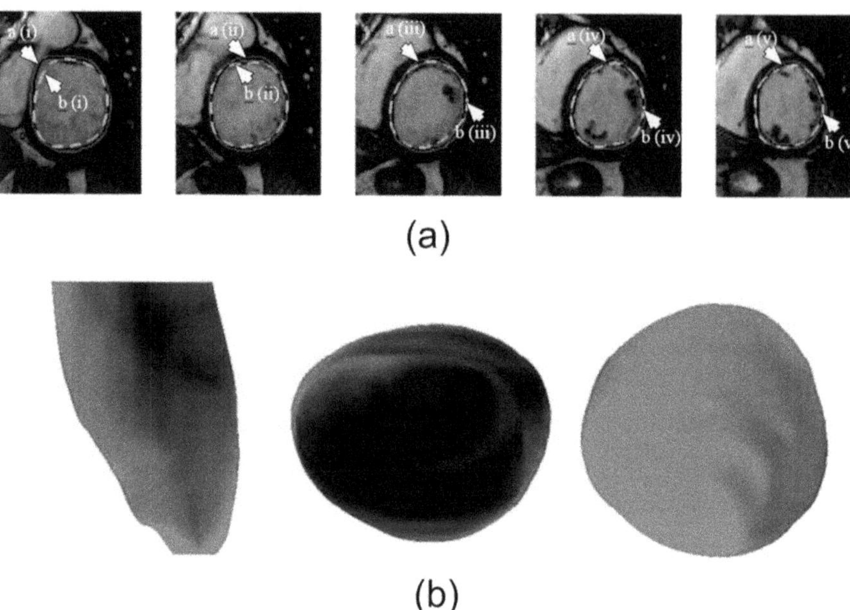

Figure 2.11. (a) Example of outputs from the DL based LV segmentation. The auto-detected LV contour is shown in red-black as indicated by arrow 'a', while the ground truth contour is displayed in green as indicated by arrow 'b'. (b) Inferred 3D shape from the given model with side, top, and bottom cross-sections. (Reproduced with permission from [42]. Copyright 2016 Elsevier.)

The DLC learns the position, orientation, and scale of the LV contour iteratively, by maximizing the probability of locality of the LV by using translation, orientation, and scale space. It is specified as

$$\left(\hat{\vec{T}}, \hat{\vec{R}}, \hat{\vec{S}}\right) = \arg\max_{\vec{T},\vec{R},\vec{S}}\left(\vec{T}, \vec{R}, \vec{S} \mid I\right),\tag{2.13}$$

where $T$ represents the translation, $R$ denotes the orientation, and $S$ signifies the scale space. Three DLCs were trained to predict the position, orientation, and space of the LV. The first DLC estimates the position, the second DLC estimates the orientation by taking input from the first, and the third DLC estimates scale by taking input from the first and second. In the second stage, an active shape model based on deep learning is applied for LV shape deformation. To prevent overfitting, dropout is employed. The dataset is 2891 3D volumes taken from 869 patients. The proposed DL based MSDLM model performed comparatively better than the earlier conventional system employing the same methodology but without the DL component. The comparative results are displayed in table 2.1. The corresponding process model and resulting images are displayed in figures 2.12 and 2.13, respectively.

**Table 2.1.** Comparison of the MSDLM and the conventional MSM model.

| Accuracy parameters | DL based MSDLM | Conventional MSM |
| --- | --- | --- |
| Position error | 1.47 mm | 3.12 mm |
| Corner error | 2.80 mm | 0.42 m |

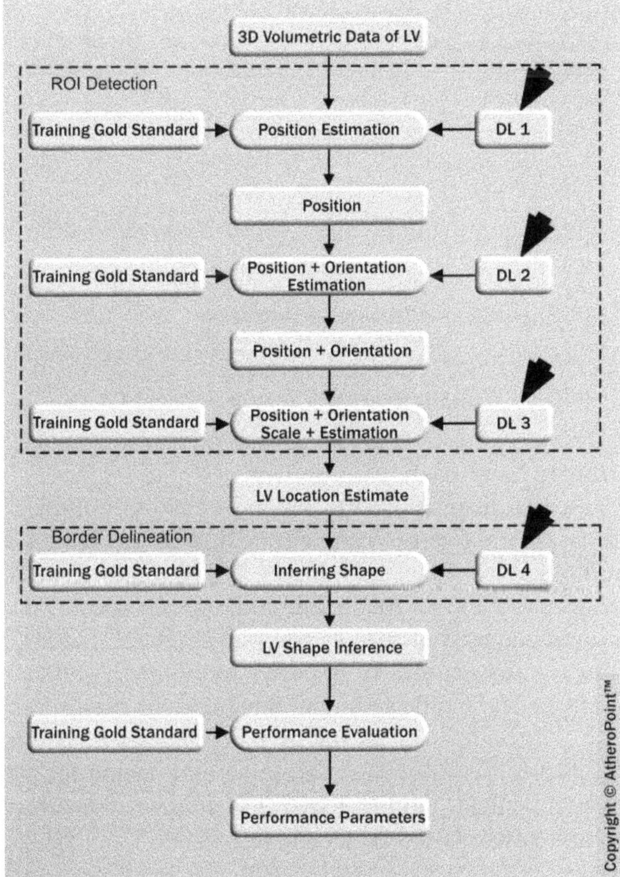


**Figure 2.12.** Deep classifier based 3D LV segmentation model.

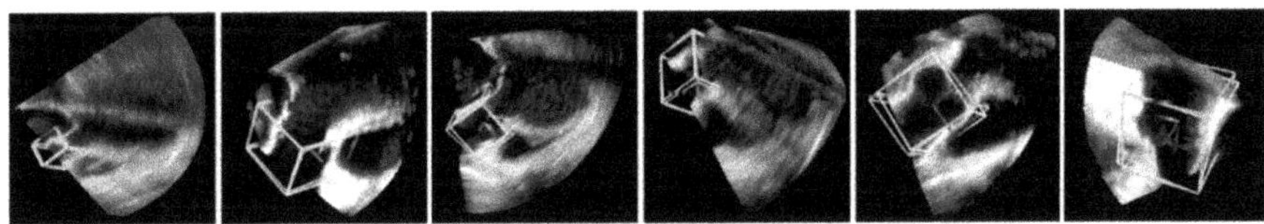

**Figure 2.13.** Deep classifier based 3D LV segmentation model outputs. The auto-detected bounding box is shown in green while the ground truth bounding box is shown in yellow. (Reproduced with permission from [43]. Copyright 2016 IEEE.)

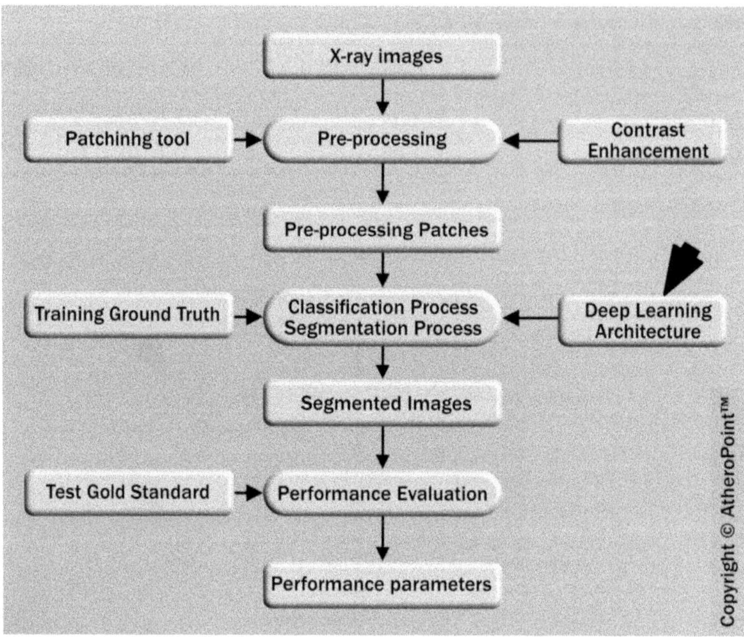


**Figure 2.14.** Segmentation of x-ray images using CNNs.

### 2.3.1.4 Vessel mining from x-ray angiograms

Globally, cardiovascular diseases (CVDs) kill more than 17 million people annually. Coronary artery disease is one of the main CVDs and a leading cause of death (figure 2.14). Traditional methods, i.e. tracking and filter based methods, are unable to provide better accuracy due to low resolution, the presence of noise, and similar echogenic organs near the ROI increasing complexity. A DL based CNN model was used for detecting vessel regions from x-ray angiography images [44]. In this experiment, 44 images were collected. The images were pre-processed to increase the contrast using the top-hat transform. A patching based method was applied, where 1 040 000 patches were mined for training and testing the CNN. The CNN (two convolutions + two max-pooling + two fully connected layers (500)) generated the probabilities for the two classes vessel and not-vessel. The CNN model generated a high accuracy of 93.5% which was comparatively better than previous methods. Moreover, the better results were generated despite the presence of noise, low-resolution, and complex backgrounds. However, the CNN model used was only six layers deep and, therefore, the dataset should be trained/tested on a deeper model for better results, reliability, and clinical use. The segmentation results are displayed in figure 2.15.

### 2.3.1.5 Plaque characterization

Vulnerable plaque identification is an important element of CVD detection and can be helpful in early risk assessment and preparation of a treatment plan. However, there are several challenges in plaque detection owing to its small size, complex

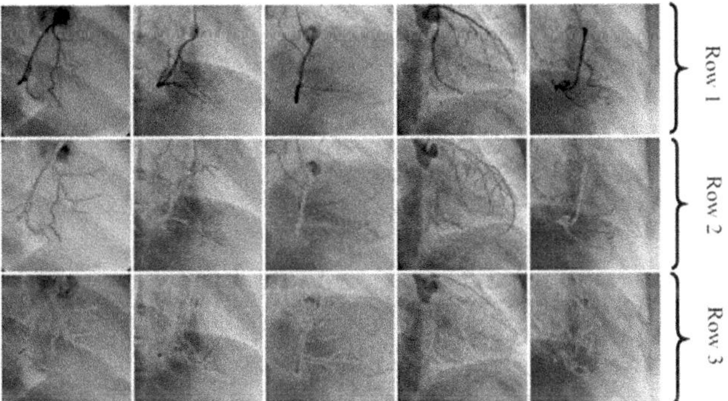


**Figure 2.15.** Row one shows five input images. Row two shows the corresponding ground truth of the row one images. Row three shows analogous images for segmentation outputs using CNN. (Reproduced with permission from [44]. Copyright 2016 IEEE.)

**Table 2.2.** Comparison of the DL and conventional models for plaque characterization [45].

| Model | Accuracy |
| --- | --- |
| DL based CNN | 78.5% |
| Single-scale SVM | 14.3% |
| Multi-scale SVM | 7.2% |

multifocal appearance, and the presence of speckle noise, making segmentation difficult and the characterization of plaque thus constituents a complicated task. In this regard, an automated DL based CNN model was proposed for plaque characterization [45]. The CNN model (four convolution layers + four ReLu + three fully connected layers) used the softmax classification technique. The cross-entropy loss function was used for training and the results of the proposed model were compared to single-scale and multi-scale support vector machines (SVMs).

For the experiment, 56 ultrasound CCA images were used. A patch based strategy was followed. Therefore, 90 000 patches were mined from the ultrasound images. They were annotated into three categories, lipid core, fibrous, and calcified, which were used as the ground truth for CNN model learning. The comparative results of the DL model and the other conventional ML model are listed in table 2.2, which shows that the DL based model outperformed the ML based models by a huge margin. The process of the model and the segmentation results are displayed in figures 2.16 and 2.17, respectively.

## 2.3.2 Neurology application

Here we discuss various DL applications in the field of neurology imaging.

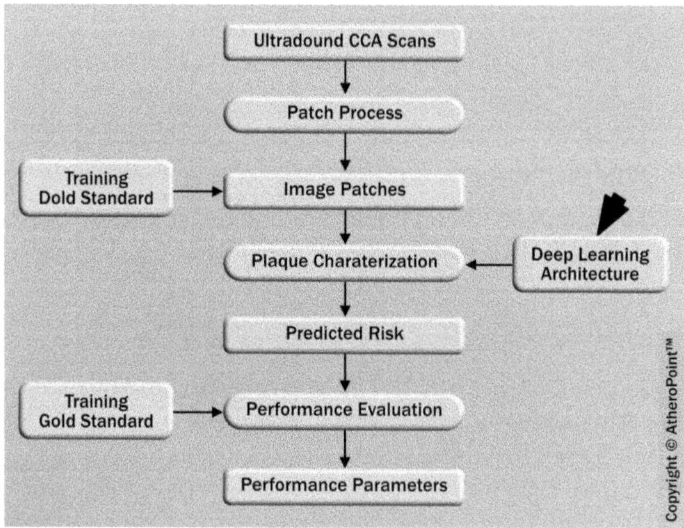


**Figure 2.16.** Global diagram for plaque characterization.

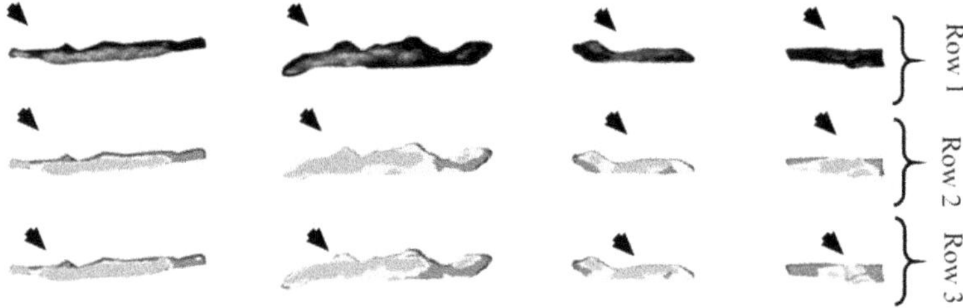


**Figure 2.17.** Row one shows the input images, row two is the corresponding ground truth, and row three is the equivalent characterization output for the given method. Here red represents the lipid core, yellow represents fibrous tissue, and green represents calcified tissue. (Reproduced with permission from [45]. Copyright 2016 IEEE.)

### 2.3.2.1 Brain lesion segmentation

In the case of an emergency, the precise localization of a brain lesion from injury is necessary for successful treatment and saving lives. However, there are several complexities to accurately estimate the location and size of a brain lesion, i.e. occurrence at multiple sites, varying shape and size, and an intensity profile matching unaffected parts of brain. Different conventional models have been developed but they have had limited performance. In this regard, a two-pathway DL model was proposed for accurately estimating and segmenting brain lesions [46]. In this model, two separate CNN architectures were proposed for each of the pathways. Each pathway processed a different dimension and scale of the same input images, which were later combined to give an accurate boundary and size estimate of the lesion. The initial pathway and the second pathway were fed the

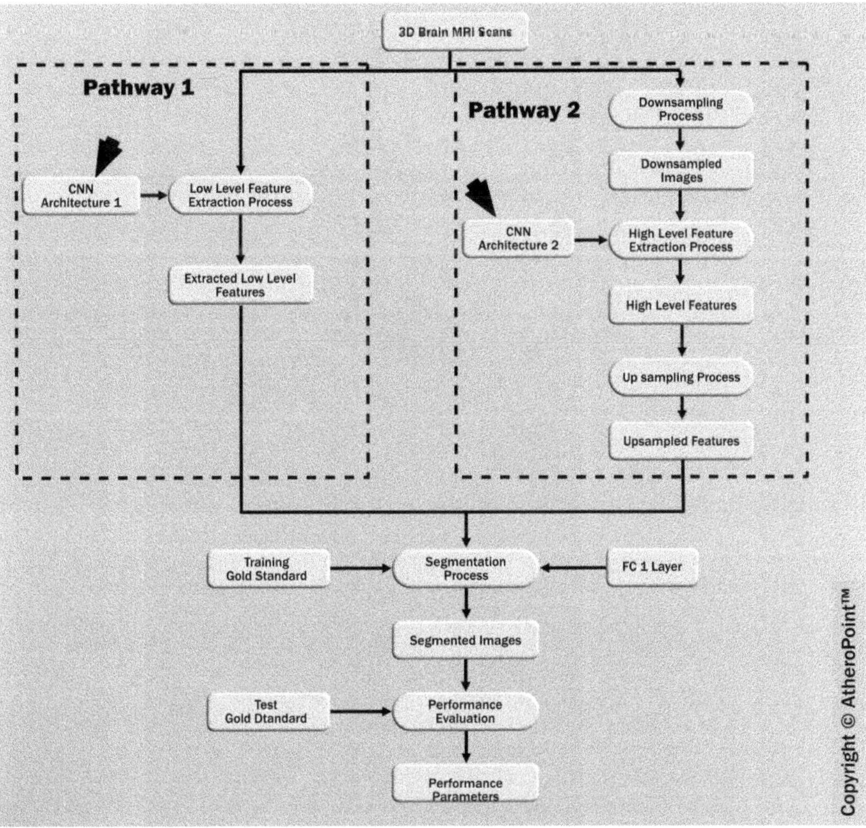



**Figure 2.18.** Deep neural network for brain lesion segmentation.

input 3D images and down-sampled images, respectively. The initial pathway worked out the detailed shape of the brain lesion, whereas the second pathway captured the location. The output feature maps of the second pathway were up-sampled to the original size and combined with the output from the first pathway to give overall lesion segmentation. The joint output was later fed into a fully connected conditional random field network for the final estimation.

The input images featuring traumatic brain injuries were collected from 66 patients and appeared in the ISLES-SISS challenge. The DS coefficient was found to be 0.59. The method outperformed the random forest method by a large margin. Although it performed better than previous methods, the performance of the DL model is considered low for clinical significance. The process of the model and segmented images are shown in figures 2.18 and 2.19, respectively.

### 2.3.2.2 *Brain tumor segmentation*

Several classification methods exist for brain tumors. Among brain tumors, gliomas have the highest fatality rate. Gliomas are further categorized into low-grade and high-grade (LGG and HGG). Between the HGG and the LGG the

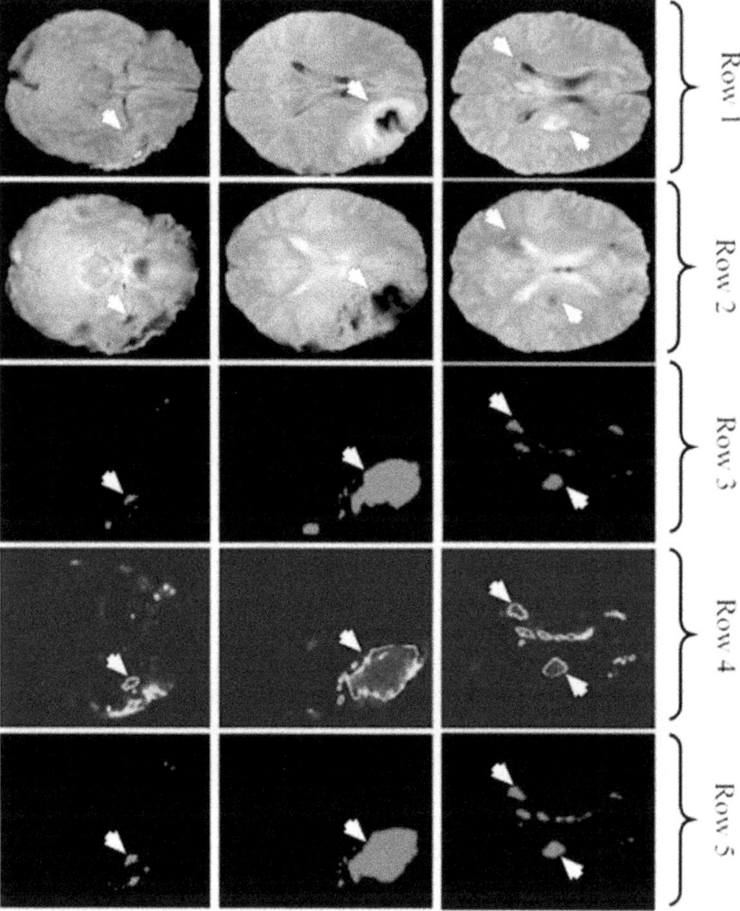

**Figure 2.19.** Row one and two show the fluid-attenuated inversion recovery and diffusion weighted imaging sequences from the dataset, row three shows the ground truth, row four shows the corresponding output of DeepMedic, and row five shows the analogous output of DeepMedic + CRF (conditional random field). (Reproduced with permission from [46]. Copyright 2017 Elsevier.)

former is more aggressive. The segmentation and classification of gliomas into LGGs and HGGs are critically important in planning the treatment, follow-up, and monitoring of patients. Due to its invasive nature biopsy is risky and dangerous. Therefore, the non-invasive procedure of imaging is preferred. However, the segmentation of gliomas manually is error-prone and tedious. MR images present a separate set of challenges, including intensity, homogeneity, and faulty acquisition scanners. In this scenario, automated techniques of brain tumor segmentation are preferred over other methods. In [47] a DL based CNN model was proposed for tumor segmentation from MR images. The tumors were categorized into necrosis, edema, non-enhancing, and enhancing classes by experts. Prior to training the CNN model, the pre-processed images were used for normalizing the dataset.

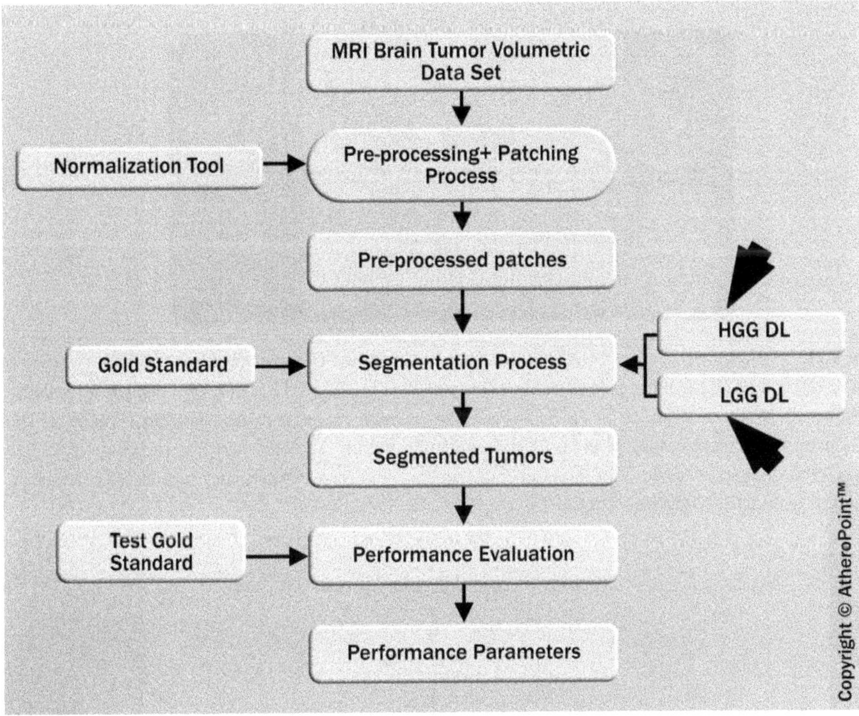


**Figure 2.20.** DL based system for brain tumor segmentation.

Two separate CNN models of dissimilar depths were used separately for LGGs and HGGs. The LGG deep learning model consisted of four convolutional, two max-pooling, and three fully connected layers. The HGG CNN model consisted of six convolutional, two max-pooling, and three fully connected layers. It can be seen that the shallower LGG deep learning model's performance did not improve by increasing depth. Two datasets were used: the first dataset consisted of 65 MR scans while the second dataset consisted of 274 MR scans. Patch based segmentation was followed. Therefore, 335 000 and 450 000 patches were mined for the LGG and HGG images and the corresponding CNN models were trained and tested. Three types of segmentation were followed: complete, core, and enhancing. The Dice similarity metrics for the three schemes were 0.88, 0.83, and 0.77, respectively. The process model and segmentation output are shown in figures 2.20 and 2.21, respectively.

### 2.3.3 Mammography application

This section provides a glimpse of a DL based application in the area of mammography. We discuss two recent DL applications in the area of breast cancer research.

#### 2.3.3.1 Mitosis detection in breast cancer

The occurrence of mitotic cells in histology images is a precursor to breast cancer. These mitosis processes represent the complex transformations that the cell nucleus

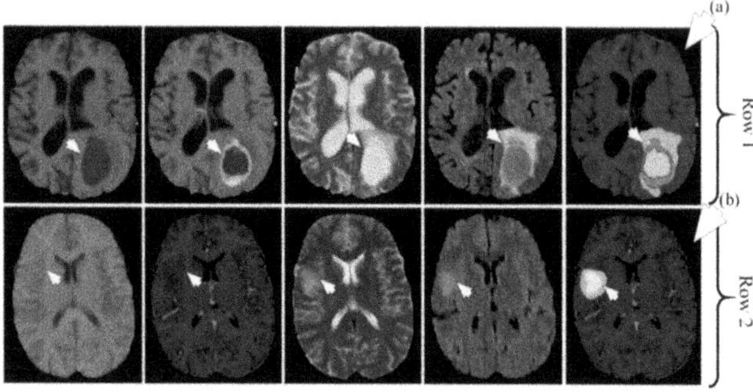



**Figure 2.21.** The segmentation of HGGs is shown in row one and the segmentation of LGG is shown in row two. The different MRI sequences from left to right are T1, T1c, T2, and FLAIR. The right-most image in row one indicated by biomedical arrow 'a' is a HGG segmented image and the right-most image in row two indicated by biomedical arrow 'b' is an LGG segmented image. The colors represent tumor types: green—edema; blue—necrosis; yellow—non-enhancing tumor; and red—enhancing tumor. (Reproduced with permission from [47]. Copyright 2016 IEEE.)

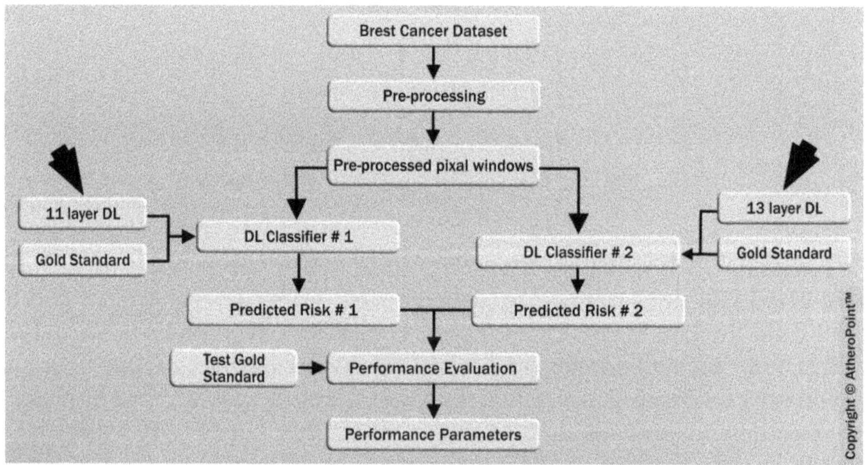



**Figure 2.22.** DL based breast cancer detection.

undergoes. Although similar structures of same intensity, shape, and size exist, only a very few are mitotic cells and therefore require extensive evaluation using deep learning models. In this study, DL based CNN models were applied to characterize mitotic and non-mitotic cells [48]. An ensemble classifier based on two separate CNN models was developed. The first CNN consisted of 13 layers (input + five convolution + five max-pooling + two fully connected layers) and the second CNN consisted of 11 layers (input + four convolution + four max-pooling + two fully connected layers). The outputs were later pooled in the ensemble version. The ensemble DL based classifier was trained and tested using 50 images taken from the MITOSIS dataset. The performance parameters of precision and F1 score were 0.88 and 0.78, respectively. The process model and the outputs are shown in figures 2.22

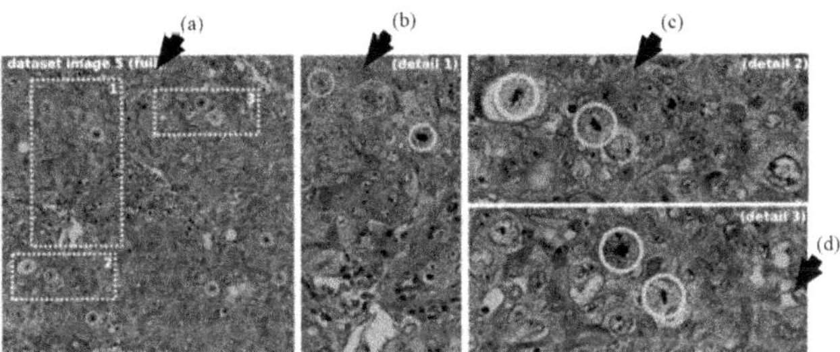


**Figure 2.23.** The left-most image show by arrow (a) represents the image with three dotted areas. The three dotted areas are displayed in detail as indicated by arrows (b), (c), and (d). Mitosis detected by the system is shown in green, red represents a false positive, and cyan signifies mitosis not detected by the given approach. (Reproduced with permission from [48]. Copyright 2013 Springer.)

and 2.23, respectively. Although the results were better compared to others in the ICPR2012 challenge, a greater number of training images would have led to better generalization.

### 2.3.3.2 Digital mammographic tumor classification

The conventional model of breast cancer diagnosis involves feature mining and application of statistical modeling for predicting malignancy. In [49] CNN transfer learning is used for categorization of breast lesions from mammography images. Further, this study also compared the features mined from a deep learning model and hand crafted lesion features. The CNN model used in the work was first trained in general object recognition, which is entirely different from breast cancer data. Finally, the trained CNN model was tested on the mammography images. It is hypothesized that models trained in a general setting are able to recognize minute details of specialized biomedical images such as breast cancer. Here, the dataset constituted 219 breast lesions from digital mammography images. A total of 607 ROIs were mined. AlexNet, a popular CNN model, was used for feature mining. The AlexNet model consisted of five convolution layers and three fully connected layers. Another set of features was mined using analytically mined handmade features. Two SVM classifiers were trained separately on the two sets of features: deep features (process P1) and handmade features (process P2). The classification outputs were combined from both P1 and P2 to make the ensemble classifier (P1 + P2). It can be seen that the performance of the ensemble classifier AUC was better than that of P2 (an AUC of 0.86 versus an AUC of 0.81). The process model and analogous output images are shown in figures 2.24 and 2.25, respectively.

### 2.3.4 Microscopy applications

The applications of deep learning in microscopy are discussed in here.

### 2.3.4 1 Analysis of individual cells in a live-cell imaging experiment

In cellular biology, a powerful research technique is dynamic live-cell imaging experiments. Cells are cultivated manually for hours for proper identification.

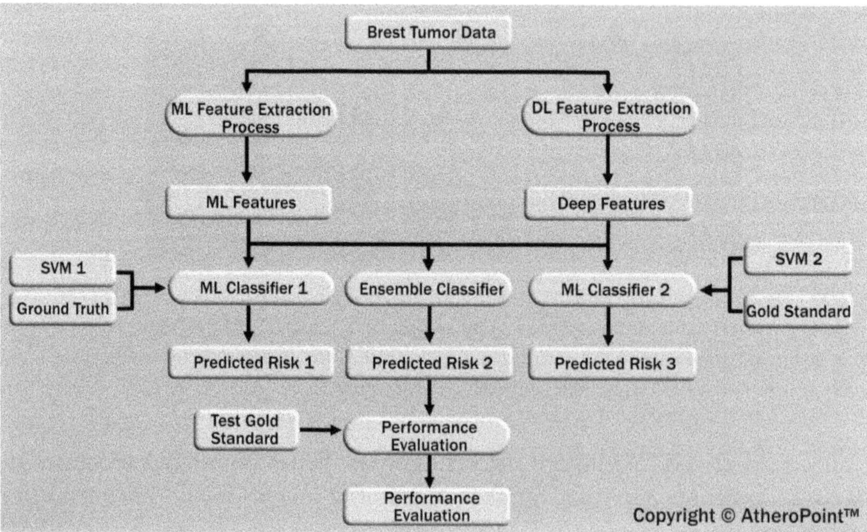


**Figure 2.24.** DL based breast tumor detection.

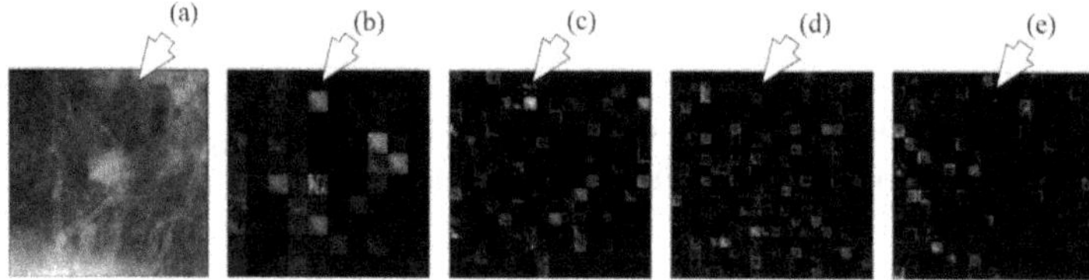


**Figure 2.25.** The outputs of different layers of AlexNet are shown: (a) the ROI input; (b) the output of pooling layer one; (c) the outputs of pooling layer two; (d) the output of convolution layer four; and (e) the output of pooling layer five. These are fed into conventional methods for characterization. (Reproduced with permission from [49]. Copyright 2016 Society of Photo-Optical Instrumentation Engineers.)

Automated techniques such as deep learning can provide faster and more accurate results in both segmentation and accuracy. In [50] a DL based model was employed for automated segmentation and characterization. Several CNNs were employed to track the growth of bacterial cells and individual mammalian nuclei without manual intervention. The patching methodology was applied, where the patches were mined from the image and annotated with their respective class. This process of patching and annotating was repeated for all the images. Thereafter, the classifier was trained and tested to characterize each patch. After classification all the labelled patches were combined to give the segmented image. The cost function applied for the function is

$$C = -\sum_i \log\left(\frac{e^{\text{label}}}{\sum_l e^l}\right) + \tau \sum_{w \in W} w^2, \tag{2.14}$$

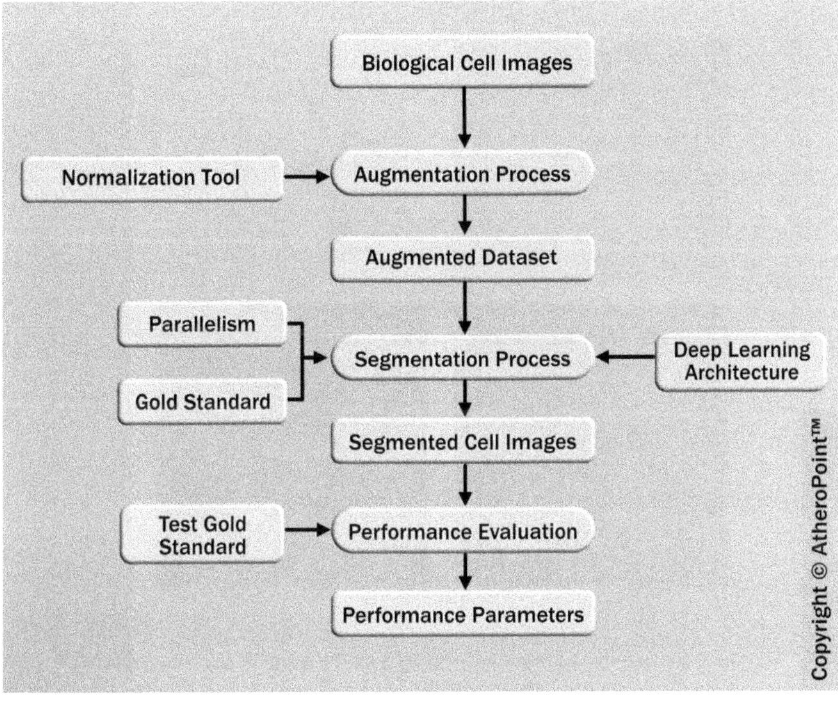


**Figure 2.26.** Model depicting the segmentation of individual cells using DL.

where $C$ is the cost function, $i$ is the images, $l$ is the labels, $w$ is the weights, $\tau$ is the regularization parameter, and $W$ represents all the weights. The weights are updated as per the formula

$$w_{i+1} = w_i - \eta \frac{\partial C}{\partial w}, \tag{2.15}$$

where $i$ denotes the iterations, $\eta$ represents the learning rate, and $\frac{\partial C}{\partial w}$ is the gradient of cost with respect to weight. The mammalian cells were categorized into five types. In each image there were approximately 300 nuclei, 500 nuclei, and 100 mammalian cells. The images were pre-processed and artificially augmented using rotation. The Jaccard index and Dice metric for the operation were 0.89 and 0.94, respectively. Model parallelism was achieved using GPUs. The process model and segmented output are shown in figures 2.26 and 2.27, respectively.

### 2.3.4.2 Classification and segmentation of microscopy images

In [51] an approach for multi-instance learning (MIL) was combined with deep learning CNN for object detection. Instead of a single instance, multi-instance learning learns from a group of instances. In this work the similarity of the CNN pooling layer and the aggregation function of multi-instance learning was used, where features from the CNN corresponded to instance features of the

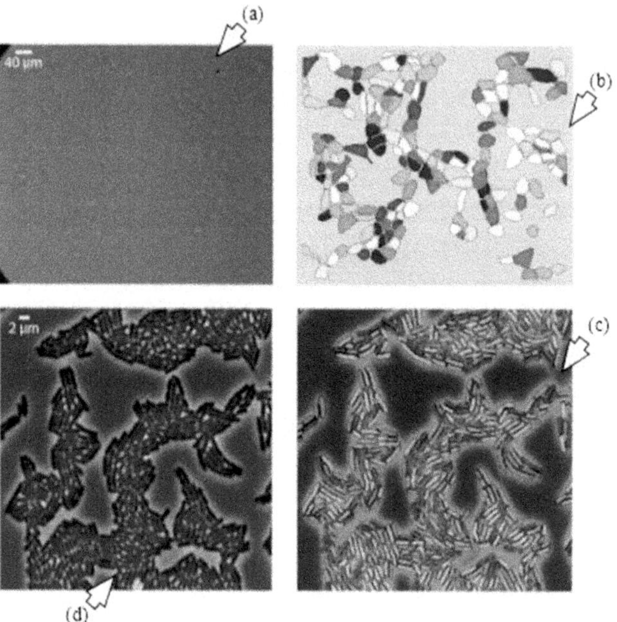

**Figure 2.27.** The phase contrast input images are displayed on the left-most side and indicated by (a) and (d). The analogous segmentation output is indicated by (b) and (c). (Reproduced with permission from [50].)

multi-instance learning [51]. The CNN produced image characterization of different shapes and sizes through multi-instance pooling layers. The softmax activations of output nodes were backpropagated through the CNN network to generate Jacobian maps for class predictions. Then thresholding was applied on the sum of Jacobian maps along the network input channels to generate the segmentation masks. A loopy belief network was employed to improve the localization prospects (reference [10] in [51]). The CNN model consisted of seven convolutions, four pooling layers, one multi-instance learning pooling layer, and one fully connected layer. The dataset consisted of microscopic MFC-7 breast cancer cell images, of which 300 were used for training and 40 for testing. Another dataset consisting of green fluorescent protein yeast data was used out of which 2200 images were used for training and 280 images were used for testing. For validation purposes, the MNIST dataset was used where 50 and 10 images were used for training and testing, respectively. The accuracy for the MFC-7 and yeast datasets were 0.97 and 0.96, respectively. The test error was 0%. The process of the model and the segmentation output are shown in figures 2.28 and 2.29, respectively.

### 2.3.5 Dermatology application

A significant percentage of people are affected by skin cancer globally. This subsection provides a guide to DL application in melanoma recognition.

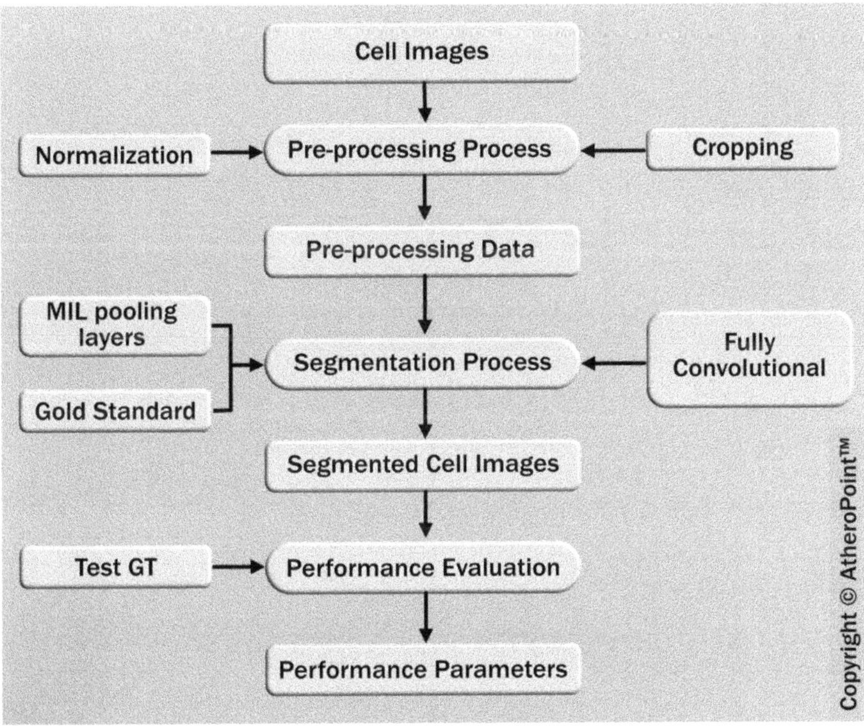



**Figure 2.28.** Model depiction for deep multiple instance learning.

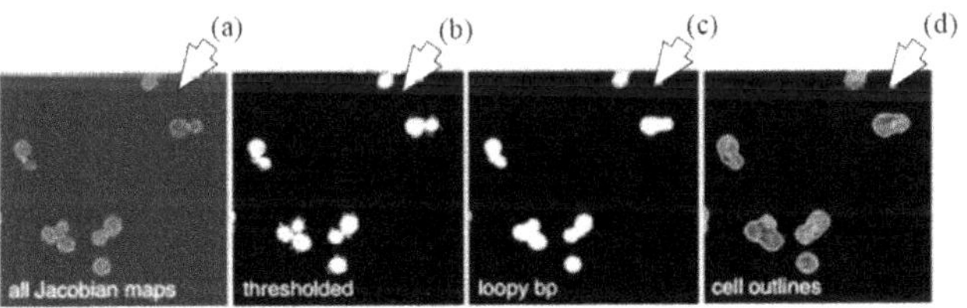



**Figure 2.29.** Segmentation using the FCN-MIL pooling methodology. (a) The Jacobian maps, (b) thresholding, (c) denoising using loopy bp, and (d) segmentation outlines. (Reproduced with permission from [51].)

## 2.3.5.1 DL based melanoma recognition

Among all types of skin cancers, melanoma has the highest fatality rate. If detected early melanoma can be cured. The ML based automated techniques for melanoma detection from biomedical images have limited performance, as they are dependent on an intermediate manual feature mining stage. In [52] DL was applied to detect melanoma from biomedical images. The DL architecture drew inspiration from UNet, CNN, and ResNets [53]. The two primary elements of the model were segmentation and classification. The network consisted of three layers of convolution and pooling followed by three layers of deconvolution and upsampling. Skip operations were employed to merge the feature maps in the intermediate layers of convolution and deconvolution, to recover spatial information lost during

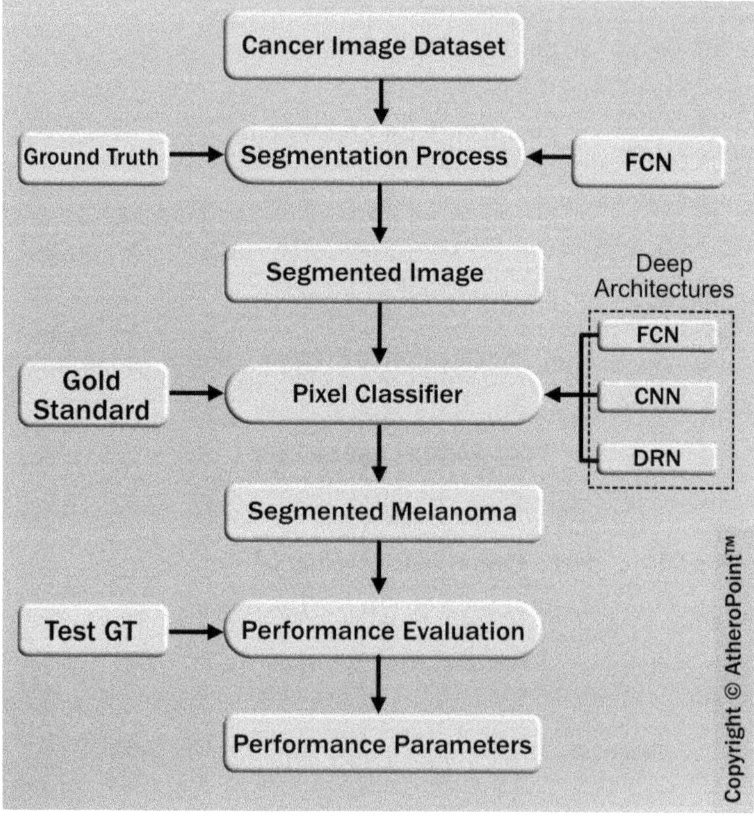



**Figure 2.30.** Segmentation and classification of melanoma.

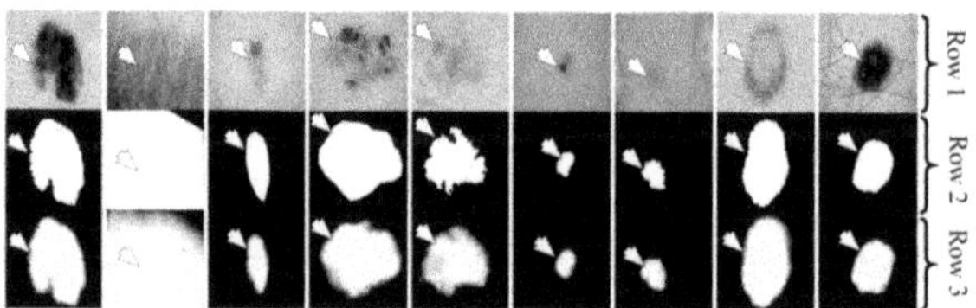



**Figure 2.31.** Segmentation of melanoma of nine patients using FCN-UNet. Row one shows the input images, row two is the corresponding ground truth, and row three is the analogous segmentation output. (Reproduced with permission from [52].)

down-sampling. The dataset was derived from the ISIC2016 challenge, of which 900 images and 379 images were used for training and testing, respectively. Finally, the segmented output was segregated using an SVM. The system performed with an accuracy of 76.0% whereas the average performance of eight dermatologists was 70.5%. This proved that the DL based automated diagnosis system performed better. The process model and segmentation outputs are displayed in figures 2.30 and 2.31, respectively.

### 2.3.6 Gastroenterology applications

DL has been applied to liver tumor detection and risk characterization for diagnosis and the preparation of treatment plans for patients from biomedical CT images.

#### 2.3.6.1 Segmentation of liver tumors

In a recent report by WHO, liver cancer was the second main cause of death among all forms of cancer. The appearance of liver lesions in CT images is fuzzy and heterogeneous, with varying densities, shapes, and sizes, hence making detection difficult. In [54] DL was applied for segmentation of tumor lesions from liver CT images. A patch based method was applied for characterization. The CNN model comprised two convolution layers, two max-pooling layers, and one softmax layer. A 30 CT image dataset was used for the experiment. Initially, the patches were pre-processed and annotated as positive if they comprised >50% lesion, otherwise they were annotated as negative. Three ML techniques were also applied for benchmarking, i.e. AdaBoost, random forest, and an SVM. The results show that the DL method outperformed the other techniques with a DM coefficient of 80.06% (SVM = 79.78%, random forest = 79.47%, and AdaBoost = 75.67%). The process of the model and segmentation outputs are shown in figures 2.32 and 2.33, respectively.

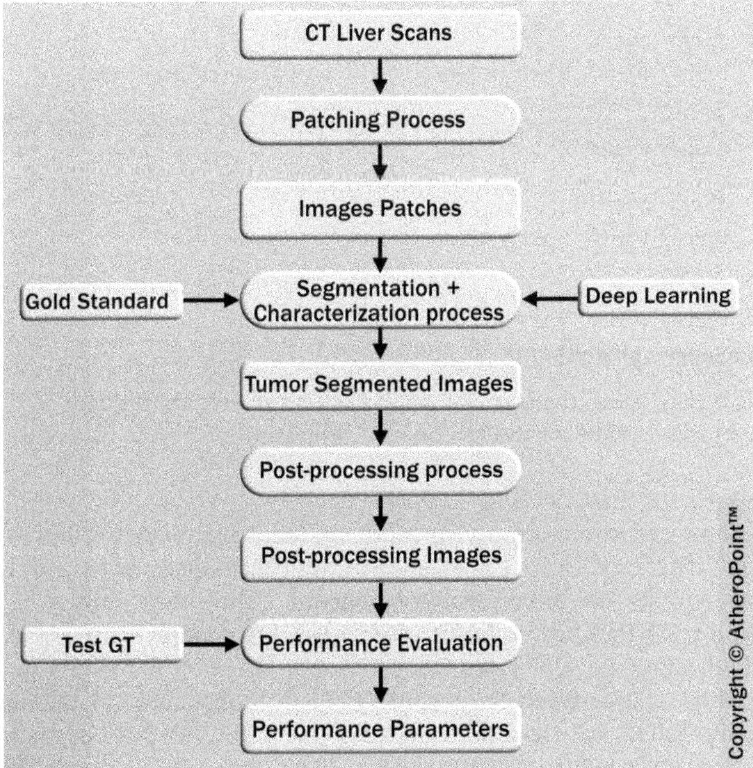


**Figure 2.32.** Segmentation of CT liver tumor.

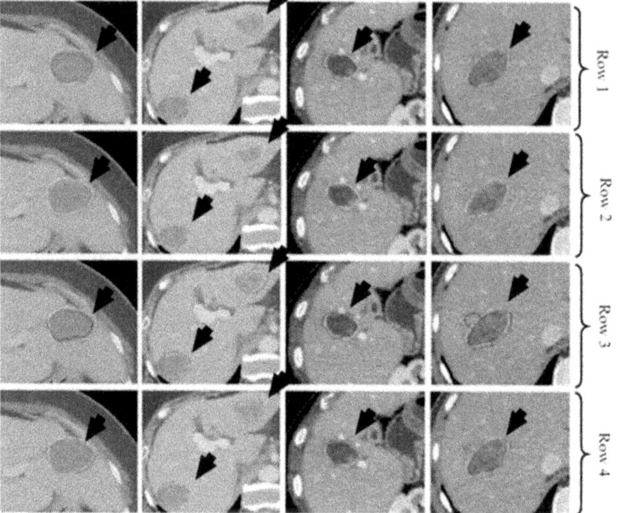


**Figure 2.33.** Segmentation of CT liver tumor for four patients using AdaBoost (purple) in row one, random forest (green) in row two, an SVM (blue) in row three, and a CNN (red) in row four. The ground truth is displayed in yellow in all rows. (Reproduced with permission from [54].)

**Table 2.3.** Result analysis of the DL application for lung cancer detection [55].

| Parameter | CNN | DBN |
| --- | --- | --- |
| Sensitivity | 73.3% | 73.4% |
| Specificity | 78.7% | 82.2% |

## 2.3.7 Pulmonary application

Lung cancer appears in the form of small nodules which are difficult to detect and diagnose. In this section we discuss one DL application of lung cancer detection.

### 2.3.7.1 Characterization of lung disease

Malignant lung cancer has a five-year survival rate of less than 10% and therefore is deadly if not diagnosed early. Lung cancer is difficult to detect even with biopsy, as small lung nodules are infrequently malignant. Also, they cannot be reliably categorized by CT-PET scans. In this regard two DL models were applied to lung cancer classification, i.e. CNN and DBN [55]. The shape of pulmonary nodules can give an initial characterization regarding their malignancy. Additionally, the internal composition, such as fluid, calcification, and fat, can provide an indication. CT scans were collected from 1010 patients. The results are given in table 2.3. The process of the model is shown in figure 2.34.

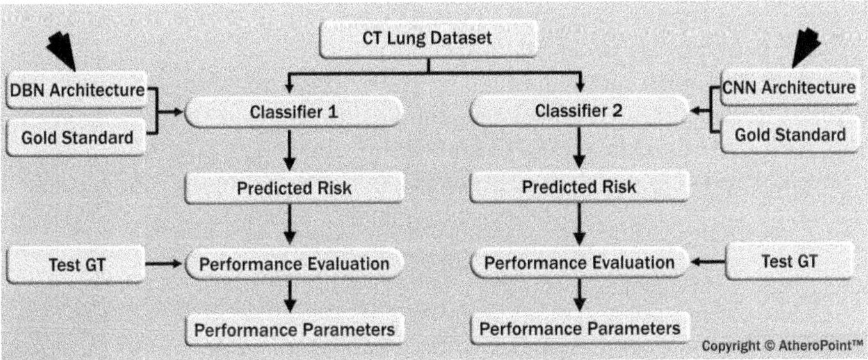


**Figure 2.34.** Characterization of CT lung cancer.

## 2.4 Discussion

The advent of computing systems led scientists to develop algorithms to emulate intelligent behavior [56]. The early algorithms were simple task-specific knowledge tools based on propositional calculus. Therefore, if the task changed the full algorithm had to be adapted. A demand for generalized algorithms for all tasks grew, which led to the development of ML. ML is a two-stage process where the first stage is feature mining and the second stage is the actual statistical model for characterization and segmentation. The issue remained the same since the feature were mined using manual tools and therefore were also task-specific. Thus the performance was limited and could not achieve the necessary generalization for widespread use. This led to the evolution of DL based tools which were single stage and had inbuilt feature mining capability. The depth of the DL networks allowed them to learn more intrinsic details of the pattern relationships. The DL network can mine images and perform characterization in any imaging modality, such as ultrasound, MR, CT, x-ray etc. Therefore, the generalization as well as scalability of the DL models can be achieved in a faster way than other intelligence methods. In this chapter we have displayed different applications in cardiography, neurology, mammography, microscopy, dermatology, gastroenterology, and lung imaging. DL has been also applied in brain lacunes [57], white matter segmentation [58], fatty liver disease detection [59], protein structure prediction [66–74], gene expression regulation [75–79], protein classification [80–82], anomaly classification [83, 87–89], and brain decoding [84–86].

### 2.4.1 Graphics processing unit and open-source software for deep learning

Deep learning modules consist of millions of parameters for training. These parameters represent the memory of the learning process and continuously evolve until convergence. The current multicore CPU systems are not suitable for DL computation. GPUs provide a better alternative. GPUs consist of thousands of cores capable of processing in parallel. Therefore, the operating system can schedule thousands of computations among these cores using multiple filter threads. The speed of GPUs is tens to thousands of times faster than that of multicore CPUs and can achieve convergence faster [8].

Python is the most accepted and popular programming language for DL application developers. Most packages for DL are derived from some version of Python, such as TensorFlow [60], Theano [61], Keras [62], CAFFE [63], and Torch [64]. Deeplearning4j [65] is derived from the Java programming language for Java developers.

# References


[1] LeCun Y, Bengio Y and Hinton G 2006 Deep learning *Nature* **521** 436–44

[2] Ciregan D, Meier U and Schmidhuber J 2012 Multi-column deep neural networks for image classification *Computer Vision and Pattern Recognition* (Piscataway, NJ: IEEE) pp 3642–9

[3] Wawrzynski P 2015 Control policy with auto correlated noise in reinforcement learning for robotics *Int. J. Mach. Learn. Comput.* **5** 91–5

[4] Mnih V, Kavukcuoglu K, Silver D, Graves A, Antonoglou I, Wierstra D and Riedmiller M 2013 Playing Atari with deep reinforcement learning, arXiv: 1312.5602

[5] Du T and Vijay K S 2009 Deep learning for natural language processing *Eecis. Udel Edu.* 1–7 https://www.eecis.udel.edu/~vijay/fall13/snlp/lit-survey/DeepLearning.pdf

[6] Hadsell R, Erkan A, Sermanet P, Scoffier M, Muller U and LeCun Y 2008 Deep belief net learning in a long-range vision system for autonomous off-road driving *Proc. Intelligent Robots and Systems* pp 628–33

[7] Rosenblatt F 1958 The perceptron: a probabilistic model for information storage and organization in the brain *Psychol. Rev.* **65** 386

[8] Raina R, Madhavan A and Ng A Y 2009 Large-scale deep unsupervised learning using graphics processors *Proc. of the 26th Annual Int. Conf. on Machine Learning* (New York: ACM) pp 873–80

[9] Najafabadi M M, Villanustre F, Khoshgoftaar T M, Seliya N, Wald R and Muharemagic E 2015 Deep learning applications and challenges in big data analytics *J. Big Data* **2** 1

[10] McCulloch W S and Pitts W 1943 A logical calculus of ideas immanent in nervous activity *Bull. Math. Biophys.* **5** 115–33

[11] Bebis G and Georgiopoulos M 1994 Feed-forward neural networks *IEEE Potentials* **13** 27–31

[12] Hopfield J J 1984 Neurons with graded response have collective computational properties like those of two-state neurons *Proc. Natl Acad. Sci. USA* **81** 3088–92

[13] Hinton G E 2009 Deep belief networks *Scholarpedia* **4** 5947

[14] Smolensky P 1986 Information processing in dynamical systems: foundations of harmony theory *Research Paper* NO. CU-CS-321-86 Department of Computer Science, Colorado University at Boulder

[15] Hinton G E, Osindero S and Teh Y-W 2006 A fast learning algorithm for deep belief nets *Neural Comput.* **18** 1527–54

[16] Hinton G, Simon O, Max W and Teh Y-W 2006 Unsupervised discovery of nonlinear structure using contrastive backpropagation *Cogn. Sci.* **30** 725–31

[17] Sutskever I and Geoffrey H 2007 Learning multilevel distributed representations for high-dimensional sequences *Proc. Mach. Learn. Res.* **2** 548–55

[18] Taylor G W, Hinton G E and Rowels S 2007 Modeling human motion using binary latent variables *Advances in Neural Information Processing Systems* vol 19 (Cambridge, MA: MIT Press) pp 1345–52

[19] Bengio Y 2009 Learning deep architectures for AI *Found. Trends Mach. Learn.* **2** 1–127

[20] Krizhevsky A, Sutskever I and Hinton G E 2012 ImageNet classification with deep convolutional neural networks *Advances in Neural Information Processing Systems* vol 12 (Cambridge, MA: MIT Press) pp 1097–105

[21] Szegedy C, Liu W, Jia Y, Sermanet P, Reed S, Anguelov D, Erhan D, Vanhoucke V and Rabinovich A 2015 Going deeper with convolutions *Proc. of the IEEE Conf. on Computer Vision and Pattern Recognition* pp 1–9

[22] Krizhevsky A, Sutskever I and Hinton G E 2012 ImageNet classification with deep convolutional neural networks *Commun. ACM* **60** 84–90

[23] Russakovsky O, Jia D, Hao S, Jonathan K, SanjeevSatheesh, Sean M and Zhiheng H 2015 ImageNet large scale visual recognition challenge *Int. J. Comput. Vision* **115** 211–52

[24] He K, Xiangyu Z, Ren S and Sun J 2016 Deep residual learning for image recognition *Proc. of the IEEE Conf. on Computer Vision and Pattern Recognition* pp 770–8

[25] Herman G T 2009 *Fundamentals of Computerized Tomography: Image Reconstruction from Projections* (Berlin: Springer)

[26] Rinck P A 2014 *Magnetic Resonance in Medicine: A Critical Introduction* (Norderstedt: Books on Demand) p 21

[27] Sanches J M, Laine A F and Suri J S 2012 *Ultrasound Imaging: Advances and Applications* (Berlin: Springer)

[28] Wu D H, Shaffer A D, Thompson D M, Yang Z, Magnotta V A, Alam R and Mayr N 2008 Iterative active deformational methodology for tumor delineation: evaluation across radiation treatment stage and volume *J. Magn. Reson. Imaging* **28** 1188–94

[29] Molinari F, Kristen M, Meiburger U, Acharya R, Zeng G, Paulo S, Rodrigues, Saba L, Nicolaides A and Suri J S 2011 CARES 3.0: a two stage system combining feature-based recognition and edge-based segmentation for CIMT measurement on a multi-institutional ultrasound database of 300 images *Annual Int. Conf. of the IEEE Engineering in Medicine and Biology Society* (Piscataway, NJ: IEEE) pp 5149–52

[30] Acharya U R, Sree S V, Kulshreshtha S, Molinari F, Koh J, Saba L and Jasjit S S 2014 GyneScan: an improved online paradigm for screening of ovarian cancer via tissue characterization *Technol. Cancer Res. Treat.* **13** 529–39

[31] Acharya U R, Sree S V, Krishnan M M R, Molinari F, Ÿnik W Z, Bardales R H, Witkowska A and Suri J S 2014 Computer-aided diagnostic system for detection of Hashimoto thyroiditis on ultrasound images from a Polish population *J. Ultrasound Med.* **33** 245–53

[32] Saba L *et al* 2014 Semi-automated analysis of carotid artery wall thickness in MRI *J. Magn. Reson. Imaging* **39** 1457–67

[33] Suri J S, Liu K, Singh S, Laxminarayan S N, Zeng X and Reden L 2002 Shape recovery algorithms using level sets in 2-D/3-D biomedical imagery: a state-of-the-art review *IEEE Trans. Inf. Technol. Biomed.* **6** 8

[34] Powell S, Magnotta V A, Johnson H, Jammalamadaka V K, Pierson R and Andreasen N 2008 Registration and machine learning-based automated segmentation of subcortical and cerebellar brain structures *Neuroimage* **39** 238–47

[35] Nair V and Hinton G E 2010 Rectified linear units improve restricted Boltzmann machines *Proc. of the 27th Int. Conf. on Machine Learning* pp 807–14

[36] Baldi P 2012 Autoencoders, unsupervised learning, and deep architectures *Proc. of ICML Workshop on Unsupervised and Transfer Learning* pp 37–49

[37] Ng A 2011 Sparse autoencoder *Lecture Notes* CS294A Stanford University, CA pp 1–19

[38] Hubel D H and Torsten N W 1962 Receptive fields, binocular interaction and functional architecture in the cat's visual cortex *J. Physiol.* **160** 106–54

[39] Carneiro G and Jacinto C N 2013 Combining multiple dynamic models and deep learning architectures for tracking the left ventricle endocardium in ultrasound data *IEEE Trans. Pattern Anal. Mach. Intell.* **35** 2592–607

[40] Georgescu B, Zhou X S, Comaniciu D and Gupta A 2005 Databased-guided segmentation of anatomical structures with complex appearance *Conf. Computer Vision Pattern Rec.*

[41] Nascimento J C and Marques J S 2008 Robust shape tracking with multiple models in ultrasound images *IEEE Trans. Imag. Proc* **17** 392–406

[42] Avendi M R, Kheradvar A and Hamid J 2016 A combined deep-learning and deformable-model approach to fully automatic segmentation of the left ventricle in cardiac MRI *Biomed. Image Anal.* **30** 108–19

[43] Ghesu F C, Krubasik E, Georgescu B, Singh V, Zheng Y, Hornegger J and Comaniciu D 2016 Marginal space deep learning: efficient architecture for volumetric image parsing *IEEE Trans. Biomed. Imaging* **35** 1217–28

[44] Nasr-Esfahani E, Samavi S, Karimi N, Soroushmehr S M R, Ward K, Jafari M H, Felfeliyan B, Nallamothu B and Najarian K 2016 Vessel extraction in X-ray angiograms using deep learning *38th Annual Int. Conf.Engineering in Medicine and Biology Society* (Piscataway, NJ: IEEE) pp 643–6

[45] Lekadir K, Galimzianova A, Betriu À, del Mar Vila M, Igual L, Rubin D L, Fernández E, Radeva P and Napel S 2016 A convolutional neural network for automatic characterization of plaque composition in carotid ultrasound *IEEE J. Biomed. Health Inform.* **21** 48–55

[46] Kamnitsas K, Ledig C, Newcombe V F J, Simpson J P, Kane A D, Menon D K, Rueckert D and Glocker B 2017 Efficient multi-scale 3D CNN with fully connected CRF for accurate brain lesion segmentation *Biomed. Image Anal.* **36** 61–78

[47] Pereira S, Pinto A, Alves V and Carlos A S 2016 Brain tumor segmentation using convolutional neural networks in MRI images *IEEE Trans. Biomed. Imaging* **35** 1240–51

[48] Cireşan D C, Giusti A, Gambardella L M and Schmidhuber J 2013 Mitosis detection in breast cancer histology images with deep neural networks *Int. Conf. on Biomedical Image Computing and Computer-assisted Intervention* (Berlin: Springer) pp 411–8

[49] Huynh B Q, Li H and Maryellen L G 2016 Digital mammographic tumor classification using transfer learning from deep convolutional neural networks *J. Biomed. Imaging* **3** 034501

[50] Van Valen *et al* 2016 Deep learning automates the quantitative analysis of individual cells in live-cell imaging experiments *PLoS Comput. Biol.* **12** e1005177

[51] Kraus O Z, Ba J L and Frey J F 2016 Classifying and segmenting microscopy images with deep multiple instance learning *Bioinformatics* **32** i52–9

[52] Codella N C F, Nguyen Q, Pankanti S, Gutman D, Helba B, Halpern A and Smith J R 2016 Deep learning ensembles for melanoma recognition in dermoscopy images *CoRR abs/1610* **04662** 153–65

[53] von Eicken T, Basu A, Buch V and Vogels W 1995 U-Net: a user-level network interface for parallel and distributed computing *ACM SIGOPS Operating Syst. Rev.* **29** 40–53

[54] Li W, Jia F and Hu Q 2015 Automatic segmentation of liver tumor in CT images with deep convolutional neural networks *J. Comp. Commun.* **3** 146

[55] Hua K-L, Hsu C-H, Hidayati S C, Cheng W-H and Chen Y-J 2015 Computer-aided classification of lung nodules on computed tomography images via deep learning technique *OncoTargets Ther.* **8** 2015–22

[56] Giraldi G, Rodrigues P, Suri J and Singh S 2011 Dual active contour models for biomedical image segmentation *Image Segmentation* (Rijeka: InTech)

[57] Ghafoorian M, Karssemeijer N, Heskes T, Bergkamp M, Wissink J, Obels J and Platel B 2017 Deep multi-scale location-aware 3D convolutional neural networks for automated detection of lacunes of presumed vascular origin *NeuroImage: Clin.* **14** 391–9

[58] Ghafoorian M, Karssemeijer N, Heskes T, Uden I W, Sanchez C I, Litjens G and Platel B 2017 Location sensitive deep convolutional neural networks for segmentation of white matter hyperintensities *Sci. Rep.* **7** 5110

[59] Biswas M, Kuppili V, Edla D R, Suri H S, Saba L, Marinhoe R T, Sanches J M and Suri J S 2018 Symtosis: liver ultrasound tissue characterization and risk stratification in optimized deep learning paradigm *Comp. Methods Prog. Biomed.* **155** 165–77

[60] Abadi M, Ashish A, Paul B, Eugene B, Zhifeng C, Craig C and Greg S C 2016 Tensorflow: large-scale machine learning on heterogeneous distributed systems, arXiv:1603.04467

[61] Bergstra J, Breuleux O, Lamblin P, Pascanu R, Delalleau O, Desjardins G, Goodfellow I, Bergeron A, Bengio Y and Kaelbling P 2011 Theano: deep learning on GPUs with Python *NIPS 2011, BigLearning Workshop, (Granada, Spain)* 3

[62] Chollet F *et al* 2015 Keras: deep learning library for Theano and Tensorflow **7** T1

[63] Jia Y, Shelhamer E, Donahue J, Karayev S, Long J, Girshick R, Guadarrama S and Darrell T 2014 Caffe: convolutional architecture for fast feature embedding *Proc. of the 22nd Int. Conf. on Multimedia* (New York: ACM) pp 675–8

[64] Collobert R 2008 Torch *NIPS Workshop on Machine Learning Open Source Software*

[65] https://deeplearning4j.konduit.ai/

[66] Lyons J, Dehzangi A and Heffernan R 2014 Predicting backbone Cα angles and dihedrals from protein sequences by stacked sparse auto-encoder deep neural network *J. Comput. Chem.* **35** 2040–6

[67] Heffernan R, Paliwal K and Lyons J 2015 Improving prediction of secondary structure, local backbone angles, and solvent accessible surface area of proteins by iterative deep learning *Sci. Rep.* **5** 11476

[68] Spencer M, Eickholt J, Cheng J and Deep A 2015 Learning network approach to *ab initio* protein secondary structure prediction *IEEE/ACM Trans. Comput. Biol. Bioinform.* **12** 103–12

[69] Nguyen S P, Shang Y and Xu D 2014 DL-PRO: A novel deep learning method for protein model quality assessment *Int. Joint Conf. on Neural Networks* (Piscataway, NJ: IEEE) pp 2071–8

[70] Baldi P, Brunak S and Frasconi P 1999 Exploiting the past and the future in protein secondary structure prediction *Bioinformatics* **15** 937–46

[71] Baldi P, Pollastri G and Andersen C A 2000 Matching protein beta-sheet partners by feedforward and recurrent neural networks *Proc. of the 2000 Conf. on Intelligent Systems for Molecular Biology (La Jolla, CA)* pp 25–36

[72] Sønderby S K and Winther O 2014 Protein secondary structure prediction with long short term memory networks, arXiv:1412.7828

[73] Lena P D, Nagata K and Baldi P 2012 Deep architectures for protein contact map prediction *Bioinformatics* **28** 2449–57

[74] Baldi P and Pollastri G 2003 The principled design of large-scale recursive neural network architectures–DAG-RNNs and the protein structure prediction problem *J. Mach. Learn. Res.* **4** 575–602

[75] Leung M K, Xiong H Y and Lee L J 2014 Deep learning of the tissue-regulated splicing code *Bioinformatics* **30** i121–9

[76] Lee T and Yoon S 2015 Boosted categorical restricted Boltzmann machine for computational prediction of splice junctions *Int. Conf. on Machine Learning (Lille, France)* pp 2483–92

[77] Zhang S, Zhou J and Hu H 2015 A deep learning framework for modeling structural features of RNA-binding protein targets *Nucl. Acids Res.* **44** gkv1025

[78] Mainak B, Venkatanareshbabu K, Luca S, Damodar R E, Elisa C G, Tato M R and Andrew N 2019 State-of-the-art review on deep learning in medical imaging *Front. Bioscience-Landmark* **24** 380–406

[79] Denas O and Taylor J 2013 Deep modeling of gene expression regulation in an erythropoiesis model *Int. Conf. on Machine Learning Workshop on Representation Learning (Atlanta, GA)*

[80] Asgari E and Mofrad M R 2015 Continuous distributed representation of biological sequences for deep proteomics and genomics *PLoS One* **10** 0141287

[81] Hochreiter S, Heusel M and Obermayer K 2007 Fast model-based protein homology detection without alignment *Bioinformatics* **23** 1728–36

[82] Sønderby S K, Sønderby C K and Nielsen H 2015 Convolutional LSTM networks for subcellular localization of proteins, arXiv:1503.01919

[83] Fakoor R, Ladhak F and Nazi A 2013 Using deep learning to enhance cancer diagnosis and classification *Proc. of the Int. Conf. on Machine Learning*

[84] Freudenburg Z V, Ramsey N F and Wronkeiwicz M 2011 Real-time naive learning of neural correlates in ECoG electrophysiology *Int. J. Mach. Learn. Comput.* **1** 269

[85] An X, Kuang D, Guo X and Deep A 2014 Learning method for classification of EEG data based on motor imagery *Intelligent Computing in Bioinformatics* (Berlin: Springer) pp 2013–210

[86] Li K, Li X and Zhang Y 2013 Affective state recognition from EEG with deep belief networks *Int. Conf. on Bioinformatics and Biomedicine* (Piscataway, NJ: IEEE) pp 305–10

[87] Huanhuan M and Yue Z 2014 Classification of electrocardiogram signals with deep belief networks *IEEE 17th Int. Conf. on Computational Science and Engineering* pp 7–12

[88] Wulsin D, Gupta J and Mani R 2011 Modeling electroencephalography waveforms with semisupervised deep belief nets: fast classification and anomaly measurement *J. Neural Eng.* **8** 036015

[89] Turner J, Page A and Mohsenin T 2014 Deep belief networks used on high resolution multichannel electroencephalography data for seizure detection *2014 AAAI Spring Symp. Series*

# Part II

## Deep learning in brain imaging

**IOP** Publishing

# Multimodality Imaging, Volume 1
### Deep learning applications
**Mainak Biswas and Jasjit S Suri**

# Chapter 3

# A review of artificial intelligence in brain tumor classification and segmentation

**Mainak Biswas and Jasjit S Suri**

The Asian continent sees the highest number of fatalities due to cancer in the brain or the central nervous system according to the World Health Organization (WHO). Brain cancer is curable through prompt diagnosis and proper treatment if detected in its earliest stage. One of the critical aspects of a brain cancer treatment plan is the proper and accurate grading of the type of tumor. The normal grading process involves the surgically invasive procedure of biopsy which is risky, expensive, and time-consuming. Therefore, there is a necessity for the development of automated non-invasive methods for diagnosing brain tumors. This review addresses three areas. The first discusses brain cancer pathophysiology. The second covers the different imaging modalities used in scanning the brain, i.e. magnetic resonance (MR) imaging, computed tomography (CT) etc. At the end this review chapter discusses the existing automated techniques in brain cancer diagnosis and grading, specifically in the area of deep learning (DL). Furthermore, this research compares the association between other brain anomalies such as Parkinson's disease and Alzheimer's disease in the perspective of medical imaging and automated approaches such as ML and DL.

## 3.1 Introduction

In Asia, the brain cancer mortality is the highest in the world, according to a World Health Organization (WHO) report of 2018 [1]. Headaches, mood swings, slurred speech, seizures, and memory loss are all signs of brain cancer. The brain tumor is the most commonly type of cancer detected in the central nervous system (the brain and spinal cord) [2]. The origin, rate of growth, and stage of progression of brain tumors are well documented and classified into multiple categories [3]. Brain tumors are categorized into two groups based on their origin: if it appears directly in brain, i.e. primary, and if it appears due to cancer occurrence in other organs such as the stomach and lungs, i.e. secondary or metastasis. Based on their pace of growth the brain tumors are classified into benign and malignant. The benign cells have slow progression, are generally non-invasive to healthy cells, and have distinct boundaries, e.g. pituitary tumors and astrocytomas. On the other hand, malignant cells actively attack neighboring healthy cells in the central nervous system, are rapidly progressing, and have fuzzy boundaries (also known as oligodendrogliomas or high-grade astrocytomas). The WHO grades brain tumors into four classes (grades I–IV) based on the rate of growth [2, 5, 9]. They are also categorized based on the progression stage (stages 0–4). At stage 0, the tumor cells are abnormal but do not attack nearby cells. In the advanced stages (1, 2, and 3) the cancer cells spreads rapidly. In stage 4 the cancer spreads all over the body. In this stage survivability rates are negligible and therefore it is prerogative to detect a cancer in the early stages.

The standard procedure of cancer detection is biopsy, where a small invasive surgery is carried out in the brain or spinal cord to collect a tumor specimen for examination in the lab for diagnosis. Non-invasive procedures involve a check-up of the body and the study of brain scans using imaging techniques, i.e. computed tomography (CT), magnetic resonance (MR), and their variants. The study of brain scans forms an important part of surgical planning. Radiologists provide their conclusive findings on shape, size, progression, and other critical aspects [10], and also carry out inter-variability tests before the actual surgery [11]. The decreasing cost of computer hardware and new innovations in artificial intelligence have given rise to the growth of computer-assisted tools (CAT) in the domain of medical diagnosis. CATs are increasingly based on smart learning techniques, such as machine learning (ML) and deep learning (DL). In the next few sections we will look into various technologies developed in the area of brain tumor diagnosis, segmentation, characterization, and grading using such methods.

The structure of the chapter is as follows: brain tumor pathophysiology is discussed in section 3.2, imaging modalities in section 3.3, tumor grading in section 3.4, tests in section 3.5, characterization methods in section 3.6, different brain anomalies in section 3.7, and the discussion and conclusion are in section 3.8.

## 3.2 Brain cancer pathophysiology

The brain cancer pathophysiology is covered in the next two subsections. Section 3.2.1 discusses the cellular level architecture and section 3.2.2 discusses the link between brain tumor and genes.

### 3.2.1 Architecture at the cellular level

The basic unit of the human body is the cell, performing all of the organ's essential activities such as waste material management, oxygen flow, and so on. There are 23 pairs of chromosomes in each cell's nucleus, containing millions of genes. Each gene's instructions are encoded in its DNA structure, which determines how it behaves. The gene's protein serves as a messenger between cells as well as within them. The 3D structure of the communicated message represents the gene [12, 13]. Genes play a crucial role in the replication of healthy cells as well as the demise of worn-out cells. Any abnormality in this system's operation can result in cancer. Any change in DNA sequence due to mutation can cause gene dysfunction, which is the primary cause of the uncontrolled cell development that eventually leads to cancer.

The responsible genes are classified into three groups. Each group is defined as follows:

a. *Tumor suppressors* are the class of genes that control the cell death cycle, i.e. apoptosis [14]. Two signaling routes exist. The first channel instructs the cell to kill itself, whereas the second pathway instructs the cell to receive death signals from nearby cells. If a mutation slows down one of these routes, cell death slows down and eventually stops; if the mutation occurs in both pathways, uncontrolled cell proliferation occurs [14, 15]. The most common tumor suppressor genes are RB1, PTEN etc [16].

b. *Repair genes* restore the DNA. Any fault in their working will lead to cancer. The most common repair genes are MGMT and p53 protein.

c. *Proto-oncogenes* are responsible for the cell division and growth process, and thus inhibit the standard cell death cycle [17, 18]. Protein signals are used by proto-oncogenes to govern the cell division cycle. These signals follow a pathway known as a signal transduction cascade, which constitutes a sequence of steps. Figure 3.1 depicts this pathway. The signal is carried from the cell to the nucleolus

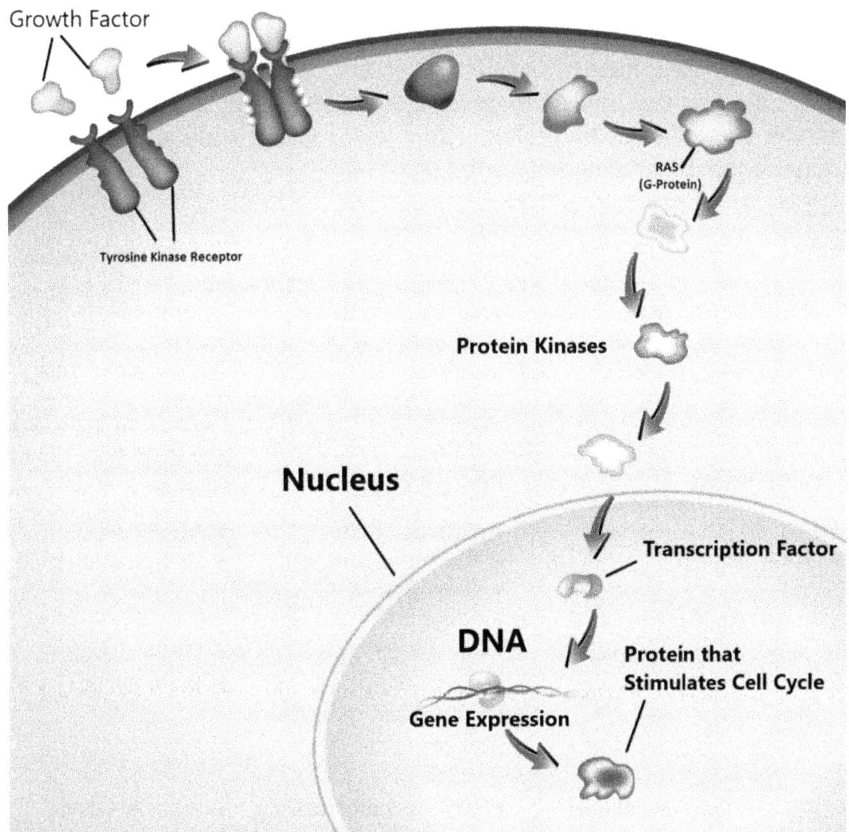


**Figure 3.1.** Cell cycle proliferation. (Image courtesy of AtheroPoint™. Reproduced with permission from [109].)

through the cytoplasm by proteins. When the cell nucleus receives the signal, it activates the transcription genes necessary for cell division. RAS is a proto-oncogene that toggles between 'on' and 'off' during cell division [19]. Any mutation may lead to a change to the oncogene, where it might not be able to turn 'off' its cell division process, leading to unstoppable growth and finally cancer.

If a cancer follows from any of these three causes in the central nervous system, then it is a primary tumor, which attacks its neighboring cells. However, if the cancer begins through the blood vessels, then it is a secondary tumor/metastasis [20]. Metastasis tumors require oxygen, nutrients, and a blood supply to grow, which is detected by some genes that are responsible for establishing a vascular network for it, a process which is called angiogenesis [21]. The details of the genes and their functionality are provided in table 3.1.

Around 15% of cancers worldwide are caused by viruses, through infecting the cells and mutating the DNA, converting proto-oncogenes to oncogenes [22]. Viruses are responsible for 15% of malignancies, yet only a few have been identified. In addition, these viruses are divided into two classes based on the type of gene infection they cause: DNA viruses and RNA viruses. Epstein–Barr, hepatitis B, human herpes virus, and human papillomavirus are the four cancer-causing DNA viruses. Hepatitis C and human T lymphotropic type 1 virus are RNA cancer-causing viruses. Other environmental and behavioral factors, such as x-rays, UV radiation, tobacco use, pollution, and daily use goods that are carcinogenic agents can also cause cancer. For example, sunlight can cause cancer by mutating suppressor genes, while tobacco smoke can cause lung cancer by altering lung cells [23].

**Table 3.1.** The significance of genomics in brain tumors. RB1: retinoblastoma 1; EGFR: epidermal growth factor receptor; 1p and 19: co-deletion; ATRX: $\alpha$-thalassemia-mental X-linked mental disability syndrome; GBM: glioblastoma; RKT: receptor tyrosine kinase; BRAF: B-Raf proto-oncogene; TP53(p53): tumor protein 53; PTEN: phosphatase and tensin homolog; HGG: high-grade glioma; IDH1/DH2: isocitrate dehydrogenase 1/2; MGMT: O6-methylguanine DNA methyltransferase. (Reproduced with permission from [109].)

| Gene | Task | Effect of mutation | Connection between genes and brain tumor (mutation degree) |
|---|---|---|---|
| TP53 (p53) [26] | DNA repair Initiating apoptosis | • Genetic instability<br>• Reduced apoptosis<br>• Angiogenesis | • More relevant to HGG<br>• Brain tumor (80%) |
| RB1 [26] | Tumor suppressor | • Blocks cell cycle progression<br>• Unchecked progression of cell cycle | • GBM relevant<br>• Brain tumor (75%) |
| EGFR [27] | Trans-membrane receptor in RTK | • Increased proliferation<br>• Increased tumor cell survival | • Primary GBM (approx. 40%) |

| PTEN [27] | Tumor suppressor | • Increased proliferation<br>• Reduced cell death | cell • Primary GBM (15–40%)<br>• GBM (up to 80%) |
|---|---|---|---|
| IDH1 and DH2 [28] | Control citric acid cycle | • Inhibits the function of enzymes | **IDH1**<br>• Primary GBM (5%)<br>• GBM grade II–III (70%–80%)<br>• IDH1 longer survival<br><br>**IDH2**<br>• Relevant to oligodendroglial tumors |
| 1p and 19q [29] | Prognosis of the disease or treatment assessment | • Poor prognosis | • Oligodendrogliomas (80%)<br>• Anaplastic oligodendrogliomas (60%)<br>• Oligoastrocytomas (30%–50%)<br>• Anaplastic oligoastrocytomas (20%–30%) |
| MGMT [30] | DNA repair predicts patient survival | • Cell proliferation | • GBM (35%–75%) |
| BRAF [26] | Proto-oncogene | • Cell proliferation<br>• Apoptosis | • Pilocyticastrocytomas (65%–80%)<br>• Pleomorphic xanthoastrocytomas and gangliogliomas (25%) |
| ATRX [26] | Genomic repeats deposition | • Genital anomalies<br>• Hypotonia<br>• Intellectual disability<br>• Mild-to-moderate anemia<br>• Secondary to α-thalassemi | • Relevant to oligodendroglial tumors |

The tumor cells have distinctive molecular uniqueness at different stages [24]. Metaplasia, hyperplasia, anaplasia, dysplasia, and neoplasia are the different phases. Aberrant growth begins during hyperplasia, yet the cells still appear regular. The cells begin to seem aberrant during metaplasia. During anaplasia the cells lose their distinguishing morphological traits and during dysplasia the cells become aberrant and aggressive. The anaplasia is the main risky stage, in which tumor cells quickly infect nearby healthy cells or enter the bloodstream, resulting in metastasis [25]. These physical changes of cells can be visualized through the high resolution imaging modalities discussed in the next section.

### 3.2.2 Links between brain tumors and genes

In the previous section we discussed the categories of brain tumor and the degree of gene mutation, specified in table 3.1. The protein TP53 is occupied by the repair of DNA and the initiation of apoptosis. A high-grade glioma brain tumor has an abnormal TP53 level. Also, mutations are found in 80% of tumors. The tumor suppression gene retinoblastoma (RB1) mutation is common in 75% of brain tumors with respect to glioblastoma [26]. The EGFR gene belongs to the tyrosine kinase family and is a trans-membrane receptor. The mutation in EGFR leads to augmented cell cycle propagation and tumor survivability. EGFR is associated with glioblastomas and 40% of mutations within it. Similarly, PTEN, a tumor suppressor gene, accounts for 15%–40% of mutations within glioblastomas [27] and maybe up to 80%. Mutations in enzymes IDH1 and IDH2, which manage the citric acid cycle, inhibits their activity.

The IDH1 mutation is more common in high-grade glioblastomas (70%–80%) than primary glioblastomas (5%). IDH2 mutations are more common within oligodendroglial tumors [28]. The deletion of the 1p and 19q chromosomes, which are helpful in treatment and prognosis assessment, indicates a tumor of oligoden-droglial nature and is more commonly seen in oligodendrogliomas (80%), oligoas-trocytomas (30%–50%), anaplastic oligodendrogliomas (60%), and anaplastic oligoastrocytomas (20%–30%) [29]. An abnormality 35%–75% in MGMT protein (responsible for DNA repair), leads to glioblastoma tumors [30]. Another proto-oncogene is BRAF, which is concerned in cell propagation, treatment assessment, and apoptosis. Mutation in BRAF leads to pleomorphic xanthoastrocytomas (80%), pilocyticastrocytomas (65%–80%), and gangliogliomas (about 25%) [26]. Another gene $\alpha$-thalassemia X-linked mental disability syndrome (ATRX) encodes a protein that is involved with TP53 and IDH1 mutation. It is generally used as prognostic indication of the presence of IDH1 mutation, and also used to distinguish between oligodendroglial tumors [26].

## 3.3 Imaging modality

MR, CT, x-ray, and ultrasound imaging are non-invasive ways to visualize the internal organs and study their behavior without invasive surgery. Different imaging modalities have different effects in terms of the detection, diagnosis, and grading of a tumor. The image scans also help in making decisions on treatment response, patient diagnosis, and surgery planning. Functional and structural imaging are the two techniques of visualizing the cancer behavior [31, 32]. Structural imaging concerns the structure, location, injuries, and other anomalies related to a brain tumor. Functional imaging is related to the detection of metabolic changes, lesions, and brain activities, which is made possible as the changes are reflected on the scan plates. The two general imaging modalities related to brain tumors are CT and MR [33, 34].

### 3.3.1 Computed tomography imaging

The CT scans are advised by doctors to study various anomalies of the brain such as tumors, hemorrhages, and blood clots. A CT scan uses x-rays concentrated around a specific part of the skull, and captures a sequence of images from different angles. Each angle is used to create a 2D cross-section image, which are later combined to provide a 3D image of the infected organ. In another variation of CT, called the positron emission tomography (PET), contrast agents are introduced into the body and then CT is performed to highlight the infected organs. One of the problems of CT scans is the ionizing radiation, which harms healthy cells and can cause cancer. In a study it was shown that the radiation from a CT scan is hundred of times stronger than a normal x-ray [35].

### 3.3.2 Magnetic resonance imaging

MRI is a better alternative to CT since it is radiation free. MRI uses the axial, sagittal, and coronal planes to image the brain. Due to its higher contrast, MRI produces clearer images of the vascular autonomy and central nervous system. The different variations of MRI are T1-weighted, T2-weighted, and FLAIR [36]. T1-weighted MRI produces high quality contrast images distinguishing the gray and white matter of the brain. T2-weighted MRI is water content sensitive. Therefore T2-weighted MR is used to capture images of brain diseases where there is water accumulation within the tissues. Both T1-weighted and T2-weighted are used to detect cerebrospinal fluid, where it appears dark in the former and brighter in the latter. FLAIR is similar to T2-weighted, except for the acquisition protocol. FLAIR is generally used to distinguish between brain and cerebrospinal fluid (CSF) anomalies. FLAIR locates an edema region from attenuating water signals and thereby the periventricular hyperintense lesion becomes clearly visible.

A visual comparison of the three methods is provided in figure 3.2. Another type of MRI sequence is diffusion weighted imaging (DWI) for detecting the random movement of water molecules inside the brain. DWI is indicative of acute stroke since it is able to produce a bright signal when water movement becomes restricted. For blood flow detection perfusion-weighted MRI (PWI) is commonly used. Diffusion-tensor MRI (DTMRI) uses a microscopic image to detect water molecules, which aids in the surgical removal of a brain tumor. The brain is occasionally more active in some areas than others, consuming more oxygen as a result. By noting the neuronal action of the brain and monitoring the oxygenation level in the brain, functional MRI (FMRI) can correlate these mental processes and find these continuous activities [38]. As MRI's motion artifact effect is inferior to CT's, it is unable to identify brain damage and acute haemorrhage with precision. When compared to other imaging modalities, MRI takes longer to acquire data.

### 3.3.3 Biopsy

The standard gold standard protocol to diagnose brain tumor is biopsy. The cell nucleus of the tissue sample is examined for color, shape, and size. Hematoxylin and

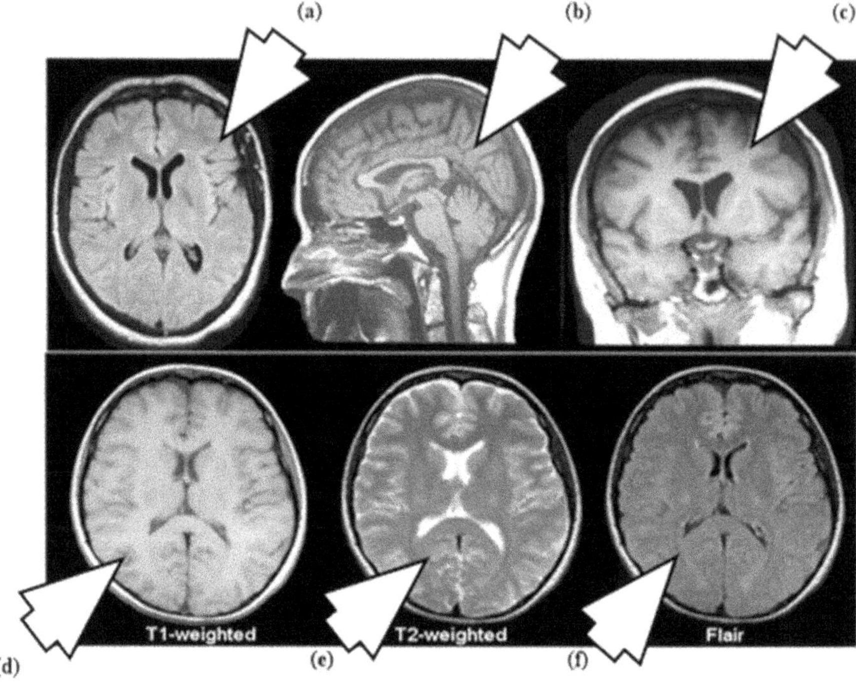



**Figure 3.2.** (a) Axial, (b) sagittal, and (c) coronal views, and (d) T1, (e) T2, and (f) FLAIR images of MRI. Reproduced with permission from http://casemed.case.edu/clerkships/neurology/webpercent20neurorad/mri-percent20basics.htm.

eosin staining is employed for biopsy sample analysis in this case. Cytopathology is a branch of pathology that studies the structure, function, and chemistry of cells. To better comprehend tissue proteins, immunofluorescence imaging is performed. Human examination of tiny cells is time-consuming and error-prone, thus computing tools are needed for scalability and precision. Better outcomes can be achieved by combining the pathologist's experience and expertise with computerized pathological picture analysis [39].

### 3.3.4 Hyperstereoscopy imaging

A brain tumor in the advanced stages actively invades neighboring cells and is increasingly complex to differentiate using the naked eye. Inaccuracy in a resection procedure can harm healthy tissues leading to a lower survivability rate amongst patients [40, 41]. Hyperspectral imaging (HSI) is used in this regard as its invasiveness is negligible, it uses non-ionizing sensing, and can examine color variation in a wider range. This allows HSI to capture more distinguishable features of the affected area [41, 42].

Intelligence based methods in combination with HSI have been used to delineate tumor boundaries in order to assist surgeons [40]. These models are a combination of supervised as well as unsupervised algorithms. The supervised algorithms consist of an SVM and kNN, and the unsupervised algorithm consists of $k$-means clustering. These algorithms perform pixel-to-pixel classification between normal and cancerous tissues, and blood vessels. The process is divided into training and testing phases. In the training stage, the intelligent model is taught using labeled samples. In the testing stage, the trained model is tested on unlabeled samples. The HIS images are obtained directly for training and testing purposes. The SVM and kNN are trained to generate a supervised classification map and a spatial–spectral map, respectively. Both these maps are fused together. Another map is produced using $k$-means clustering. Further, majority voting is performed to fuse all the maps. Principle component analysis (PCA) was performed in the method for dimensionality reduction.

In another work the application of an intelligence based deep learning (DL) model was based on the hyperspectral paradigm [43]. Data from 50 patients were used to train/test the CNN network (six layers of convolution + three fully connected layers), which included head and neck tissue, squamous-cell carcinoma, and thyroid cancer. A batch size of 250 was used to train the DL network for 25 000 iterations. The system achieved accuracy, specificity, and sensitivity of 80%, 78%, and 81%, respectively. This was better than all contemporary models.

### 3.3.5 MR spectroscopy

The detection of metabolic changes in the brain is the focus of magnetic resonance spectroscopy (MRS). Each tumor type is linked to a different set of metabolites. This phenomenon is beneficial for identifying tumors, strokes, and epilepsy in brain tissue. Amino acids, lactate, alanine, lipids, and other metabolites have distinct signatures that are utilized to characterize them [44]. As a result, each tumor grade has its own distinct hallmarks. The neurologist uses these changes to distinguish between malignant and healthy tissue by mapping the frequency of each metabolite. MRS images were used for training/testing a deep learning model for brain tumor diagnosis [45]. Three classes of normal, low-grade, and high-grade tumor samples were used. In a different study, the authors used brain tumor grading models with MRS [46]. The study showed that superior accuracy can be achieved using short and long echo times (TEs). Another ML based model was used for classifying tumors into benign and malignant [47]. Four ML models, an SVM, random forest, local weighted learning (LWL), and multi-layer perceptron, were used. The random forest AUC was 0.91 and LWL sensitivity was 86.1%.

Sometimes, different imaging modalities are combined to reduce the redundancy and randomness of individual medical data. Statistical models are applied on these fused dataset to obtain better diagnosis. The image modalities are fused based on different parameters such as morphology, wavelets, fuzzy logic, and different knowledge areas [48].

## 3.4 Guidelines for tumor grading by the WHO

A series of guidelines was supplied by the WHO with regards to the grading of brain tumors. Medical practitioners use these guidelines to prepare treatment plans and increase the survivability rate of patients. Since 1979 there have been five amendments which are listed in table 3.2. The 1979 amendment included mitotic activity,

**Table 3.2.** WHO recommendations on tumor assessment in different editions. (Reproduced with permission from [109].)

| Edition | Year | Recommended parameters for tumor assessment |
| --- | --- | --- |
| I | 1979 | Miotic activity, necrosis and infiltration |
| II | 1993 | Immunohistochemistry (IHC) |
| III | 2000 | Genetic profile |
| IV | 2007 | Genetic profile and histological variation |
| V | 2016 | Molecular features and histology |

infiltration, and necrosis for tumor grading. The 1993 amendment included immune histochemistry, the 2000 amendment included genetic profile, and the 2007 amendment included histological variation and genetic profile. Finally, the 2016 amendment defined the tumor cells precisely which helped in better characterization [49].

Studies have conclusively proven that the tumor grades and groups can be precisely defined by molecular signatures [50]. The four fundamental parameters for grading of the WHO are microvascular proliferation, necrosis, nuclear atypia, and mitoses [51]. The grading is assigned from grade I–IV, in terms of increasing aggression [49–52], with grade I being the most normal looking and slow spreading. Grade 2 tumor cells are abnormal, progress gradually, and may invade neighboring healthy cells. Grade I and grade II cells can be healed by treatment. Grade III cells readily attack neighboring healthy cells. Grade IV cells are the most aggressive and grow in size faster than those of the other grades. The grading of cells microscopically is extremely difficult and challenging to a pathologist due to their fuzzy structure and boundaries

## 3.5 Brain tumor tests

During the medical examination of a patient, the doctor checks for vision, hearing, muscle strength, reflexes, and alertness. The eyes may also be checked to look for signs of swelling. After that, scans, biopsy, and biomarkers are used to confirm and grade the cancer. Based on the results and symptoms, the physician will declare the malignancy of the tumor. The details of the tests are given in the following sections.

### 3.5.1 Biomarker test

As the mutation of genes is the root cause of cancer, the degree of mutation is calculated through biomarker tests. The genes responsible are already discussed in table 3.1. The biomarker tests confirm or decide the type and grade of cancer, which may help in estimating the tumor growth, treatment therapy, and also the response [53].

### 3.5.2 Biopsy

The fundamental test for brain cancer diagnosis is biopsy [54]. It is an intrusive operation. During surgery, a section of the brain tumor is taken. A number of laboratory tests are performed on the biopsy sample to determine the grade based on

**Table 3.3.** Overview of some open challenges in digital pathology image analysis worldwide. (Reproduced with permission from [109].)

| Year | Challenges | References |
|---|---|---|
| 2012 | ICPR Mitosis Detection Competition | http://www.icpr2012.org/contests.html |
| 2012 | EM Segmentation Challenge 2012, 2D segmentation of neuronal processes | https://imagej.net/ Segmentation_of_neuronal_structures_in_ EM_stacks_challenge_-_ISBI_2012 |
| 2013 | MICCAI Grand Challenge on Mitosis Detection | https://warwick.ac.uk/fac/sci/dcs/research/tia/ glascontest/ |
| 2014 | MICCAI Brain Tumor Digital Pathology Challenge | |
| 2014 | MICCAI Brain Tumor Digital Pathology Challenge | |
| 2015 | MICCAI Gland Segmentation Challenge Contest | |
| 2016 | Tumor Proliferation Assessment Challenge 2016 | https://tupac.grand-challenge.org/ |
| 2017 | CAMELYON17 Challenge | https://camelyon17.grand-challenge.org/ |
| 2018 | Medical Imaging with Deep Learning (MIDL-2018) | https://2018.midl.io/ |

cellular patterns. It is difficult to make an accurate diagnosis because low and high-grade gliomas have similar structures and are difficult to identify. A proper diagnosis of the tumor's grade and type leads to an appropriate treatment plan. As these studies are so essential, they are submitted to inter-reader variability testing based on histopathologists' knowledge [55].

Human interpretation of the grade of tumor is challenging since it is difficult to ascertain using the naked eye. The use of computer aided tools can partially circumvent these shortcomings [56]. There are several intermediate steps for computerized image analysis, i.e. registration, pre-processing, feature extraction, ROI identification, segmentation, and characterization. The challenges for computer-assisted diagnosis are the paucity of images for training, large sized single images, and complex clinical features. The Association of Medical Image Computing and Computer Assisted Intervention (MICCAI) offers open challenges and open conferences on digital histopathology image analysis in order to promote the advancement of computer-assisted tools. Table 3.3 lists a few of these challenges.

### 3.5.3 Imaging test

The different non-invasive imaging modalities are CT, MRI, PET, and SPECT. The most commonly used modality for confirmation of the tumor diagnosis is MRI, due to its good contrast and lack of radiation. The challenges of automated tumor diagnosis are overlapping intensities, inconsistencies in shape, size, and orientation, noise, and low contrast [57]. The different open challenges are displayed in table 3.4. In the next few sections different intelligence based methods such as ML and DL are proposed for brain tumor image characterization, detection, and segmentation.

**Table 3.4.** Overview of open challenges for brain image analysis worldwide. (Reproduced with permission from [109].)

| Challenge | Objective | Modality | References |
|---|---|---|---|
| BraTS 2012 | Segmentation of brain tumors. | MRI | http://www2.imm.dtu.dk/projects/BRATS2012/index.html |
| BraTS 2013 | Segmentation of brain tumors. | MRI | https://sites.google.com/site/miccaibrats2014 |
| BraTS 2014 | Segmentation of brain tumors. | MRI | https://sites.google.com/site/miccaibrats2014/ |
| BraTS 2015 | Segmentation of brain tumors. | MRI | https://sites.google.com/site/braintumorsegmentation/home/brats2015 |
| BraTS 2016 | Quantify longitudinal changes: compare the volumetric changes between any two time points for accuracy. | MRI | https://sites.google.com/site/braintumorsegmentation/home/brats_2016 |
| BraTS 2017 | Glioma segmentation in pre-operative imaging. Patient overall survival (OS) prediction from pre-operative scans. | MRI | http://www.miccai2017.org/ |
| BraTS 2018 | Glioma segmentation in pre-operative MRI scans. Prediction of patient overall survival (OS) from pre-operative scans. | MRI | https://www.med.upenn.edu/sbia/brats2018.html |
| MICCAI 2018 | Multi-sequence brain MR images with and without (large) diseases, the segmentation of gray matter, white matter, cerebrospinal fluid, and other structures. To assess the effect of (large) pathologies on segmentation and volumetry. | MRI | http://mrbrains18.isi.uu.nl/ |
| HC-18 | To create an algorithm that can measure the fetal head circumference automatically using a 2D ultrasound image. | Ultrasound images | https://hc18.grand-challenge.org/ |

## 3.6 Classification methods

The definition of ML can be summarized as follows: the machine performance improves with experience for a given a task [58]. The learning procedure is categorized into two groups: supervised and unsupervised [59, 61]. The supervised model learns from labeled samples, while the unsupervised model tries to understand the relationships between data from unlabeled instances. Machine learning models are a two-stage process while deep learning models have a single stage. The two stages of ML models are feature extraction and statistical model application. The process model of ML is shown in figure 3.3.

The feature extraction stage employs purpose-built mathematical models to extract features based on texture, contrast, brightness, and other image properties. Different features from diverse extraction algorithms are combined to make for more effective characterization. If the number of features is huge, dimension reduction algorithms are applied to use only the most relevant features. The different statistical models of learning are k-nearest neighbors (kNN) [63], support vector machines (SVMs) [64], artificial neural networks (ANNs) [65] etc. kNN is a lazy learning classification algorithm which assigns the most frequent class of its k-nearest neighbors to the unlabeled instance. An SVM draws the largest separating hyperplane between two classes. If the classes are not separable, it uses kernels to map the classes to higher dimensions where they are linearly separable. ANNs are hierarchical networks of nodes that are capable of learning. There are different variations of ANN learning models based on the hidden layer numbers, activation function, architecture etc. Some of the most common ANN based models are extreme learning machines (ELMs) [66], recurrent neural networks (RNNs) [67], restricted Boltzmann machines (RBMs) [68], etc. ELMs are single-hidden-layer feed-forward neural networks, while RNNs apply a feedback mechanism and RBNs use stochastic learning rules.

DL networks are an adaptation of ANNs with a huge number of hidden layers. Since the number of layers is greater, the network learns more complex relationship within the data. The network in this case is capable of high performance and

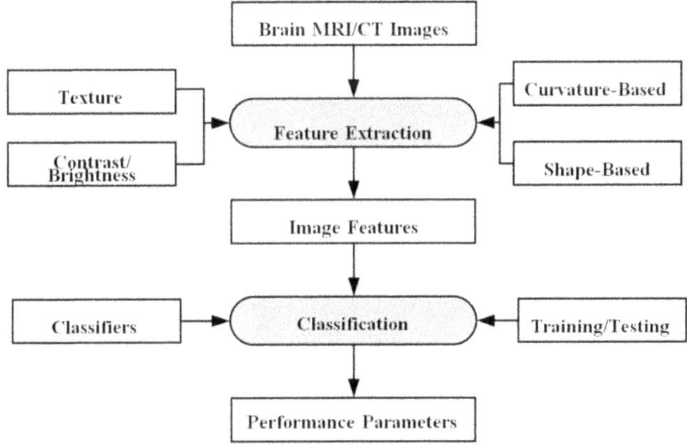


**Figure 3.3.** Working of ML based systems. (Reproduced with permission from [109].)

robustness. These networks are called deep learning models [69]. The most common DL models are CNNs [70], autoencoders [71], and deep belief networks [72].

### 3.6.1 Machine learning

The most common ML models are kNN, SVMs, decision trees, naïve-Bayes classifiers,, random forest etc. These techniques have been used previously in the analysis of brain images either individually or in combination, and are discussed in table 3.4. Some of the applications are discussed next.

#### 3.6.1.1 ANN based MRI brain tumor classification using genetic features
An ANN based characterization of brain images was proposed to classify the brain tumor into three groups: normal, benign, and malignant [57], as shown in figure 3.4.

A dataset of 100 MR images were collected (35 normal + 35 benign + 30 malignant). The ROI was extracted using a semi-automated process. A feature extraction algorithm based on wavelets was applied for the features. Three feature selection algorithms were applied, i.e. genetic, PCA, and sequential algorithms. Finally, a three-layer ANN was applied for characterization. It can be seen that the 29 features extracted by the genetic feature selection algorithm gave the highest accuracy at 98%. The process model is shown in figure 3.5.

#### 3.6.1.2 A brain tumor hybrid characterization system
Two separate ML models were used, i.e. kNN and ANN, for characterization of brain MR images [73]. For this purpose, 70 MR images were collected (60 diseased and 10 normal). DWT and PCA were used for feature extraction [74] and selection [75], respectively. The ANN algorithm used back-propagation learning for updating the network parameters [76]. The ANN performed with an accuracy of 97% and the kNN performed with an accuracy of 98%. The process model is shown in figure 3.6.

#### 3.6.1.3 An AI based brain tumor characterization and grading system
In [77] brain tumor characterization was proposed using general MR images and rCBV maps obtained from perfusion MRI. The model characterizes gliomas into three grades (II, III, and IV). For this purpose 102 MR images were used. On the

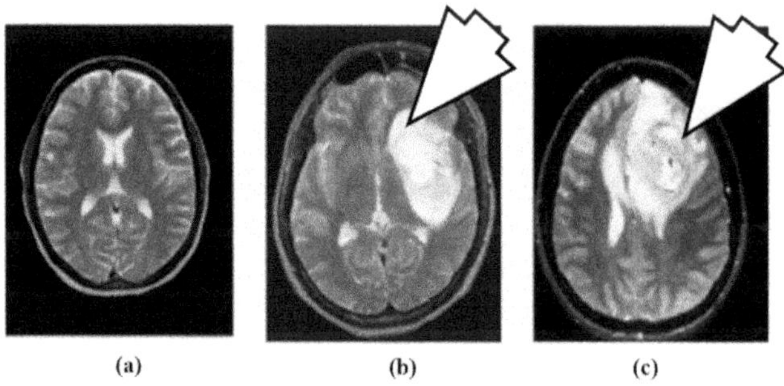

(a)        (b)        (c)

**Figure 3.4.** Brain MR images: (a) normal brain, (b) benign tumor (seven o'clock arrow), and (c) malignant tumor (seven o'clock arrow). Reproduced with permission from [57].

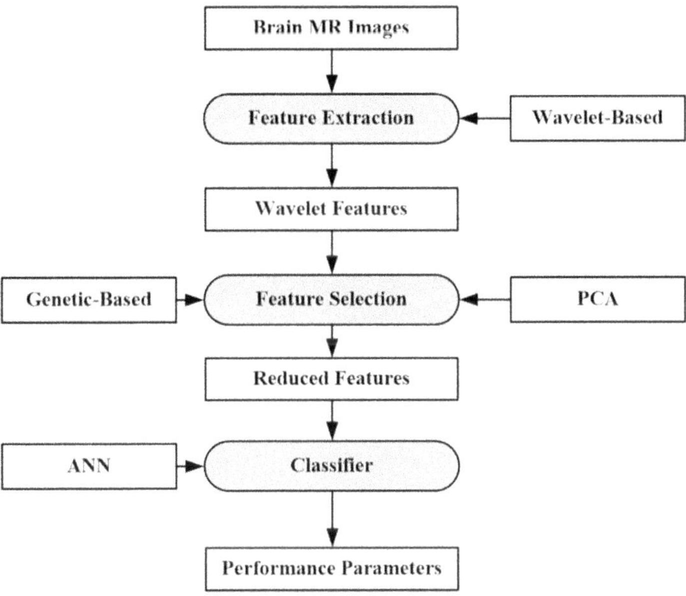


**Figure 3.5.** Process model of the ANN based classification model. (Reproduced with permission from [57].)

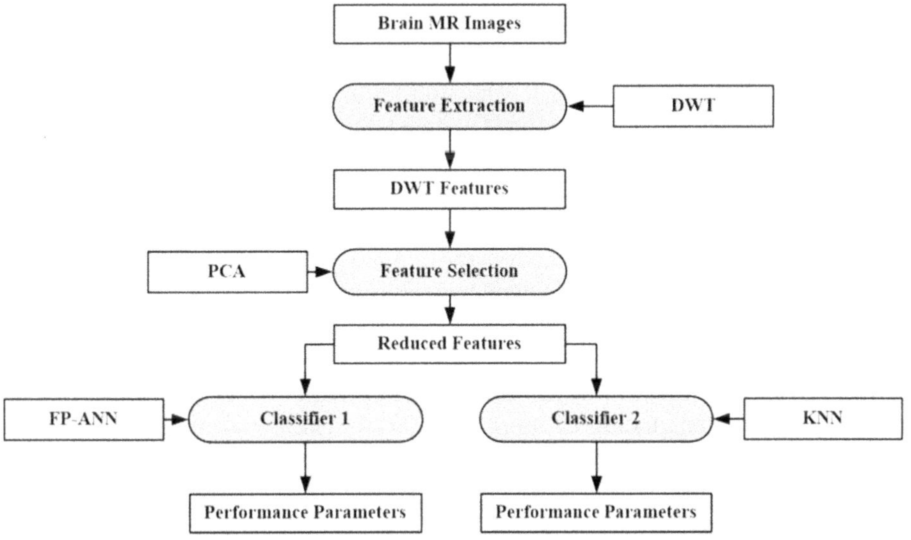


**Figure 3.6.** Hybrid characterization system for brain cancer characterization. (Reproduced with permission from [73].)

same dataset, linear discriminant analysis (LDA) based on principle component regression had been applied previously [78], with limited accuracy. In this work the authors extracted features using tumor shape, image intensity, and Gabor features. Thereafter, ranking-based and SVM-recursive feature selection was applied to

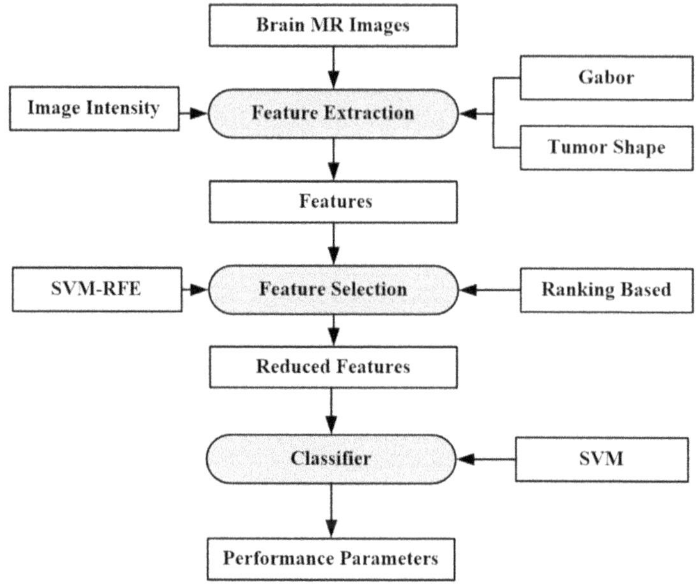


**Figure 3.7.** Process model using an SVM based grade estimation method. (Reproduced with permission from [77].)

choose the important features. Thereafter, an SVM was applied for characterization. For high-grade gliomas and low-grade gliomas, the accuracies were 91.7% and 90.0%, respectively. An accuracy of about 97.8% was obtained for characterizing grade II gliomas from metastasis. The minimum accuracy was obtained differentiating between grade II and grade III gliomas (acc: 75%). The model is shown in figure 3.7.

### 3.6.1.4 Brain neoplasm multi-parametric tissue characterization system

In this current work, ML based models were applied for characterization of neoplastic tissue, healthy tissue, tumor, and edema components [79]. In this regard, voxel-wise intensity features were extracted from the data, which were taken from 14 patients. The SVM and Bayesian classifiers was used for the purpose [80–82]. The Bayesian classifier had the highest accuracy in classifying edema (97.03%), enhancing neoplasm (96.39%), and non-enhancing neoplasm (93.05%). The SVM obtained the highest accuracy for characterizing CSF (91.34%). The process model is given in figure 3.8.

### 3.6.1.5 Extreme learning machine

This section discusses the extreme learning machine (ELM), which is lightweight when compared to more computation intensive algorithms. The ELM is a single layer feedforward neural network. The input-to-hidden layers weights are initialized randomly. The hidden-to-output layer weights are computed using the Moore–Penrose inverse form [82]. The nature of this network allows the weights to be trained in a single pass, thereby reducing the complexity and the time required, and increasing accuracy [83, 84]. A model of the ELM is shown in figure 3.9.

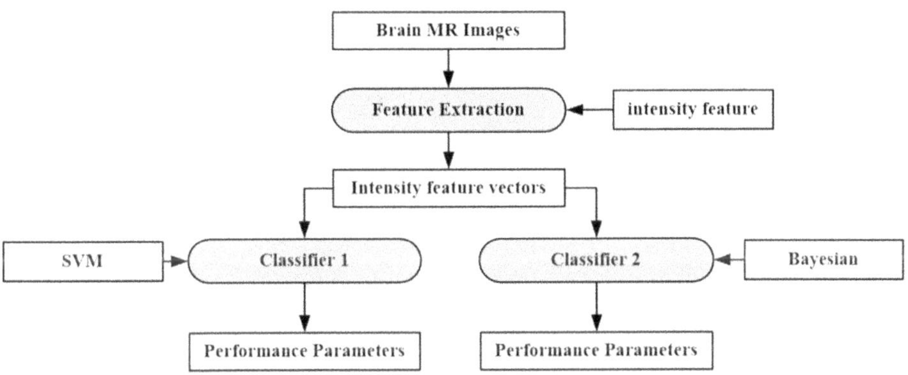


**Figure 3.8.** Process model of SVM based grade estimation method. (Reproduced with permission from [77].)

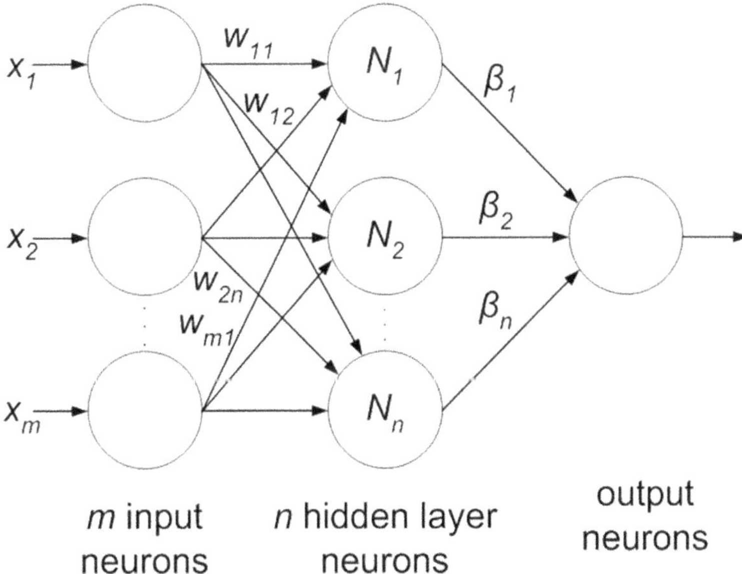


**Figure 3.9.** Extreme learning machine. (Image courtesy of AtheroPoint™. Reproduced with permission from [109].)

### 3.6.2 Deep learning

The advent of DL has revolutionized the medical imaging industry. DL has a single stage, unlike ML. It is able to extract features directly from the images and characterize them in a single stage. The CNN is the most popular amongst all the DL models and is used for both characterization and segmentation. The CNN uses three distinctive operations repeatedly to extract features: convolution, ReLu [85], and pooling (described in previous chapters). At the end it appends fully connected layers for characterization or up-sampling layers for segmentation. A CNN network with a fully connected network is shown in figure 3.10.

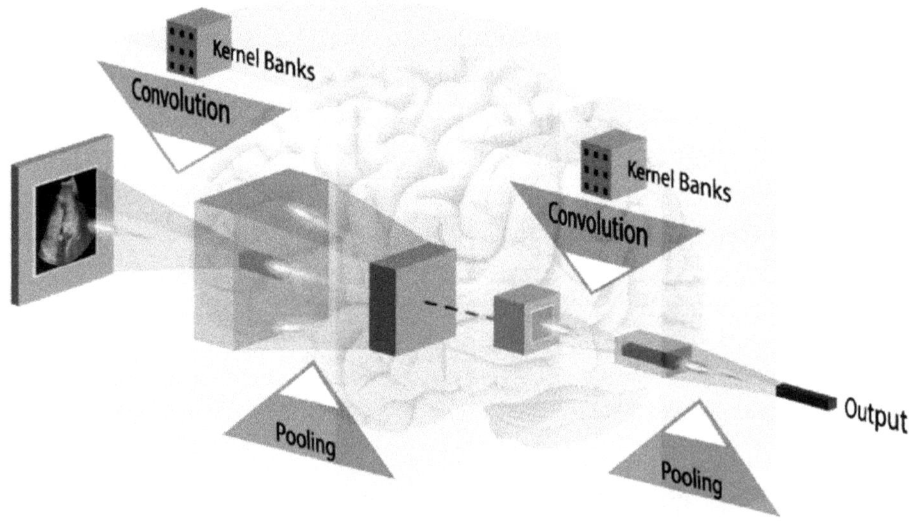


**Figure 3.10.** CNN architecture. (Image courtesy of AtheroPoint™. Reproduced with permission from [109].)

### 3.6.3 Brain image analysis using deep learning

In brain imaging, DL has been instrumental in accurately characterizing brain tumors, and segmentation of edema, enhancing and non-enhancing tumor regions, Alzheimer's disease, and stroke lesions [86]. In the next few subsections we will learn about applications of DL in brain tumor detection, characterization, and segmentation.

*3.6.3.1 DL based inter-institutional brain tumor segmentation*
In [87] an automated brain segmentation model was proposed based on CNN. Three separate CNNs were used on the data from multiple institutions. Each CNN consisted of six layers (four convolution + two fully connected layers). A total of 68 medical images were collected. Patching was used as a segmentation approach. The patching process consists of extracting small equal sized patches. They are annotated as normal, enhancing, abnormal, FLAIR abnormal, and necrotic regions by an experienced radiologist. The ground truth images and descriptions of all the classes are given in figure 3.11. The initial CNN was applied in a dataset from institute 1 and the subsequent CNN was applied on a dataset from institute 2. The cross-validation protocol followed for training/testing was ten-fold cross-validation. The performance parameters were the Dice similarity coefficient (DSC) and Hausdorff distance. It can be seen that the performance decreased when CNN models were trained and tested for different datasets instead of the same dataset. The DSC for data from different institutes (Dice coefficients: $0.68 \pm 0.19$ and $0.59 \pm 0.19$) and data from the same institution (Dice coefficients: $0.72 \pm 0.17$ and $0.76 \pm 0.12$) are different by large margin. The entire process model is shown in figure 3.12.

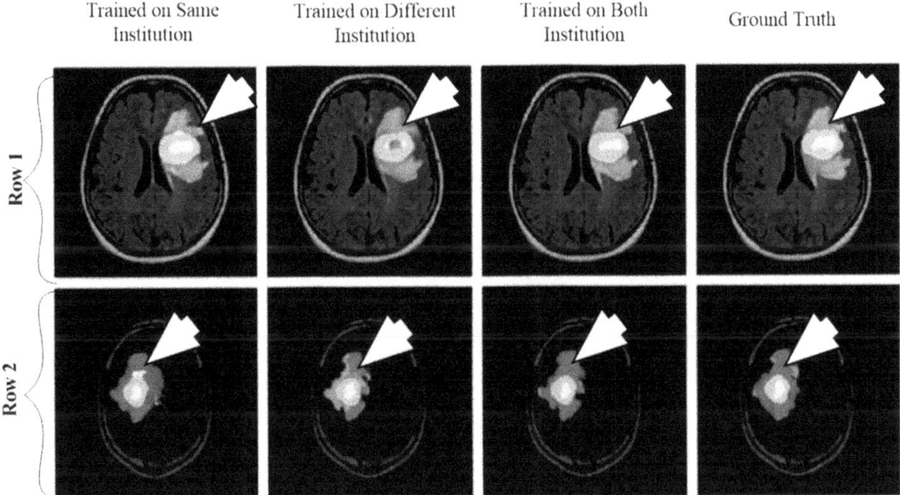


**Figure 3.11.** Two separate patients contributed to the segmentation. Class 1 is the ground truth; class 2 (enhancing area) is in green; class 3 (necrotic region) is in yellow; class 4 (abnormality-hypo intensity region on T1 excluding enhancing and necrotic regions) is in red; and class 5 (abnormality on FLAIR excluding classes 2–4) is in blue. Reproduced with permission from [87].

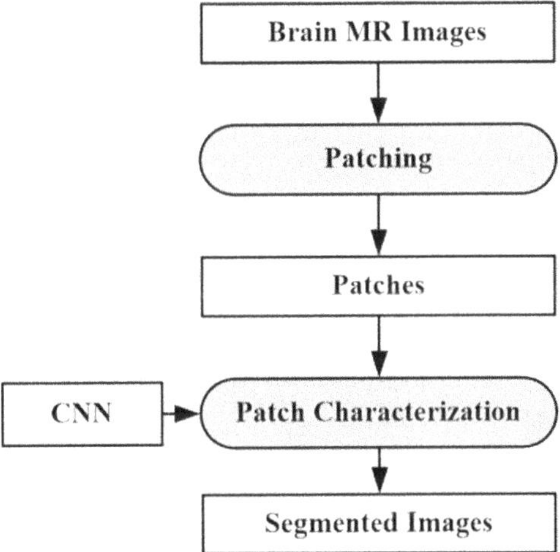


**Figure 3.12.** Process model for segmentation. (Reproduced with permission from [87].)

### 3.6.3.2 Two-pathway CNN for brain tumor segmentation

In this section we learn about the two-pathway model for the segmentation of brain tumors [88]. The brain tumors are segmented into low-grade and high-grade gliomas from MR images. The pathways are divided into local and global segmentation for localization and orientation of brain tumors. The local pathway uses small convolution filters while the global pathway uses large filters. In the final stage, the segmentation features from the pathways are fused to give the overall

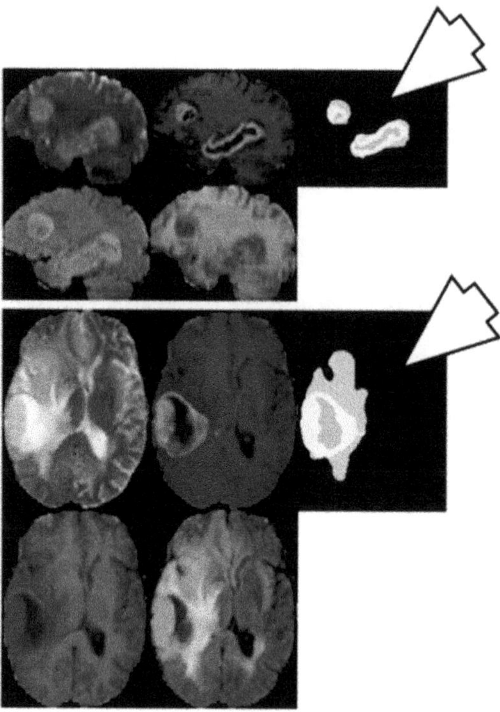

**Figure 3.13.** Two separate patients contributed to the segmentation. Green indicates edoema, yellow indicates an enhanced tumor, pink indicates necrosis, and blue indicates a non-enhanced tumor. Reproduced with permission from [88].

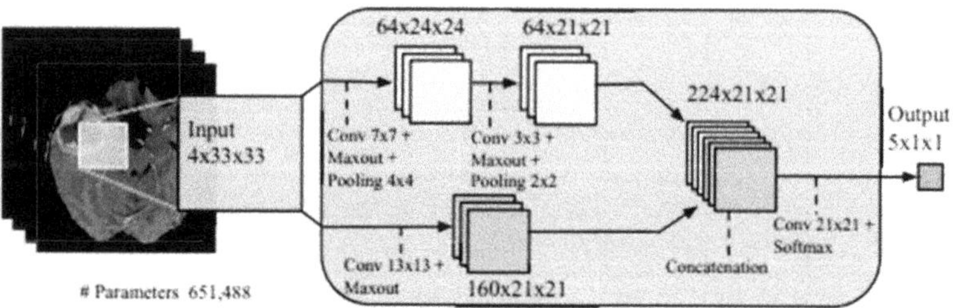


**Figure 3.14.** Model architecture. Reproduced with permission from [88].

segmentation map. For this aim, three DL networks were created and cascaded: MF cascade CNN, input cascade CNN, and local cascade CNN. The input cascade CNN gave a better performance with a Dice similarity of 0.89. The process models and segmented output are shown in figures 3.13 and 3.14.

### 3.6.4 A plausible solution for brain cancer classification

The most common brain tumors are gliomas. Gliomas are further categorized into low-grade and high-grade. The low-grade gliomas are further divided into grade I and II and

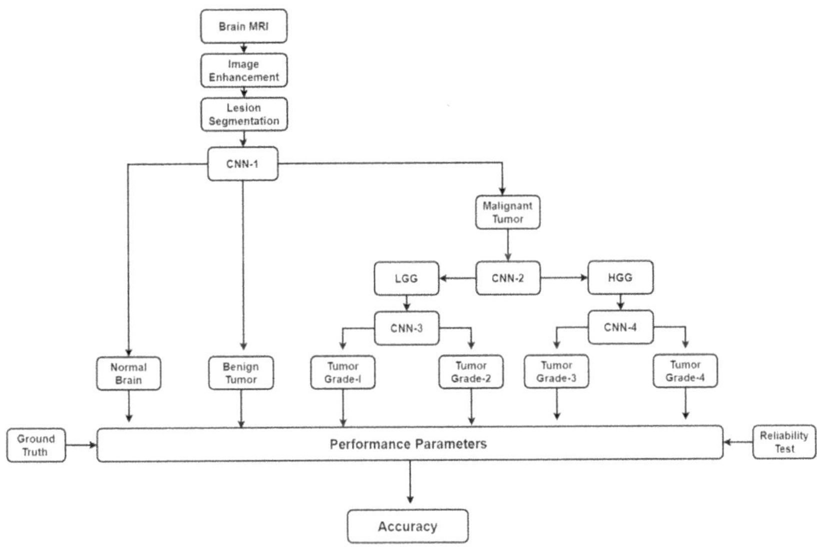


**Figure 3.15.** A plausible solution for brain tumor grading. (Reproduced with permission from [109].)

high-grade gliomas into grade III and IV by the WHO. Although the classification of low-grade and high-grade gliomas is easy, further categorization into their respective grades is difficult. Therefore, physical properties such as shape, size of nuclei, and distribution of cells are used to estimate the degree of malignancy or the respective grades. This is essential since it determines the approach for targeted therapy and prognosis assessment. Although biopsy is the ideal standard, medical practitioners prefer imaging modalities such as MRI (FLAIR, T1, and T2), for localization of tumors because of the better quality of contrast and non-radiation imaging. The aggressiveness of the brain tumor is estimated in most examinations. The effectiveness of detection and estimation depends on the expertise and experience of the medical practitioners and results are thus subjected to inter-variability tests to achieve better results. Computer-assisted tools are slowly gaining prominence in achieving better accuracy.

Several automated models for brain tumor characterization from MR images have been proposed previously based on the intelligence paradigm [77, 89, 90]. DL is increasingly performing well, as suggested by the volume of literature published in the field. A plausible solution for tumor grading using ML and DL paradigms is shown in figure 3.15 (discussed later).

## 3.7 Brain cancer and other brain disorders

### 3.7.1 Stroke

The two types of stroke are hemorrhagic and ischemic [91]. Hemorrhagic stroke is caused by artery injury or an irregular vascular structure, whereas ischemic stroke is caused by a disruption in blood flow. Some experts have looked at the connection between brain cancer and stroke, despite the fact that they are completely distinct in type and origin. The occurrence of brain cancer development among stroke patients was investigated in one such study [92]. In this study 35 malignant tumor patients with and without a history of stroke were considered. Stroke patients had a high chance of acquiring brain cancer over time, with a hazard ratio of 3.09 and a 95% confidence interval of 1.80–5.30. Another significant finding was that females between the ages of 40 and 50, as well as elderly stroke patients, have a higher risk of acquiring brain cancer.

### 3.7.2 Alzheimer's disease

Alzheimer's disease (AD) is a neurodegenerative illness that affects the brain. It mainly affects elderly people who develop increased symptoms, such as memory loss, stuttered speech, and failure in self-care [92]. Studies suggest that AD and brain cancer have an inverse relationship. In one such study over an average follow up of 10 years, cancer survivors had a 33% reduced risk of AD, while AD patients had a 61% decreased risk of cancer.

### 3.7.3 Parkinson's disease

Parkinson's disease (PD) is a motor neuron disease that mainly causes rigidity, tremors, and delayed movement in the limbs, although there are few instances where procedural and behavioral changes are observed. The link between PD and brain tumors is positive [93]. In a study of eight groups of 329 276 patients, the incidence of brain cancer was found to be relatively higher among diagnosed PD patients (1.55 odds ratio, 95% CI 1.18 ± 2.05), which is more statistically significant than for non-PD diagnosed patients (1.21 odds ratio, 95% CI 0.93 ± 1.58).

### 3.7.4 Leukoaraiosis

Leukoaraiosis is the modification of the white matter near the left ventricle, mainly affecting elderly people. It marks the beginning of Binswanger's disease [94]. There is no direct link between brain cancer and leukoaraiosis.

### 3.7.5 Multiple sclerosis

Multiple sclerosis (MS) is related to the spinal cord and brain, where the immune system of the body attacks the myelin cover of the nerve fibers, hampering communication between the brain and the rest of the body. The degree of nerve damage determines the severity of the MS. There are many symptoms of MS, such as slurred speech, tingling, immobility etc. In recent studies, it was found that the patients with MS are vulnerable to brain cancer [95, 96].

### 3.7.6 Wilson's disease

Wilson's disease (WD) is a hereditary illness that is passed down through the generations. The brain and liver are both affected by the build-up of copper in the body. Tremors, personality changes, anxiety, and other brain-related symptoms occur. Weakness, fluid build-up, tremors, stiffness, yellow skin, and itching are all

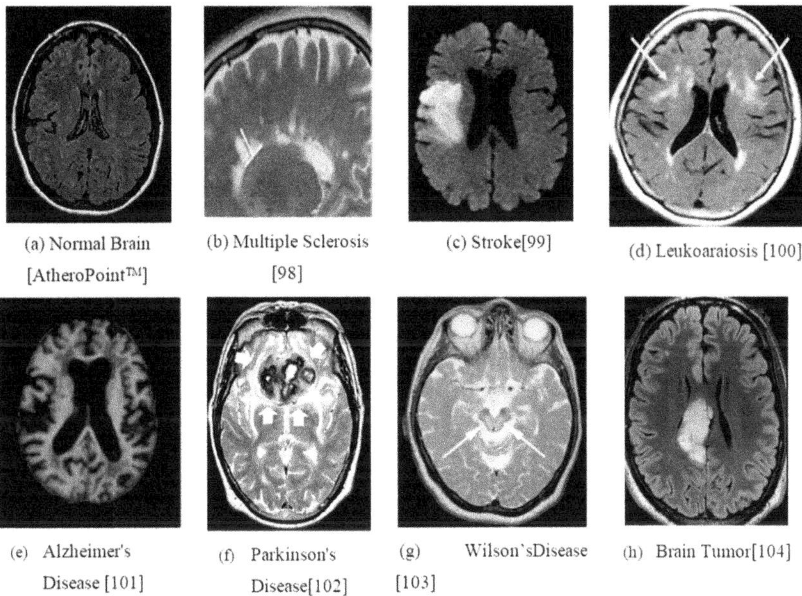



**Figure 3.16.** Comparison of a brain tumor with other brain disorders. Reproduced with permission from (a) AtheroPoint™; (b) [98]; (c) Wikipedia; (d) [100]; (e) [101]; (f) [102]; (g) [103]; (h) [104].

indications of a diseased liver [97]. The different brain diseases are compared in figure 3.16.

## 3.8 Discussion

The computer-assisted study of a brain tumor is difficult and time-consuming. Pre-processing, categorization, and post-processing are the three major stages. Each phase introduces its own set of complications and obstacles, further complicating the work. There are currently no techniques that can accurately establish the presence of malignancy and the grade of a brain tumor. As a result, the optimal test for cancer is an intrusive biopsy, which is dangerous and time-consuming [54, 55]. Pathologists examine the biopsy tissues under a microscope for any abnormalities. Inter-variability tests are also performed on the results. Digital pathological image (DPI) analysis of cells has been a growing topic of research because the laboratory analysis of samples is a difficult undertaking. In order to quantify the degree of malignancy, the DPI study evaluates the size and shape of cells and nuclei, and the cell distribution. The distinction between benign and malignant cells is easier to make. However, due to the asymmetrical formation of the cell, the classification of sub-groups is much more complex, and therefore a challenge. External variables including noise, the use of different scanners, and color changes can also lead to incorrect interpretation. Another difficulty is that the DPI is based on whole slide image (WSI) scanners, which produce only a 2D image and ignore depth, which is a crucial factor in establishing tissue class. This problem will be addressed by the predicted development of 3D WSI scanners [105].

Registration, augmentation if necessary, feature extraction, selection, and eventually characterization are all part of brain image analysis. Picture analysis quality is determined by a number of aspects, including similarity measurements, transformation, image content, optimization, and implementation. Image registration is a crucial stage of analysis since any anomalies here will spread to the rest of the process. Noise, blur owing to organ dynamics, and other irregularities are examples. Low contrast also degrades the quality of the image.

Some common image enhancement techniques include Gaussian filters, histogram equalization, and contrast increase, among others. Characterization becomes computationally difficult when a huge bank of characteristics is created. In this case, the selection of key traits is required. PCA, linear discriminant analysis, and genetic algorithms are the most frequently used feature selection techniques. ML models such as kNN, Bayes classifiers, and ANNs are utilized for characterization in the characterization phase. These are statistical models which work on the features selected in the previous stage. The challenges of this stage are the different variabilities that exist within the data. DL models are gaining popularity, as they give better accuracy and output than ML models. The reason behind this is that the feature maps generated by DL systems are translation invariant and more stable to deformations, and thus robust to small variations. Therefore, the characterization and segmentation of DL models are far better than those of ML models. In this regard, we discuss the model displayed in figure 3.15. For tumor characterization and grading, the suggested DL model consists of four CNN models. The tumor is segmented and the brain images are pre-processed. CNN1 classifies the tumor as malignant, benign, or normal. If the tumor is malignant, the second CNN classifies it as low-grade or high-grade. CNN3 classifies low-grade tumors as grade I or II, and CNN4 classifies high-grade tumors as grade III or IV. As the DL models need a large amount of computation, they are not suited for use on desktop computers. These computations have a lot of storage requirements, which adds to the overhead. In a GPU context, DL models perform better because GPUs allow for parallel computing over thousands of cores. Furthermore, it has been discovered that increasing the depth of DL models does not always result in high accuracy.

### 3.8.1 A note on cancer detection biomarkers

CSF from the spinal cord can also be used to detect the spread of cancer in the neurological system. The extraction procedure is known as a lumbar puncture/spinal tap [107]. Several biomarkers of the brain are detected within the CSF [108]. Tests are performed on the tumor sample to identify proteins, genes, and cells associated with the tumor. These biomarkers are used by doctors to confirm the type and grade of tumors and to choose therapy options. Based on the results of the biomarkers, doctors can begin the initial treatment. The application of ML and DL techniques for biomarker evaluation can increase the accuracy and reduce costs. The benchmarking of ML based systems for brain tumor characterization and segmentation is given in table 3.5.

**Table 3.5.** Overview of brain tumor classification methods. (Reproduced with permission from [109].)

| Reference | Tissue classes | MRI subtype | Data size | Feature processing | Feature reduction | Architecture for classification | Highest performance |
|---|---|---|---|---|---|---|---|
| Sasikala et al (2008) [57] | N, ABN, B, M | T2W | 100 (N = 35, B = 35, M = 30) | DWT | GA | ANN | ACC = 98%, SEN = NA, SPC = NA, AUC = NA |
| Verma et al (2008) [79] | Neoplasms, edema, and healthy tissue | DWI, B0, FLAIR, T1, and GAD | 14 (G III = 8, G IV = 7) | | | Bayesian and SVM | ACC = NA, SEN = 91.84, SPC = 99.57, AUC = NA |
| Zacharaki et al (2009) [77] | Metastasis, meningiomas, gliomas (G II–III), GBM | T1W, T2W, FLAIR, rCBV | 102 (metastasis = 24, meningiomas = 4, gliomas G II = 22, gliomas G III = 18, GBM = 34) | SVM, RFE | Feature ranking | LDA, kNN, NL-SVM | ACC = 97.8%, SEN = 100%, SPC = 95%, AUC = 98.6% |
| El-Dahshan et al (2010) [73] | N, ABN | T2W | 60 (N = 60, ABN = 10) | DWT | PCA | FP-ANN, kNN | ACC = 98.6%, SEN = 100, SPC = 90, AUC = NA |
| Ryu et al (2014) [106] | Glioma (G II, III, IV) | DWI, ADC | 42 gliomas (G II N = 8, G III N = 10, and G IV N = 22) | GLCM | | Entropy, histogram | ACC = 84.4%, SEN = 81.8%, SPC = 90%, AUC = 94.1% |
| Skogen et al (2016) [90] | LGG (G II), HGG (G III–IV) | T1W, T2W, FLAIR | 95 (LGG = 27 (G II); HGG = 68 (G III = 34 and G IV = 34) | Statistical Analysis | | Standard deviation | ACC = 84.4%, SEN = 93%, SPC = 81%, AUC = 91% |

GLCM: gray level co-occurrence matrix; NL-SVM: nonlinear SVM; MDF: most discriminant factor; LDA: linear discriminant analysis; ADC: apparent diffusion coefficient; GA: genetic algorithm; DWT: discrete wavelet transform; SVM: support vector machine; RFE: recursive feature elimination; N: normal; ABN: abnormal; GBM: glioblastoma; LGG: low-grade glioma; HGG: high-grade glioma; B: benign; M: malignant; T1W: T1-weighted; T2W: T2-weighted; FLAIR: fluid-attenuated inversion recovery; rCBV: relative cerebral blood volume; G: grade; ANN: artificial neural network; DWT: discrete wavelet transform; FP-ANN: feed-forward, back-propagation ANN; ACC: accuracy; SEN: sensitivity; SPC: specificity; AUC: area under curve; ROC: receiver operating characteristic.

## 3.9 Conclusion

In this review we have focused on various facets of brain tumor analysis and characterization. We have given details on the pathophysiology of brain tumors, different imaging modalities, WHO guidelines on tumor grading, biopsy, and computer-assisted tools such as ML and DL. Different brain anomalies and their relationships with brain cancer were discussed. We have also stated that DL with its automated feature extraction techniques and better accuracy is becoming more popular than ML models. Intelligence based methods such as ML and DL can provide non-invasive, fast, cost effective, and better diagnosis, helping to save countless lives.

**Funding:** This research received no external funding.

**Conflicts of interest:** The authors declare no conflict of interest.

## References


[1] International Agency for Research on Cancer *Global Cancer Observatory* https://gco.iarc.fr/

[2] Brain Tumor Basics https://thebraintumorcharity.org/

[3] American Cancer Society *Cancer A–Z* https://www.thebraintumourcharity.org/

[4] American Society of Clinical Oncology *Brain Tumor: Diagnosis* (https://cancer.net/cancer-types/brain-tumor/diagnosis

[5] World Health Organization *WHO Statistics on Brain Cancer* http://who.int/cancer/en/

[6] Shah V and Kochar P 2018 Brain cancer: implication to disease, therapeutic strategies and tumor targeted drug delivery approaches *Recent Pat. Anti-Cancer Drug Discov.* **13** 70–85

[7] Ahmed S, Iftekharuddin K M and Vossoug A 2011 Efficacy of texture, shape, and intensity feature fusion for posterior-fossa tumor segmentation in MRI *IEEE Trans. Inf. Technol. Biomed.* **15** 206–13

[8] Behin A, Hoang-Xuan K, Carpentier A F and Delattre J 2003 Primary brain tumors in adults *Lancet* **361** 323–31

[9] Deorah S, Lynch C F, Sibenaller Z A and Ryken T C 2006 Trends in brain cancer incidence and survival in the United States: surveillance, epidemiology, and end results program, 1973 to 2001 *Neurosurg. Focus* **20** E1

[10] Mahaley M S Jr, Mettlin C, Natarajan N, Laws E R Jr and Peace B B 1989 National survey of patterns of care for brain-tumor patients *J. Neurosurg.* **71** 826–36

[11] Hayward R M, Patronas N, Baker E H, Vézina G, Albert P S and Warren K E 2008 Inter-observer variability in the measurement of diffuse intrinsic pontine gliomas *J. Neuro-Oncol.* **90** 57–61

[12] Griffiths A J F, Wessler S R, Lewontin R C, Gelbart W M, Suzuki D T and Miller J H 2005 *An Introduction to Genetic Analysis* (New York: Macmillan)

[13] Shinoura N, Chen L, Wani M A, Kim Y G, Larson J J, Warnick R E, Simon M, Menon A G, Bi W L and Stambrook P J 1996 Protein and messenger RNA expression of connexin43 in astrocytomas: implications in brain tumor gene therapy *J. Neurosurg.* **84** 839–45

[14] Evan G I and Vousden K H 2001 Proliferation, cell cycle and apoptosis in cancer *Nature* **411** 342

[15] Burch P R 2012 *The Biology of Cancer: A New Approach* (New York: Springer)

[16] Song M S, Salmena L and Pandolfi P P 2012 The functions and regulation of the PTEN tumor suppressor *Nat. Rev. Mol. Cell Biol.* **13** 283

[17] Rak J, Filmus J, Finkenzeller G, Grugel S, Marme D and Kerbel R S 1995 Oncogenes as inducers of tumor angiogenesis *Cancer Metastasis Rev.* **14** 263–77

[18] Yarden Y, Kuang W-J, Yang-Feng T, Coussens L, Munemitsu S, Dull T J, Chen E, Schlessinger J, Francke U and Ullrich A 1987 Human proto-oncogene c-kit: a new cell surface receptor tyrosine kinase for an unidentified ligand *EMBO J.* **6** 3341–51

[19] Greenberg M E, Greene L A and Ziff E B 1985 Nerve growth factor and epidermal growth factor induce rapid transient changes in proto-oncogene transcription in PC12 cells *J. Biol. Chem.* **260** 14101–10

[20] Sneed P K *et al* 2002 A multi-institutional review of radiosurgery alone vs radiosurgery with whole brain radiotherapy as the initial management of brain metastases *Int. J. Radiat. Oncol. Biol. Phys.* **53** 519–26

[21] Bertram J S 2000 The molecular biology of cancer *Mol. Asp. Med.* **21** 167–223

[22] Liao J B 2006 Cancer issue: viruses and human cancer *Yale J. Biol. Med.* **79** 115

[23] Golemis E A, Scheet P, Beck T N, Scolnick E M, Hunter D J, Hawk E and Hopkins N 2018 Molecular mechanisms of the preventable causes of cancer in the United States *Genes Dev.* **32** 868–902

[24] Swartling F J, Čančer M, Frantz A, Weishaupt H and Persson A I 2015 Deregulated proliferation and differentiation in brain tumors *Cell Tissue Res.* **359** 225–54

[25] Montes-Mojarro I, Steinhilber J, Bonzheim I, Quintanilla-Martinez L and Fend F 2018 The pathological spectrum of systemic anaplastic large cell lymphoma (ALCL) *Cancers* **10** 107

[26] Mabray M C, Barajas R F and Cha S 2015 Modern brain tumor imaging *Brain Tumor Res. Treat.* **3** 8–23

[27] Hegi M E, Murat A, Lambiv W L and Stupp R 2006 Brain tumors: molecular biology and targeted therapies *Ann. Oncol.* **17** 191–7

[28] Yan H *et al* 2009 IDH1 and IDH2 mutations in gliomas *N. Engl. J. Med.* **360** 765–73

[29] Hu N, Richards R and Jensen R 2016 Role of chromosomal 1p/19q co-deletion on the prognosis of oligodendrogliomas: a systematic review and meta-analysis *Interdiscip. Neurosurg.* **5** 58–63

[30] Lee E, Yong R L, Paddison P and Zhu J 2018 Comparison of glioblastoma (GBM) molecular classification methods *Seminars in Cancer Biology* (New York: Academic)

[31] Amyot F *et al* 2015 A review of the effectiveness of neuroimaging modalities for the detection of traumatic brain injury *J. Neurotrauma* **32** 1693–721

[32] Pope W B 2018 Brain metastases: neuroimaging *Handb. Clin. Neurol.* **149** 89–112

[33] Morris Z *et al* 2009 Incidental findings on brain magnetic resonance imaging: systematic review and meta-analysis *Brit. Med. J.* **339** b3016

[34] Lagerwaard F, Levendag P C, Nowak P J C M, Eijkenboom W M H, Hanssens P E J and Schmitz P M 1999 Identification of prognostic factors in patients with brain metastases: a review of 1292 patients *Int. J. Radiat. Oncol. Biol. Phys.* **43** 795–803

[35] Smith-Bindman R, Lipson J, Marcus R, Kim K, Mahesh M, Gould R, de González A B and Miglioretti D L 2009 Radiation dose associated with common computed tomography examinations and the associated lifetime attributable risk of cancer *Arch. Intern. Med.* **169** 2078–86

[36] Dong Q, Welsh R C, Chenevert T L, Carlos R C, Maly-Sundgren P, Gomez-Hassan D M and Mukherji S K 2004 Clinical applications of diffusion tensor imaging *J. Magn. Reson. Imaging* **19** 6–18

[37] Khoo M M Y, Tyler P A, Saifuddin A and Padhani A R 2011 Diffusion-weighted imaging (DWI) in musculoskeletal MRI: a critical review *Skelet. Radiol.* **40** 665–81

[38] Savoy R L 1999 Functional magnetic resonance imaging (fMRI) *Encyclopedia of Neuroscience* (Charlestown, MA: Elsevier)

[39] Gurcan M N, Boucheron L, Can A, Madabhushi A, Rajpoot N and Yener B 2009 Histopathological image analysis: a review *IEEE Rev. Biomed. Eng.* **2** 147

[40] Fabelo H *et al* 2018 An intraoperative visualization system using hyperspectral imaging to aid in brain tumor delineation *Sensors* **18** 430

[41] Petersson H, Gustafsson D and Bergstrom D 2016 Hyperspectral image analysis using deep learning—a review *Proc. of the IEEE 2016 6th Int. Conf. on Image Processing Theory Tools and Applications (Oulu, Finland, 12–15 December)* pp 1–6

[42] Lu G and Fei B 2014 Medical hyperspectral imaging: a review *J. Biomed. Opt.* **19** 010901

[43] Halicek M, Lu G, Little J V, Wang X, Patel M, Griffith C C, El-Deiry M W, Chen A Y and Fei B 2017 Deep convolutional neural networks for classifying head and neck cancer using hyperspectral imaging *J. Biomed. Opt.* **22** 060503

[44] Nelson S J 2003 Multivoxel magnetic resonance spectroscopy of brain tumors *Mol. Cancer Ther.* **2** 497–507

[45] Olliverre N, Yang G, Slabaugh G, Reyes-Aldasoro C C and Alonso E 2018 Generating magnetic resonance spectroscopy imaging data of brain tumors from linear, non-linear and deep learning models *Int. Workshop on Simulation and Synthesis in Medical Imaging* (Cham: Springer) pp 130–8

[46] Hamed S A I and Ayad C E 2017 Grading of brain tumors using MR spectroscopy: diagnostic value at short and long *IOSR J. Dental Med. Sci.* **16** 87–93

[47] Ranjith G, Parvathy R, Vikas V, Chandrasekharan K and Nair S 2015 Machine learning methods for the classification of gliomas: initial results using features extracted from MR spectroscopy *Neuroradiol. J.* **28** 106–11

[48] James A P and Dasarathy B V 2014 Medical image fusion: a survey of the state of the art *Inf. Fusion* **19** 4–19

[49] Louis D N, Perry A, Reifenberger G, Von Deimling A, Figarella-Branger D, Cavenee W K and Ellison D W 2016 The 2016 World Health Organization classification of tumors of the central nervous system: a summary *Actaneuropathologica* **131** 803–20

[50] DeAngelis L M 2001 Brain tumors *N. Engl. J. Med.* **344** 114–23

[51] Louis D N, Ohgaki H, Wiestler O D, Cavenee W K, Burger P C, Jouvet A, Scheithauer B W and Kleihues P 2007 The 2007 WHO classification of tumors of the central nervous system *Actaneuropathologica* **114** 97–109

[52] Collins V P 2004 Brain tumors: classification and genes *J. Neurol. Neurosurg. Psychiatry* **75** ii2–11

[53] Ludwig J A and Weinstein J N 2005 Biomarkers in cancer staging, prognosis and treatment selection *Nat. Rev. Cancer* **5** 845–56

[54] Sharma H, Alekseychuk A, Leskovsky P, Hellwich O, Anand R S, Zerbe N and Hufnagl P 2012 Determining similarity in histological images using graph-theoretic description and matching methods for content-based image retrieval in medical diagnostics *Diagn. Pathol.* **7** 134

[55] Bardou D, Zhang K and Ahmad S M 2018 Classification of breast cancer based on histology images using convolutional neural networks *IEEE Access* **6** 24680–93

[56] Xu Y, Jia Z, Wang L B, Ai Y, Zhang F, Lai M and Chang C 2017 Large scale tissue histopathology image classification, segmentation, and visualization via deep convolutional activation features *BMC Bioinform.* **18** 281

[57] Sasikala M and Kumaravel N 2008 A wavelet-based optimal texture feature set for classification of brain tumors *J. Med. Eng. Technol.* **32** 198–205

[58] Haykin S S 2009 *Neural Networks and Learning Machines* vol 3 (Upper Saddle River, NJ: Pearson)

[59] Nasrabadi N M 2007 Pattern recognition and machine learning *J. Electron. Imaging* **16** 049901

[60] Wernick M N, Yang Y, Brankov J G, Yourganov G and Strother S C 2010 Machine learning in medical imaging *IEEE Signal Process Mag.* **27** 25–38

[61] Erickson B J, Korfiatis P, Akkus Z and Kline T L 2017 Machine learning for medical imaging *Radiographics* **37** 505–15

[62] Vasantha M, Subbiah Bharathi V and Dhamodharan R 2010 Medical image feature, extraction, selection and classification *Int. J. Eng. Sci. Technol.* **2** 2071–6

[63] Altman N S 1992 An introduction to kernel and nearest-neighbor nonparametric regression *Am. Stat.* **46** 175–85

[64] Cortes C and Vapnik V 1995 Support vector machine *Mach. Learn.* **20** 273–97

[65] Yegnanarayana B 2009 *Artificial Neural Networks* (New Delhi: PHI Learning)

[66] Huang G-B, Zhu Q-Y and Siew C-K 2006 Extreme learning machine: theory and applications *Neurocomputing* **70** 489–501

[67] Grossberg S 2013 Recurrent neural networks *Scholarpedia* **8** 1888

[68] Hinton G E 2012 A practical guide to training restricted Boltzmann machines *Neural Networks: Tricks of the Trade* (Berlin: Springer) pp 599–619

[69] LeCun Y, Bengio Y and Hinton G 2015 Deep learning *Nature* **521** 436

[70] Krizhevsky A, Sutskever I and Hinton G E 2012 ImageNet classification with deep convolutional neural networks *Advances in Neural Information Processing Systems (Nevada)* vol 25 pp 1097–105

[71] Hinton G E and Salakhutdinov R R 2006 Reducing the dimensionality of data with neural networks *Science* **313** 504–7

[72] Hinton G E 2009 Deep belief networks *Scholarpedia* **4** 5947

[73] El-Dahshan E S A, Hosny T and Salem A B M 2010 Hybrid intelligent techniques for MRI brain images classification *Dig. Signal Process.* **20** 433–41

[74] Yang G, Nawaz T, Barrick T R, Howe F A and Slabaugh G 2015 Discrete wavelet transform-based whole-spectral and subspectral analysis for improved brain tumor clustering using single voxel MR spectroscopy *IEEE Trans. Biomed. Eng.* **62** 2860–6

[75] Jolliffe I 2011 Principal component analysis *International Encyclopedia of Statistical Science* (Berlin: Springer) pp 1094–6

[76] Rumelhart D E, Hinton G E and Williams R J 1985 Learning internal representations by error propagation No. ICS-8506 *OCLC Number* 20472667 Defense Technical Information Center, VA, California University San Diego, La Jolla Institute for Cognitive Science

[77] Zacharaki E I, Wang S, Chawla S, Yoo D S, Wolf R, Melhem E R and Davatzikos C 2009 MRI-based classification of brain tumor type and grade using SVM-RFE *IEEE Int. Symp. on Biomedical Imaging: From Nano to Macro* pp 1035–8

[78] Barker M and Rayens W 2003 Partial least squares for discrimination *J. Chemometr. J. Chemometr. Soc.* **17** 166–73

[79] Verma R, Zacharaki E I, Ou Y, Cai H, Chawla S, Lee S, Melhem E R, Wolf R and Davatzikos C 2008 Multiparametric tissue characterization of brain neoplasms and their recurrence using pattern classification of MR images *Acad. Radiol.* **15** 966–77

[80] Murphy K P 2006 *Naive Bayes Classifiers* vol 18 (Vancouver, BC: University of British Columbia)

[81] Leung K M 2007 *Naive Bayesian Classifier* (New York: Polytechnic University Department of Computer Science/Finance and Risk Engineering)

[82] John G H and Langley P 1995 Estimating continuous distributions in Bayesian classifiers *Proc. of the Eleventh Conf. on Uncertainty in Artificial Intelligence (Montreal, QC, Canada, 18–20 August)* (San Francisco, CA: Morgan Kaufmann) pp 338–45

[83] Huang G B, Zhu Q Y and Siew C K 2004 Extreme learning machine: theory and applications. extreme learning machine: a new learning scheme of feedforward neural networks *Proc. of 2004 IEEE Int. Joint Conf. on in Neural Networks* pp 985–90

[84] Kuppili V, Biswas M, Sreekumar A, Suri H S, Saba L, Edla D R, Marinhoe R T, Sanches J M and Suri J S 2017 Extreme learning machine framework for risk stratification of fatty liver disease using ultrasound tissue characterization *J. Med. Syst.* **41** 152

[85] Biswas M, Kuppili V, Edla D R, Suri H S, Saba L, Marinho R T, Sanches J M and Suri J S 2018 Symtosis: a liver ultrasound tissue characterization and risk stratification in optimized deep learning paradigm *Comput. Methods Prog. Biomed.* **155** 165–77

[86] Litjens G, Kooi T, Bejnordi B E, Setio A A A, Ciompi F, Ghafoorian M, van der Laak J A W M, Ginneken B and Sánchez C I 2017 A survey on deep learning in medical image analysis *Med. Image Anal.* **42** 60–88

[87] AlBadawy E A, Saha A and Mazurowski M A 2018 Deep learning for segmentation of brain tumors: impact of cross-institutional training and testing *Med. Phys.* **45** 1150–8

[88] Havaei M, Davy A, Warde-Farley D, Biard A, Courville A, Bengio Y and Larochelle H 2017 Brain tumor segmentation with deep neural networks *Med. Image Anal.* **35** 18–31

[89] Erickson B J, Korfiatis P, Akkus Z, Kline T and Philbrick K 2017 Toolkits and libraries for deep learning *J. Dig. Imaging* **30** 400–5

[90] Skogen K, Schulz A, Dormagen J B, Ganeshan B, Helseth E and Server A 2016 Diagnostic performance of texture analysis on MRI in grading cerebral gliomas *Eur. J. Radiol.* **85** 824–9

[91] Kreisl T N, Toothaker T, Karimi S and DeAngelis L M 2008 Ischemic stroke in patients with primary brain tumors *Neurology* **70** 2314–20

[92] Burns A and Iliffe S 2009 Alzheimer's disease *Brit. Med. J.* **338** b158

[93] Ye R, Shen T, Jiang Y, Xu L, Si X and Zhang B 2016 The relationship between Parkinson disease and brain tumor: a meta-analysis *PLoS ONE* **11** e0164388

[94] Wardlaw J M, Sandercock P A G, Dennis M S and Starr J 2003 Is breakdown of the blood brain barrier responsible for lacunar stroke, leukoaraiosis, and dementia *Stroke* **34** 806–12

[95] Plantone D, Renna R, Sbardella E and Koudriavtseva T 2015 Concurrence of multiple sclerosis and brain tumors *Front. Neurol.* **6** 40

[96] Bahmanyar S, Montgomery S M, Hillert J, Ekbom A and Olsson T 2009 Cancer risk among patients with multiple sclerosis and their parents *Neurology* **72** 1170–7

[97] Reitan R M and Wolfson D 1985 *The Halstead-Reitan Neuropsychological Test Battery: Theory and Clinical Interpretation* vol 4 (Tucson, AZ: Reitan Neuropsychology)

[98] Cahalane A M, Kearney H, Purcell Y M, McGuigan C and Killeen R P 2017 MRI and multiple sclerosis—the evolving role of MRI in the diagnosis and management of MS: the radiologist's perspective *Irish J. Med. Sci.* **187** 781–7

[99] Terroni L *et al* 2015 The association of post-stroke anhedonia with salivary cortisol levels and stroke lesion in hippocampal/parahippocampal region *Neuropsych. Dis. Treat.* **11** 233

[100] Nakano K *et al* 2014 Leukoaraiosis significantly worsens driving performance of ordinary older drivers *PLoS One* **9** e108333

[101] Islam J and Zhang Y 2018 Brain MRI analysis for Alzheimer's disease diagnosis using an ensemble system of deep convolutional neural networks *Brain Inform.* **5** 2

[102] Heim B, Krismer F, De Marzi R and Seppi K 2017 Magnetic resonance imaging for the diagnosis of Parkinson's disease *J. Neural Transm.* **124** 915–64

[103] Bandmann O, Weiss K H and Kaler S G 2015 Wilson's disease and other neurological copper disorders *Lancet Neurol.* **14** 103–13

[104] Villanueva-Meyer J E, Mabray M C and Cha S 2017 Current clinical brain tumor imaging *Neurosurgery* **81** 397–415

[105] Madabhushi A and Lee G 2016 Image analysis and machine learning in digital pathology: challenges and opportunities *Med. Image Anal.* **33** 170–5

[106] Ryu Y J, Choi S H, Park S J, Yun T J, Kim J H and Sohn C H 2014 Glioma: application of whole-tumor texture analysis of diffusion-weighted imaging for the evaluation of tumor heterogeneity *PLoS One* **9** e108335

[107] Quincke H I 1909 *Lumbar Puncture. Diseases of the Nervous System* ed A Church (New York: Appleton) p 223

[108] Lynch H T, Lynch J F, Shaw T G and Lubiński J 2003 HNPCC (Lynch syndrome): differential diagnosis, molecular genetics and management—a review *Hered. Cancer Clin. Pract.* **1** 7

[109] Gopal S T *et al* 2019 A review on a deep learning perspective in brain cancer classification *Cancers* **11** 111

**IOP** Publishing

# Multimodality Imaging, Volume 1
### Deep learning applications
**Mainak Biswas and Jasjit S Suri**

# Chapter 4

# MRI based brain tumor classification and its validation: a transfer learning paradigm

**Luca Saba, Gopal S Tandel, Mainak Biswas, Michele Porcu, N N Khanna, Monica Turk, Christopher K Asare, Annabel A Ankrah and Jasjit S Suri**

Biopsy is considered as the gold standard for tumor confirmation and grading. However, this procedure is invasive, complicated, and life-threating for the patient. Further, for treatment planning it is of paramount importance to quickly and accurately diagnose and grade a tumor using a 3D standardized imaging modality, such as magnetic resonance imaging (MRI). In this context, a novel deep learning (DL) based framework is proposed to accurately and automatically classify a tumor into different grades in these MRI brain volumes.

The design consists of two stages: stage I involves brain segmentation, while stage II consists of tumor characterization. As part of the brain segmentation protocol, the skull is removed to extract the brain region, while in stage II, tumor characterization is implemented using an eight-layer convolution neural network (CNN) model.

The performance between the K5 and K10 cross-validation protocols were compared for T2-weighted 2D MRI axial slices. Using a database of 79 patients consisting of three kinds of tumors: 40 astrocytoma (grade II: 65% and grade III: 35%), 16 oligodendroglioma (grade II: 56% and grade III: 43%) and 23 glioblastoma (only grade IV), the DL model resulted in the following average performance: accuracy (98.33% ± 1.42%), sensitivity (97.88% ± 3.06%), specificity (98.47% ± 1.72%), positive predicted value (97.90% ± 2.27%), negative predicted value (98.71% ± 1.61%), and area under the curve (0.98, $p < 0.0001$). The system was benchmarked against various machine learning (ML) classifiers, which demonstrated an improvement of 13.86%. Further, the model was validated using a special index called the tumor separation index (TSI), which is designed via statistical analysis of learned features by CNN. The CNN was also verified using standardized biometric face data, demonstrating 99% accuracy.

# 4.1 Introduction

The International Agency for Research on Cancer (IARC) reported an alarming global cancer statistic of 18.1 million new cancer cases and 9.6 million deaths in 2018 [1]. Further, the report revealed that one in five men and one in six women worldwide suffer from cancer during their lifetime, while one in eight men and one in 11 women die from cancer. The global annual incidence of primary brain tumor has been estimated at 3.7 and 2.6 per 100 000 for men and women, respectively, with a higher rate in developed than developing countries [2]. The American Cancer Society (ACS) had recently (2019) reported that 23 820 adults in the United States of America (USA) were diagnosed with a primary brain or spinal cord tumor, comprising 13 410 men and 10 410 women [3]. Similar statistics were reported by the World Health Organization (WHO) [4].

The glioma is the most common brain tumor, which may further consist of astrocytoma, oligodendroglioma, and glioblastoma. The grade of the tumor is defined based on the rate at which the cancerous cells grow, i.e. the WHO defines gliomas from low to high and they are divided into four grades (I, II, III and IV) [5–8]. Further, brain tumors are categorized into various types based on attributes such as (i) nature of invading, (ii) origin, (iii) rate of growth and (iv) stage of progression or grading [4]. There are two main stages: benign or malignant. As regards the rate of growth, benign brain tumor cells rarely invade neighboring healthy cells and have distinct borders with a slow progression rate. In contrast, malignant cells readily attack neighboring cells in the brain or spinal cord, and have fuzzy borders and a high progression rate. Considering the place of origin as an attribute, brain tumor cells are classified into two types: primary and secondary tumors [9]. A primary tumor originates directly inside the brain, whereas a tumor emerging in the brain due to cancer somewhere else in the body, such as the lungs, stomach, etc, is known as a secondary brain tumor or metastasis [10].

The correct diagnosis of glioma is highly important for the prognosis of the disease and determining suitable treatment. Magnetic resonance imaging (MRI) is primarily preferred in the initial stage of tumor diagnosis, and gives a detailed 3D view of the brain structure and assists in locating the tumor exactly. Further, tissue biopsy is the gold standard for tumor grading, and is an invasive procedure. In a biopsy, tissue samples are first collected by making an incision, and then the samples are sent to the pathologist for microscopic examination. This procedure may take from a few days to a few weeks. Due to its invasive nature, this procedure is risk-prone and life-threating as well, in particular for a brain tumor. Vision loss, stroke, infection, brain swelling, blood clots, seizures, bleeding, and reactions to anesthesia are the common risks of brain biopsy. Likewise, the microscopic inspection of tissue structure is a tedious job and subject to inter-observer variability. It is therefore necessary to integrate computer-aided diagnosis (CAD) tools using imaging techniques for a quick and accurate diagnosis.

Since accurate diagnosis and stage estimation are essential for treatment planning for cancer [11], a fundamental question is how to best detect and classify a brain tumor in the region-of-interest, given the specified medical imaging modality such as

MRI or computed tomography (CT). The most successful approach for stage or grade estimation is the knowledge-based paradigms that point us to machine learning based solutions. The global idea in machine learning (ML) based methods is to extract the offline image-based features using training datasets, and then use these offline features to transform the online testing datasets to predict the type of cancer [12–14]. Several studies have been published which present excellent feature extraction methods [15–17], such as wavelet type [18] and genetic algorithm based methods [19, 20], as well as classification tuners [21].

It is important to note that even though ML based methods claim to have higher accuracy, there are inherent challenges in their fundamental design. Since the quest for accuracy and specificity is the primary motive in ML based methods, one is therefore interested in computing a large set of features in the training and testing paradigms [22]. The larger the number of features, the more time is required in training the ML system.

Even though one can claim that principal component analysis (PCA) provides a solution to the reduction of features, it does require the design of offline systems [23, 24]. This raises the question of whether there could be an alternative solution where feature power and feature selection both are derived simultaneously. The second challenge in ML based models is their poor ability to handle variation in the input data, leading to low sensitivity in performance.

With the invention of the deep learning (DL) paradigm and big data [25], it was recently observed that the weaknesses posed by ML models can be overcome. DL models are a variation of artificial neural networks (ANNs) with many hidden layers. ANNs with a single hidden layer can understand simple patterns such as simple lines at different angles. With an increase of hidden layers, the ANNs can understand complex relationship patterns between features of an image, very similar to the functioning of the human brain. Recently, DL techniques have shown remarkable performance in automatic tissue characterization [26–28]. Our objective is to develop a CAD tool using DL techniques to automatically detect a tumor and determine the grade (or classify) of a brain tumor with precision and consistency. A series of studies are proposed for CAD and brain tumor classification [16, 29–31] using imaging methods.

Through the comprehensive literature survey presented in the next section, we conclude that very limited work has been performe on this topic and the major focus of all the studies has been on high-grade glioma (HGG) versus low-grade glioma (LGG) classification. Except for a few studies, the existing work has not been cross-validated properly. There has been no serious attempt to verify or validate the models. Therefore, the major objective and sub-objectives of our study are as follows:

1. To develop a non-invasive brain tumor grading system (TGS) based on CAD.
2. Classify brain tumor astrocytoma and oligodendroglioma into their sub-grades along with LGG versus HGG classification based on an advanced state of DL, such as transfer learning.
3. Compare and contrast the DL method against the conventional ML techniques for TGS.

The following are the sub-objectives:

1. Validate the hypothesis using a new approach called the tumor separation index (TSI).
2. Verification of the software and system design using offline biometric face data.
3. Exhaustive data analysis and understanding different partition protocols for system optimization.

In this study we have designed a two-stage system for brain tumor classification (see figure 4.1): stage I involves brain segmentation, while stage II consists of tumor characterization. As part of the brain segmentation protocol, the skull is removed to extract the brain region, while in stage II tumor characterization is implemented using an eight-layer convolution neural network (CNN) model. For protocol optimization, we adapted K5 and K10 cross-validation protocols on T2-weighted 2D MRI axial slices. Using a database of 79 patients consisting of three kinds of tumors, 40 astrocytoma (grade II: 65% and grade III: 35%), 16 oligodendroglioma (grade II: 56% and grade III: 43%) and 23 glioblastoma (only grade IV), our DL model resulted in the following average performance: accuracy (98.33% ± 1.42%), sensitivity (97.88% ± 3.06%), specificity (98.47% ± 1.72%), positive predicted value (97.90% ± 2.27%), negative predicted value (98.71% ± 1.61%), and area under the curve (0.98; $p < 0.0001$). The system was benchmarked against ML system demonstrating an improvement of 13.86%. In order to validate the model, we designed a unique parameter named the tumor separation index (TSI) based on a statistical analysis approach. Further, the model was verified by standardized biometric face data.

The remaining sections of the paper are organized as follows. The background literature survey is presented in section 4.2. Patient demographics and data preparation is discussed in section 4.3. Section 4.4 presents the engineering methodology. In section 4.5 we present the experimental protocol, results, and performance evaluation. The validation and verification of the tumor grading system is shown in section 4.6. Detailed discussion and benchmarking is presented in section 4.7, and the paper concludes in section 4.8.

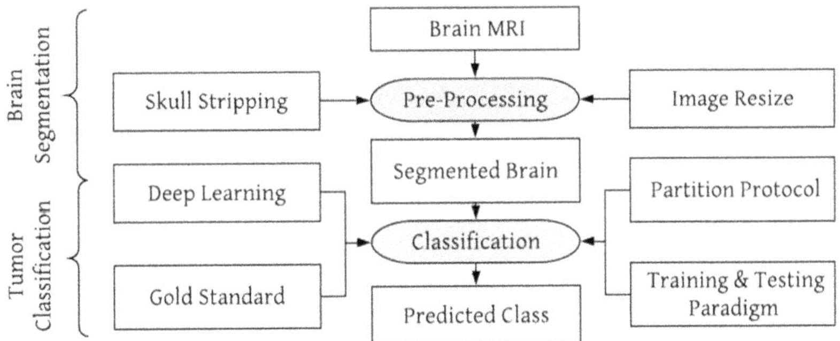


**Figure 4.1.** Global architecture showing the two-stage system: stage I (brain segmentation) and stage II (DL method for grading the tumor).

## 4.2 Background literature survey

The objective of this section is to survey the background literature in the area of tumor grading, primarily in terms of adapting machine learning and deep learning paradigms. Since the tumor grading system includes components such as pre-processing of the imaging volumes, segmentation of the region-of-interest, and finally tumor grading (so-called classification), we therefore have attempted to search the standard sources on the web (PubMed, IEEE) for works related to our objectives.

As discussed above, brain biopsy is very risk-prone and the microscopic examination of samples is time-consuming and error-prone. Automatic digital histopathology image analysis [32–34] is one solution to this problem, which alleviates the workload of pathologists. However, this technique is still risk-prone because it requires a highly invasive procedure for collecting tumor samples. Imaging techniques such as MRI provide images of the systematic 3D brain structure and help in localizing a brain tumor. Typically, a brain tumor is initially detected by doctors via general observation of the image volume. If the number of subjects is large (say above 100), then this procedure becomes tedious and there is an increased risk of human error. Thus, incorporating CAD tools not only alleviates the workload, but it also offers an alternative to automatic brain tumor grading. Therefore, brain tumor detection and grade estimation using artificial intelligence (AI) techniques plays a vital role in the workflow of for both the radiologist and the neurosurgeon.

When it comes to analysing imaging volumes, pre-processing, segmentation, and automated classification of a tumor are the three major challenges. Due to the inherent nature of MR acquisition protocols causing a low signal-to-noise ratio, low contrast-to-noise ratio, and artefacts due to motion, the image resolution can be degraded [35, 36] and, therefore, pre-processing has become a necessity to prevent propagation of errors during the segmentation and classification stages. Standard solutions such as contrast adjustment, intensity adjustment, histogram equalization, binarization, and morphological operations are adopted for pre-processing [37, 38]. This section is mainly going to present tumor segmentation and tumor classification using the ML and DL paradigms.

MRI is a routine procedure that is used for screening for many neurological disorders such as Alzheimer's disease, brain tumors, epilepsy, multiple sclerosis (MS), and schizophrenia. The region-of-interest is the brain, thus brain MRI segmentation is an important and critical process for image analysis, where the labels are assigned to each pixel in a 2D MR image (from 3D voxels). Manual segmentation is a gold standard procedure [39], where the labels are assigned to each pixel manually. Further, the classification requires predicting the correct set of labels assigned to a specific section of the image or the image itself. Assigning labels to 2D MRI slices is a highly tedious process, and is error-prone due to human error. Substantial efforts have been made by researchers to develop classical segmentation algorithms for brain MRI (e.g. gray matter, white matter, or abnormal tissue segmentation).

Recently, new paradigms incorporating intelligence have started to penetrate the field. Not long ago, conventional machine learning techniques were proposed for automatic brain tumor segmentation [40–45] and classification [6, 46–50]. Despite the intensive efforts that have been made toward automatic segmentation and classification, there is still a need for a tumor grading system which can automatically grade tumor, demonstrating both validation and verification of the system. This is required because of the morphological variation of brain images, variation in MRI acquisition protocols, and image acquisition imperfections.

In conventional ML techniques, feature selection is most crucial step. There are two main challenges in the ML paradigm: (a) robustness of features and (b) the amount of time it takes for online systems to predict the class labels or risk labels [51]. Further, the strain of the number of data samples required is one of the deep rooted problems which puts a burden on system performance [52]. Recently, the deep learning and big data innovations have opened the doors to combat the above challenges. They have penetrated into nearly all walks of life [52]. They provide the ability to carry out automatic feature extraction, self-learning, and generalization in image segmentation and classification [39, 53]. Our DL model is inspired by these innovations while focusing on generalized automated tumor grading. Before we present our system, we examine the most recently published deep learning paradigms.

Pan *et al* [54] presented a brain tumor characterization method that classified tumors into high and low grades using multi-phase MRI images. The authors developed a CNN based strategy for brain tumor grading and further compared this against the conventional neural network method. The database consisted of 190 tumor samples, 25 of which were low-grade tumor samples. Only sensitivity and specificity were presented as part of performance evaluation. Finally, the system was not validated or verified.

In another study, Mohsen *et al* [55] proposed a seven-layer deep neural network (DNN) model for brain tumor classification into four different classes: normal, glioblastoma, sarcoma, and metastatic bronchogenic carcinoma. The database consisted of 66 patients (22 normal and 44 abnormal) sourced from Harvard Medical School. The segmentation adapted a fuzzy C-means (FCM) paradigm followed by feature extraction based on a discrete wavelet transform (DWT). These features were reduced using principal component analysis (PCA) by adapting a deep neural network classifier. The highest average performance of classification for recall, precision, F-measure, and area under the curve (AUC) for all the above classes were 96.97%, 97%, 97%, 97%, and 98%, respectively. However, the system had challenges, such as the manual extraction of features, and lacked validation and verification, while the number of samples was relatively low, thus it lacked robustness.

Ari and Hanbay [16] proposed a brain MRI classification technique to classify brain MRIs into normal and abnormal brains using an extreme learning machine with local receptive fields (ELM-LRF) classifier. The study consisted of a very small dataset of 16 patients that was partitioned into training and testing sets of nine and seven patients, respectively. The MRI slices consisted of three orthogonal plains: axial, coronal, and sagittal. In this study, the ELM-LRF method (accuracy: 97.18%) outperformed both the CNN (accuracy: 96.45%) and ML methods (accuracy: 93.65%). A very similar comparison of DL versus ML versus ELM was conducted on a liver cancer dataset, showing slightly better results for DL compared to the ELM [56].

Ge *et al* [57] also carried out a study for low-grade and high-grade glioma classification using multi-sensor images (e.g. T1W, T2W, and FLAIR) via a feature aggregation method. The authors proposed a seven-layer multi-stream CNN model. The network demonstrated an accuracy of 90.87% for low versus high-grade glioma classification. Similarly, a relatively low accuracy (89.39%) was found in gliomas with/without 1p19q co-deletion classification. Although the number of patients was reasonable, the system was neither validated nor verified.

In another study, Anaraki *et al* [58] presented a study for glioma classification (low and high grade) using MRI. An optimal CNN was developed using a genetic algorithm (GA) and two studies were performed. In the first study, glioma samples were classified into three grades and achieved an accuracy of 90.9%. Similarly, in the second study, glioma, meningioma, and pituitary tumors were classified with 94.2% accuracy. This model was also neither validated nor verified.

In another study, Anaraki *et al* [58] presented an approach for glioma classification between low and high grade using MRI. An optimal CNN was developed using a genetic algorithm (GA). Two studies were performed. In the first study, glioma samples were classified into three grades and achieved an accuracy of 90.9%. Similarly, in the second study, glioma, meningioma, and pituitary tumors were classified with 94.2% accuracy. This model were also neither validated nor verified.

Yang *et al* [29] classified gliomas into high and low grades using the well-known CNN models, AlexNet and GoogLeNet with transfer learning. Public brain MRI data were used, whereas the rectangular region-of-interest (ROI) (i.e. tumor area) was extracted from 2D MRI slices. Two methods of training were adopted, training from scratch and training using pre-trained features (via transfer learning). The transfer learning method outperformed the learning from scratch in both models, in particular GoogleNet, and the highest K5 cross-validation mean performance in validation accuracy, test accuracy, and test AUC was 0.867, 0.909, and 0.939, respectively. These models were also neither validated nor verified. While the literature survey above presents the key studies in deep learning, the benchmarking subsection in the discussion presents a tabular comparison between them.

## 4.3 Demographics and data preparation

### 4.3.1 Patient demographics

The study consisted of MRI data [59, 60] taken from 79 patients with different brain tumor types (grades of tumor). The brain tumor patients were aged between 15 and 89, with a mean age of 47.5 years. The average survival rate of the patients was 47 months. The brain tumor types consist of three classes and corresponding patient numbers were for each tumor type were: (i) astrocytoma (40 patients), (ii) oligodendroglioma (16 patients), and (iii) glioblastoma-multiform (23 patients). Table 4.1 summarizes the different grades of brain tumor. Representative examples of the different grades for each tumor type are depicted in figure 4.2, where the patient number and slice number are the indices for each image. The ground truth

**Table 4.1.** Tumor types demonstrating different grades.

| Tumor type | Grades | #Patients |
|---|---|---|
| Astrocytoma (AST) | Grade II (g2) | 26 |
| | Grade III (g3) | 14 |
| Oligodendroglioma (OLI) | Grade II (g2) | 9 |
| | Grade III (g3) | 7 |
| Glioblastoma-multiform (GBM) | Grade IV (g4) | 23 |
| Total patients | **79** | |

corresponding to the tumor type was confirmed by the biopsy test. Although the complete dataset comprised different MRI protocols such as T1W, T2W, FLAIR, diffusion-weighted imaging (DWI), and their subtypes, we, however, selected T2W for our study, since the oedema region surrounding the tumor appears bright. This gives us a focused study paradigm. The slices were picked manually under the supervision of experts. A 2D brain MRI slice is considered as a positive sample if the tumor lesion is visible. Similarly, the MRI slice is considered as a negative sample if no such area exists.

## 4.3.2 Data preparation

The brain slices used the Digital Imaging and Communications in Medicine (DICOM) standard. Each brain volume was converted into the Neuroimaging Informatics Technology Initiative (NIfTI) file format (*.nii) using MRIcron software[1] [61] and then into 2D images for skull stripping. T2W MRIs in the axial view were used in this experiment due to the good contrast between white and gray matter. Skull stripping is a very important step in improving accuracy during brain image classification and involves removing the skull from each brain slice. Brain Suite18a software [62] (University of California, LA) was used for this application. The complete data preparation procedure is detailed below.

NIfTI was developed for tool-user and tool-developer communities in the field of brain image analysis. The primary goal of NIfTI is to provide coordinated and targeted services, training, and research to speed up development and enhance the utility of informatics tools related to neuroimaging. The BrainSuite tool was developed with the same objectives. BrainSuite contains brain surface extraction (BSE) tools to automatically extract the brain from the outer skull and allows the user to manually edit the parameters for skull stripping when it is unable to detect gray matter automatically. Initially, we obtained all the brain slices in DICOM format. The BrainSuite software takes input files in *.nii (NIFTI) format and we, therefore, used Micron software to convert the DICOM images into the correct format. The complete procedure of skull stripping, and gray and white mask generation is as follows:

---

[1] https://www.nitrc.org/projects/mricron

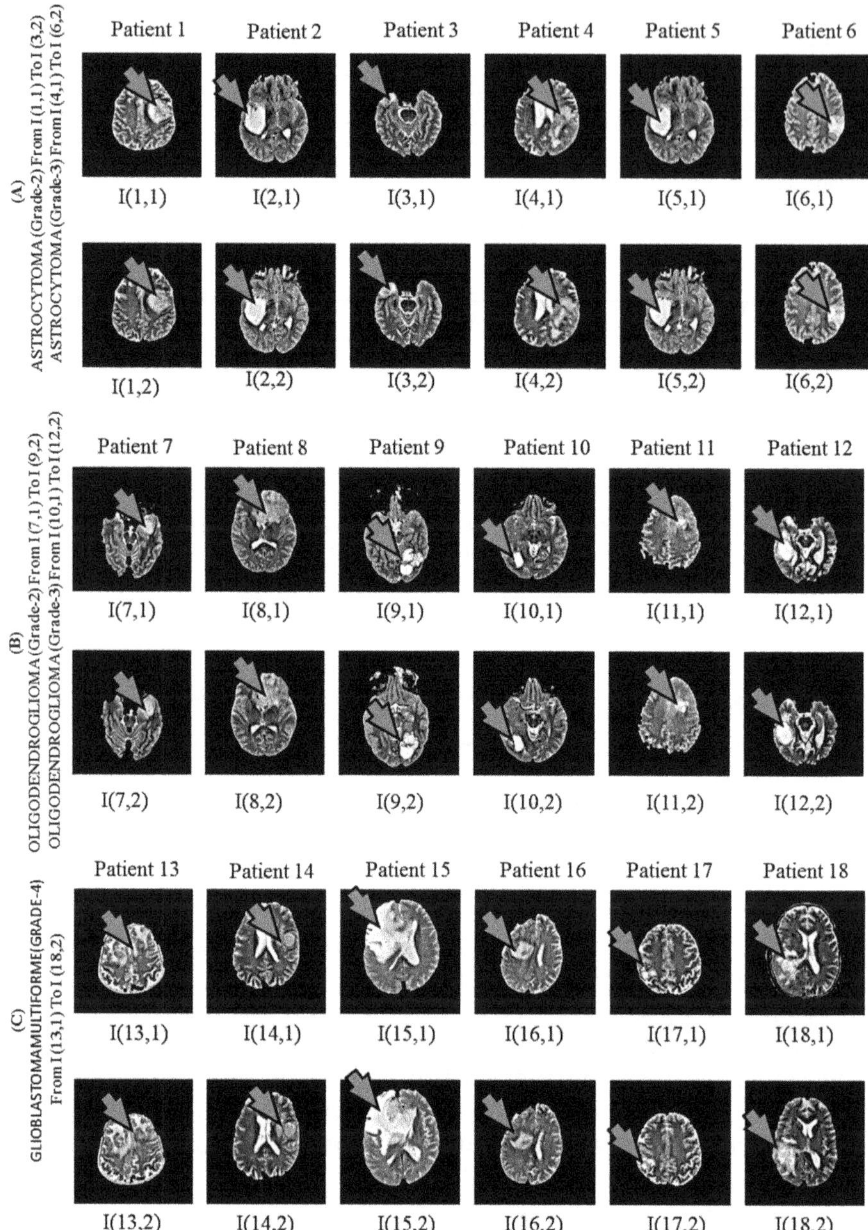


**Figure 4.2.** (a) Astrocytoma (g2) from I(1,1) to I(3,2) and (g3) from I(4,1) to I (6,2), (b) oligodendroglioma (g2) from I(7,1) to I(9,2) and (g3) from I(4,1) to I(12, 2), and (c) glioblastoma-multiform (g2) from I(13,1) to I(15,2) and (g3) from I(16,1) to I(18,2).

1. Load the brain volume (figure 4.3(a)).
2. The complete skull stripping procedure using BSE is semi-automatic: it automatically detects the brain (gray matter) in good quality brain MRIs, but fails for inferior MRIs, in which case the parameters can be adjusted

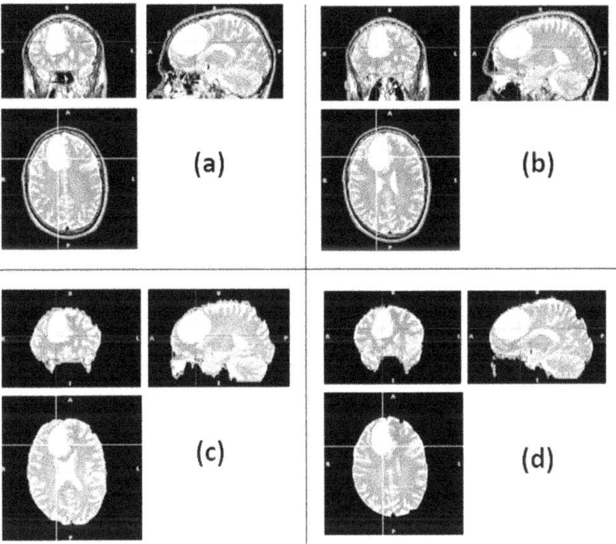

**Figure 4.3.** (a) Loading a brain volume; (b) estimating a brain boundary via gray matter detection; (c) skull stripping; and (d) final stage.

manually using the skull stripping tool. We choose skull stripping (BSE) via the cortex dialog (figure 4.4) and set the following five parameters for manual gray matter selection: (i) automatic iterations; (ii) diffusion iterations; (iii) diffusion constant; (iv) edge constant; and (v) erosion size.

3. The BSE procedure is finished when suitable parameters are found and the process terminates, as shown in figures 4.3(b) and (c). Otherwise, we can click back to change the parameters (particularly the diffusion constant and edge constant parameters) and repeat the procedure until a suitable overlay is not found. The final stage of the skull stripping procedure depicted in figure 4.3(d).

## 4.4 Methodology

DL methods are extensively used for brain image analysis in many brain-related issues [17, 63–65]. The CNN is the most popular DL model and categorizes many brain-related disorders such as brain tumor classification into normal and diseased brains [38], i.e. segmentation of a tumor region into oedema, and enhancing and non-enhancing regions [66], segmentation of stroke lesions [67], diagnosis of brains which have Alzheimer's disease [68], and autism spectrum disorder detection [69].

### 4.4.1 CNN model

The convolutional layer is the basic building block of a CNN, which learns the spatial relationship between pixels in a hierarchical manner by convolving the image using suitable filters to build a hierarchy of feature maps. This convolution function is implemented in several layers, such that the features obtained are invariant to

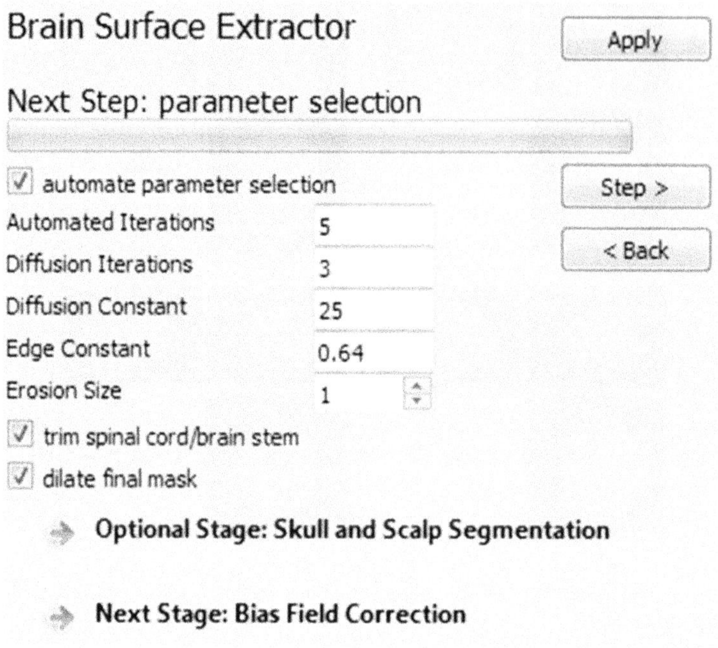

**Figure 4.4.** Skull stripping tool (BSE) dialog.

translation and distortion, resulting in a high degree of accuracy. The basic CNN architecture is shown in figure 4.5. This consists of an input signal or image, convolution, a rectified linear unit (ReLU), max-pooling, and fully connected layers.

The convolution layer (known as Conv) extracts the essential features from the image via a convolution operation between the appropriate filter and input image. The filters are an $(n \times n)$ array, which comprises parameters or weights of either 0 or 1. An ReLU is an activation function, which is used to introduce nonlinearity into the system, primarily to avoid overtraining. The ReLu is respresented by the function $Max(x, 0)$ where $x$ is the input argument. The pooling is a special operation in the network, which is generally applied after the convolution operation to diminish the number of parameters and computations during the training. The max-pooling and average pooling are two popular operations. In max-pooling a maximum value is chosen from the filter size feature vector. Similarly in average pooling, the mean of the feature vector is chosen and forwarded to the feed-forward network. The softmax classifier is used for the fully connected layer to perform probabilistic label classification using a class score of between zero and one.

Many DL networks, such as AlexNet, ResNet, GoogLeNet, etc, use the well-known CNN model. They were introduced at the annual conference of the ImageNet Large Scale Visual Recognition Challenge (ILSVRC) [70]. These networks are trained on more than a million images and are able to classify more than 1000 objects, such as a mouse, a pencil, animals, etc. It is believed that these models have been shown to have significantly improved the performance in image classification over other non-DL methods.

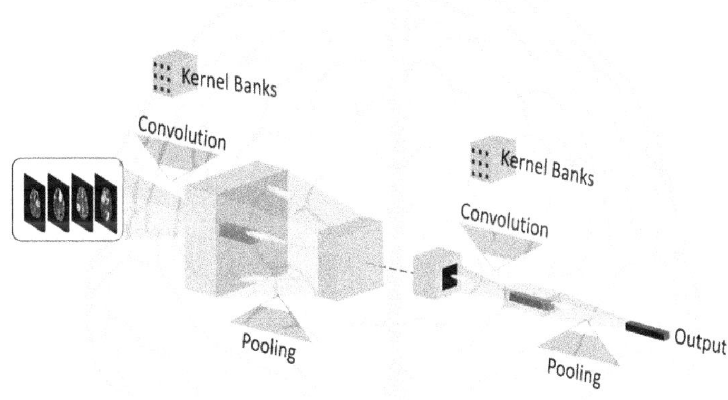


**Figure 4.5.** CNN architecture. (Image courtesy of AtheroPoint™.)

Suitable layer selection for a deep network for medical image classification is a challenging and tedious job, because it requires a lot of iterations of trial and error to find a suitable network. Transfer learning is a particular method where an existing trained deep network can be reused. In this concept, the earlier trained model, including layers and learned weights, is used as the starting point for the new task. This arrangement avoids the overheads required for suitable network design. Thus we have chosen a pre-trained (CNN) network to achieve higher classification performance in a moderate training time.

### 4.4.2 The architecture of AlexNet

AlexNet [70] was first introduced in ILSVRC 2012 and was the winner of the competition with a top-five test error rate of 15.3%, outperforming the second-best entry (26.2% top-five test error). It is an eight-layer CNN, as shown in figure 4.6, including five convolutional (Conv) and three fully connected (FC) layers. The network requires a three-channel image (red–green–blue) of size ($227 \times 227 \times 3$). The layering details of AlexNet are described in table A1 (of appendix A). The main highlights of the model are as follows.

In order to include nonlinearity in the model, an ReLU ($f(a) = \max(0,a)$) was first introduced in this model as an activation function with the objective to enhance the training speed. Previously, the tanh function ($f(a) = (1 + e^{-a})^{-1}$) was used as a popular function. The ReLU is many times faster than the tanh function and helps to diminish the top-five error rate. Further, due to limited available memory, the AlexNet architecture was designed to run on two GPUs simultaneously, whereas the convolution operations with the corresponding kernels are divided into two paths. Consequently the top-one and top-five error rates were reduced by 1.7% and 1.2%, respectively. AlexNet also used local response normalization. This speeds the training process up to the convergence and leads to reduced top-one and top-five error rates by 1.4% and 1.2%, respectively. As an alternative, the concept of

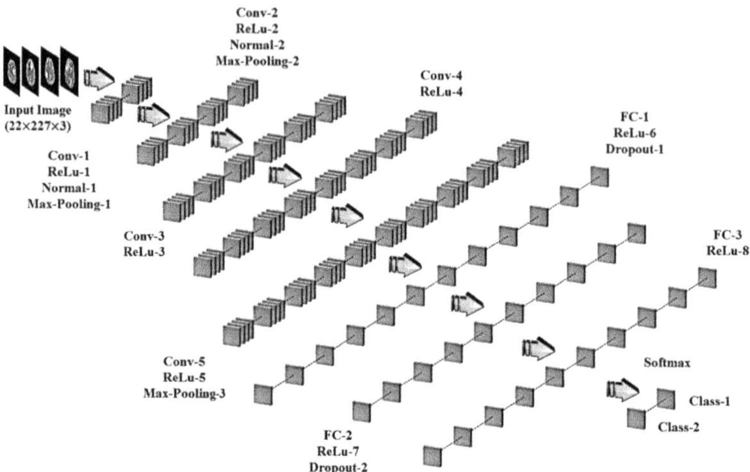


**Figure 4.6.** The AlexNet architecture showing a total of eight layers. (Image courtesy of AtheroPoint™.)

overlapping and pooling was introduced, which summarized the outputs of neighboring groups in the same kernel compared to the traditional adjacent pooling operation where overlapping with neighboring groups is not possible. Using this technique, they can manage to reduce the top-one and top-five error rates by 0.4% and 0.3%, respectively. Overfitting is a serious issue, where the network responds well for known samples but fails to recognize unknown samples. Data augmentation and dropout techniques have been introduced to overcome overfitting. Data augmentation was achieved via translation, rotation, reflection, and altering the intensities of the sample images. In the dropout technique, the selected neurons cannot forward or back-propagate the information they have. In this model, the probability of the dropout of each hidden neuron was 0.5 (the dropout rate) and applied on two fully connected (FC-6, FC-7) layers.

In a competition called ImageNet LSVRC, several deep learning networks were proposed, such as GoogleNet [71] (22 layers), ResNet [72] (18+ layers), an VGG [73] (16+ layers). The error rate was the criterion adopted to compare and contrast these networks. It was noted that even though AlexNet had a smaller number of network layers (an eight-layer architecture), it was a faster and lighter system. We therefore decided to adapt AlexNet for our classification paradigm. AlexNet had already been used successfully for medical image classification, such as breast cancer detection [74], brain hemorrhage classification [75], and diagnosis systems for diabetic retinopathy [76]. Although GoogLeNet is deeper than AlexNet, it has fewer parameters due to the inception structures, which can reduce the risk of overfitting [28] and, due to limited computational resources (a CPU), we decided to go with a less deep network.

### 4.4.3 Transfer learning and workflow

Transfer learning has recently gained attention in the field of DL, as it enables us to reuse a pre-trained model on a new problem with fewer data [77], which can

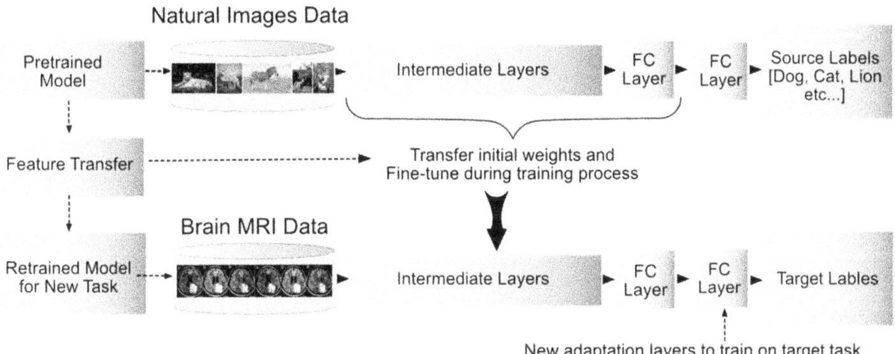


**Figure 4.7.** The transfer learning paradigm. (Image courtesy of AtheroPoint™.)

optimize the classification accuracy of the model. Transfer learning makes the model more generalized for the classification via knowledge gained by the model during the training of an earlier task [78]. It has been proven that a pre-trained model performs as well as training the model from scratch, even though the earlier learned knowledge is entirely different [29]. Therefore, the pre-trained model AlexNet, which was earlier trained on millions of natural images, is reused for brain tumor image classification. The working mechanism of transfer learning is depicted in figure 4.7. The pre-trained features/weights/knowledge are transferred to a new model and act as the initial weights of the same. The new modified model has the same layers of AlexNet except for the last fully connected layer, which is modified for two-class problems (i.e. two brain tumor classes). Although GoogLeNet is deeper than AlexNet, it has fewer parameters owing to the inception structures, which can reduce the risk of overfitting [29].

## 4.4.4 Weight optimization

Gradient descent (GD) is the most popular technique for weight optimization in a neural network [79]. The basic objective of the GD algorithm is to minimize the loss function in different iterations of the training process. The loss function is the difference between the predicted output and the actual output of the samples. As shown in figure 4.8, network weights are updated in each iterative loop, with the objective to minimize error or loss function. The loss function acts as a guide for the training process to move in the right direction and to reach global minima [80]. Typically, there are three types of GD: (a) batch gradient descent, (b) stochastic gradient descent (SGD) [81], and (c) mini-batch gradient descent.

In batch GD the whole dataset is taken during each iteration of the training. However, as all the data are required to reduce the mean error, a problem arises with large data size. In SGD, a few random samples are selected in each iteration instead of the whole dataset. Since AlexNet was previously trained on millions of natural data samples, the stochastic gradient descent algorithm [82] was selected for weight

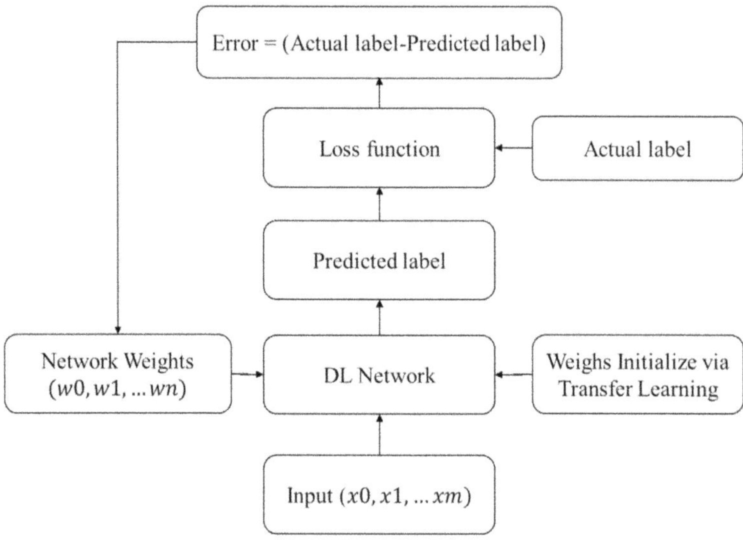


**Figure 4.8.** Weight optimization process.

optimization. The SDG estimates the error gradient for the current state of the model using examples from the training dataset, then updates the weights of the model using the back-propagation of errors algorithm, referred to simply as back-propagation [83]. The amount that the weights are updated by during training is referred to as the step size or the 'learning rate'. The learning rate controls how quickly the model is adapted to the problem. Smaller learning rates require more training epochs, and smaller changes made to the weights in each update, whereas larger learning rates result in rapid changes and require fewer training epochs. A learning rate that is too large can cause the model to converge too quickly to a suboptimal solution, whereas a learning rate that is too small can cause the process to get stuck. Therefore, the selection of the learning rate and epochs is a trade-off between time and accuracy and is challenging. Thus, in our model we kept the optimal learning rate as 0.0001, compared to other state-of-the-art methods (see benchmarking table 4.5) for better weight optimization.

### 4.4.5 A generalization system for tumor classification

The performance of the DL system is compared with various ML techniques, including SVMs and decision trees, etc, using Matlab19a. First- and second-order statistical features were used for ML classification, as shown in table A2 in appendix A. The DL/ML experimental architecture consisted of training and testing phases, as illustrated in figure 4.9. The major steps of the experiment were pre-processing, *K*-fold data partition, feature extraction, and classification. The model was trained and tested on K5 and K10 cross-validation (CV) protocols. Finally, the tumor grade was predicted based on the knowledge (weights) acquired during the training.

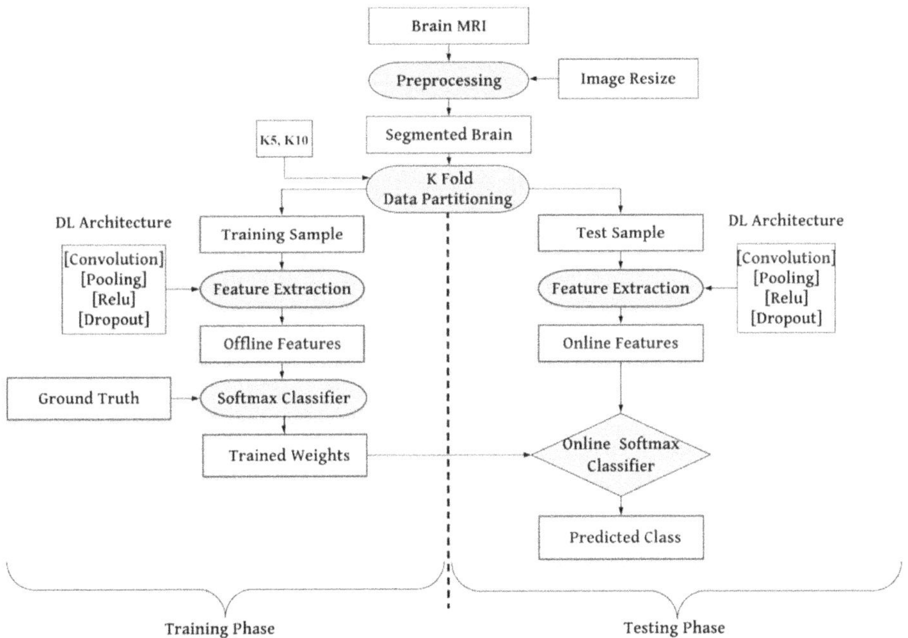

Figure 4.9. Tumor classification system using deep learning.

## 4.5 Experimental protocol, results, and performance evaluation

Well known public brain tumor data (REMBRANDT) [60] was used for the experiments. The brain tumor types, AST, OLI, and GBM were examined for sub-grading for a given ground truth using an eight-layer CNN (AlexNet). The average number of 2D MRI slices per patient in the dataset was 50. The slices (samples) were picked manually under the supervision of experts. The slice is considered as a positive sample if a tumor lesion is visible. Similarly, an MRI slice is considered a negative sample if no such lesion area exists. The true positive (TP), true negative (TN), false negative (FN), and false positive (FP) values of the brain images were evaluated after the testing. The system performance was evaluated in percentages. The performance measures calculated were accuracy (ACC), sensitivity (SE), specificity (SP), positive predictive value (PPV), negative predicted value (NPV), and the area under the curve (AUC) of the receiver operator characteristic (ROC). The average performance (AVG) and standard deviation (SD) were calculated for each set.

### 4.5.1 Parameter selection and simulation

The parameter selection is a highly important and sensitive step in fine-tuning the essential features. Parameters such as data partition protocol, epochs, learning rate, and batch size influence the model performance strongly. In comparison to existing state-of-the-art methods (see the benchmarking table 4.5), we have set the ideal parameters as follows:

- *Partition protocol.* A five- and ten-fold cross-validation (K5 and K10 CV) protocol was adopted in the experiments. The cross-validation protocol checks the ability of the model to prediction unseen data, which leads to diminishing overfitting. In the K5 CV protocol, five different sets were designed, for each set 80% of the samples for training were selected randomly and 20% for validation. Whereas in the K10 CV protocol, ten different random sets were made with 90% training and 10% validation samples in each set.
- *Epochs.* An epoch is a cycle, which refers to passing the entire dataset through all the layers in the network. At the end of each iteration, an error is calculated based on the difference between the actual and predicted labels. The error is considered as a loss and back-propagated the same in the network to update the weights of the network. In addition, the training data are shuffled randomly in each iteration, which makes the training robust. The model performance improves as the number of epochs is increased, but training time is increased as well. As a general aproach, we kept a moderate epoch rate of 1000 during the experiment, which is quite a lot higher than that for other existing state-of-the-art methods.
- *Batch size.* The batch size is a subset of the data samples, which are passed through all the layers per iteration to update the weights in the network. The possible batch size is between one oand the maximum number of samples in the dataset. The batch size is inversely proportional to the iteration per epoch. If the batch size is large the iteration per epoch is less and vice versa. During the whole experiments we decided to select the same batch size of 128, which was used by the developer of .AlexNet [70].
- *Learning rate.* Deep learning neural networks are trained using an optimization algorithm, stochastic gradient descent. In this method, the weights are updated via back-propagating the errors/losses in the network per iteration. The rate of the change of the weights is referred to as the learning rate. The learning rate is between (0–1), a small learning rate needs more epochs to make learning effective and vice versa. In our case, the learning rate was set to 0.0001.
- *Simulation.* Matlab (which is publicly available) provides a platform where we can create, modify, and visually analyze DL architectures using applications (such as Deep Network Designer). Thus, is it very convenient to achieve the goals of the project. Additionally, this automatically labels the data (including, image, video, and audio data) and accelerates the algorithm automatically using a GPU when using frameworks such as TensorFlow, PyTorch, and MxNet [84]. We used a free trial version of Matlab2019a. Further, all the experiments were performed on a Core i3, 2.5 GHz CPU machine with 4 GB RAM. The deep learning simulation was performed on MatlabR2019a using a pre-trained AlexNet convolutional neural network. This package was downloaded from https://www.mathworks.com. In order to check the significance, a *t*-test was performed using MedCalc (Version 19.1) software (Belgium).

### 4.5.2 Results

As discussed above, there are three types of brain tumor: (i) astrocytoma (AST), (ii) oligodendroglioma (OLI), and (iii) glioblastoma-multiform (GBM). AST and OLI only exist in grade II (g2) and grade III (g3), but GBM also presents as grade IV (g4). Our aim is to perform three studies for tumor characterization including: (i) AST (g2 versus g3), (ii) OLI (g2 versus g3), and (3) low-grade glioma versus high-grade glioma classification. These studies were performed on the available ground truth of brain tumor patients as described in the following.

*Study 1: Astrocytoma (g2 versus g3) classification*

In this study, 40 patients with AST participated, including 26 patients with g2 and 14 patients with g3. A total of 557 brain tumor samples (2D brain MRI slices) were used, which further consisted of 356 and 201 tumor samples of grade g2 and g3, respectively. The K5 and K10 performance are compared in figure 4.10.

*Study 2: Oligodendroglioma (g2 versus g3) classification*

This study included 16 patients with nine and seven patients of g2 and g3, respectively. In total 219 samples were manually selected including 128 samples of g2 and 91 samples of g3. The K5 and K10 performances are compared in figure 4.11.

*Study 3: Low-grade glioma versus high-grade glioma classification*

This study included all 79 patients (AST: 40 patients, OLI: 16 patients, and GBM: 23 patients) for low-grade glioma (LGG) versus high-grade glioma (HGG) classification. The LGG includes AST and OLI samples of g2. The HGG class includes AST and OLI samples of g3 and GBM of 4. A total of 1115 samples were used for the experiment including 484 samples of LGG and 631 samples of HGG. The K5 and K10 comparative performance is shown in figure 4.12.

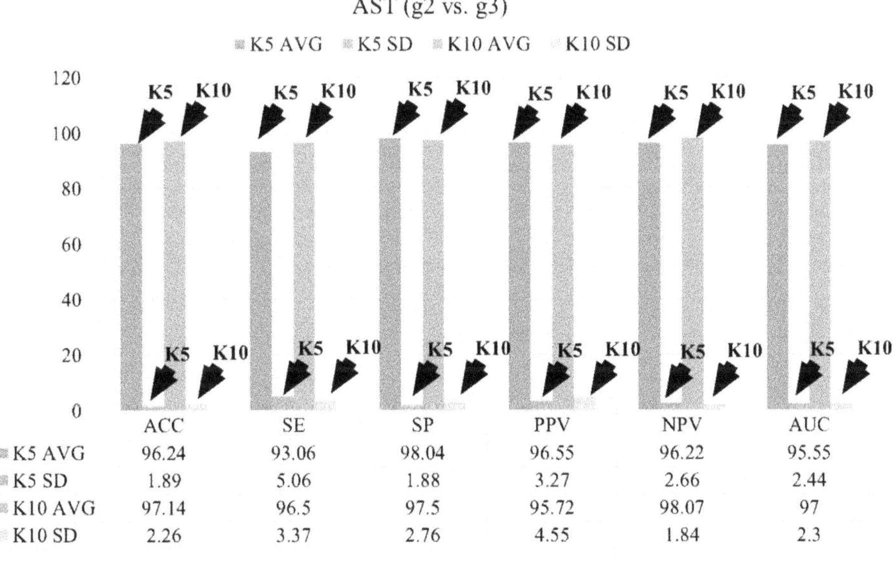



|  | ACC | SE | SP | PPV | NPV | AUC |
|---|---|---|---|---|---|---|
| K5 AVG | 96.24 | 93.06 | 98.04 | 96.55 | 96.22 | 95.55 |
| K5 SD | 1.89 | 5.06 | 1.88 | 3.27 | 2.66 | 2.44 |
| K10 AVG | 97.14 | 96.5 | 97.5 | 95.72 | 98.07 | 97 |
| K10 SD | 2.26 | 3.37 | 2.76 | 4.55 | 1.84 | 2.3 |

**Figure 4.10.** Comparison of K5 versus K10 CV protocols for AST (g2 versus g3) classification.

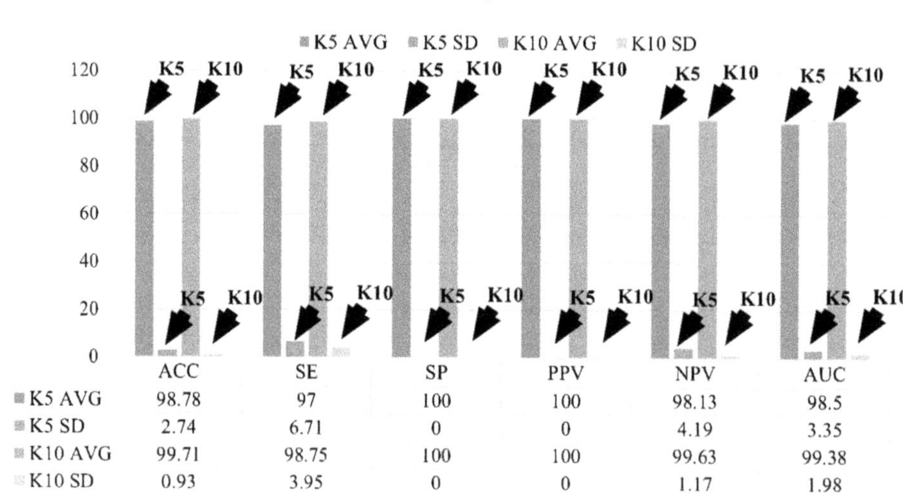



| | ACC | SE | SP | PPV | NPV | AUC |
|---|---|---|---|---|---|---|
| K5 AVG | 98.78 | 97 | 100 | 100 | 98.13 | 98.5 |
| K5 SD | 2.74 | 6.71 | 0 | 0 | 4.19 | 3.35 |
| K10 AVG | 99.71 | 98.75 | 100 | 100 | 99.63 | 99.38 |
| K10 SD | 0.93 | 3.95 | 0 | 0 | 1.17 | 1.98 |

**Figure 4.11.** Comparison of K5 versus K10 CV protocols for OLI (g2 versus g3) classification.

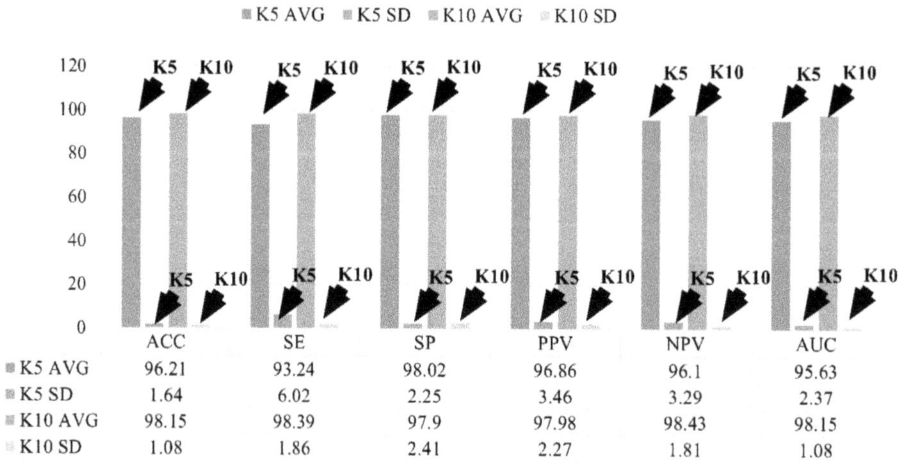



| | ACC | SE | SP | PPV | NPV | AUC |
|---|---|---|---|---|---|---|
| K5 AVG | 96.21 | 93.24 | 98.02 | 96.86 | 96.1 | 95.63 |
| K5 SD | 1.64 | 6.02 | 2.25 | 3.46 | 3.29 | 2.37 |
| K10 AVG | 98.15 | 98.39 | 97.9 | 97.98 | 98.43 | 98.15 |
| K10 SD | 1.08 | 1.86 | 2.41 | 2.27 | 1.81 | 1.08 |

**Figure 4.12.** Comparison of K5 versus K10 CV protocols for LGG versus HGG classification.

The comparative performance of all three studies for the K5 and K10 cross-validation protocols is shown in table 4.2. It is observed that the K10 CV performance is better than that for the K5 CV protocol. Additionally, the standard deviation of K10 was also much lower compared to K5. The average performance of all the above classifications are as follows: ACC (98.33% ± 1.42%), SE (97.88% ± 3.06%), SP (98.47% ± 1.72%), PPV (97.90% ± 2.27%), NPV (98.71% ± 1.61%), and AUC (0.98; $p < 0.0001$). It is also clearly visible that the classification performance is highest for the tumor type OLI (g2 versus g3) and is as follows: ACC (99.71% ± 0.93%),

**Table 4.2.** Overall performance comparison of AST (g2 versus g3), OLI (g2 versus g3), and LGG versus HGG classifications.

| | AST (g2 versus g3) | | | | OLI (g2 versus g3) | | | | LGG versus HGG | | | |
| | K5 | | K10 | | K5 | | K10 | | K5 | | K10 | |
| | AVG | SD | AVG | SD | AVG | SD | AVG | SD | AVG | SD | AVG | SD |
|---|---|---|---|---|---|---|---|---|---|---|---|---|
| ACC | 96.24 | 1.89 | 97.14 | 2.26 | 98.78 | 2.74 | 99.71 | 0.93 | 96.21 | 1.64 | 98.15 | 1.08 |
| SE | 93.06 | 5.06 | 96.50 | 3.37 | 97.00 | 6.71 | 98.75 | 3.95 | 93.24 | 6.02 | 98.39 | 1.86 |
| SP | 98.04 | 1.88 | 97.50 | 2.76 | 100.00 | 0.00 | 100.00 | 0.00 | 98.02 | 2.25 | 97.90 | 2.41 |
| PPV | 96.55 | 3.27 | 95.72 | 4.55 | 100.00 | 0.00 | 100.00 | 0.00 | 96.86 | 3.46 | 97.98 | 2.27 |
| NPV | 96.22 | 2.66 | 98.07 | 1.84 | 98.13 | 4.19 | 99.63 | 1.17 | 96.10 | 3.29 | 98.43 | 1.81 |
| AUC | 95.55 | 2.44 | 97.00 | 2.30 | 98.50 | 3.35 | 99.38 | 1.98 | 95.63 | 2.37 | 98.15 | 1.08 |
| $p$ | < 0.0001 | | < 0.0001 | | < 0.0001 | | < 0.0001 | | < 0.0001 | | < 0.0001 | |

**Table 4.3.** Performance of ML versus DL approaches.

| Cross-validation protocol | DL (CNN: AlexNet) | | ML (SVM) | | Performance improvement of mean DL over mean ML (%) |
|---|---|---|---|---|---|
| | AST (g2 versus g2) OLI (g2 versus g2) LGG versus HGG Mean ACC (%) | AST (g2 versus g2) OLI (g2 versus g2) LGG versus HGG Mean AUC | Mean ACC (%) | Mean AUC | |
| K10 CV | 98.33 ± 1.42 | 0.98 ± 0.17 ($p < 0.0001$) | 84.7 ± 6.00 | 0.89 ± 0.10 ($p < 0.0001$) | 13.86 |
| K5 CV | 97.07 ± 2.09 | 0.97 ± 0.20 ($p < 0.0001$) | 84.4 ± 5.75 | 0.89 ± 0.10 ($p < 0.0001$) | 13.05 |

SE (98.75% ± 3.95%), SP (100.00% ± 0.00%), PPV (100.00% ± 0.00%), NPV (99.63% ± 1.17%), and AUC (99.38% ± 1.98%). In order to check the significance of the experiments, the $t$-test was performed and the significance of the $p$-value was ($\leqslant$0.0001) for all the combinations.

### 4.5.3 Performance evaluation

The performance of the DL classifier is compared with the most popular ML classifiers for first- and second-order statistical features, mean, variance, standard deviation, skewness, kurtosis, contrast, energy entropy, correlation, and homogeneity. These features were extracted from the existing tumor images. The mathematical expressions for these features are given in table A2 in appendix A. Various popular ML classifiers, such as support vector machines (SVMs), linear regression, $k$-nearest neighbors, decision trees, and ensemble methods were used for classification. The inbuilt ML classification tool in MatlabR2019a was used in the experiment. The performance of all ML classifiers is given in table A3 in appendix A. The highest tumor classification accuracy was 84.4%, which was attained via an SVM classifier with a cubic kernel. The performance comparison of the ML and DL techniques is given in table 4.3, and a 13.86% improvement in accuracy was achieved in the DL classifier compared to the ML classifier. Their performance is also compared via the ROC (AUC) curve, as shown in figure 4.13.

## 4.6 Model validation and verification

### 4.6.1 Hypothesis validation

A DL network learns the features during training via suitable filters. The low level features are accrued in the network through earlier layers and the high-level features via later layers. The convolution layers feed the learned features forward to further layers after the convolution operation using suitable filters in the form of multiple

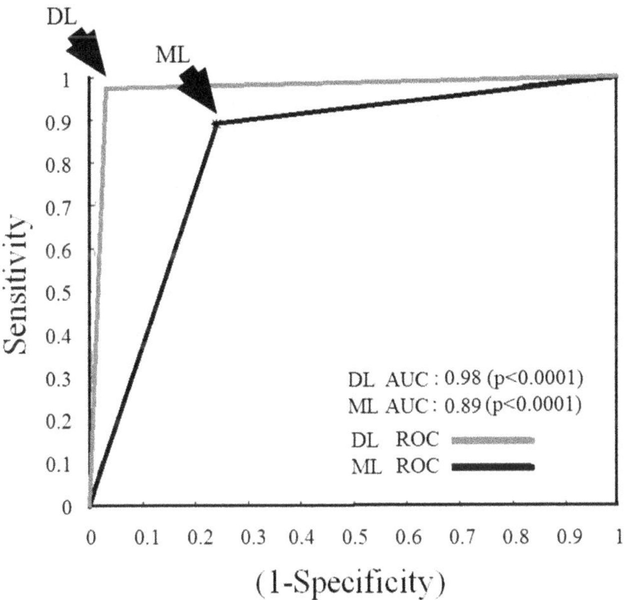


**Figure 4.13.** ROC (AUC) comparison between the ML and DL methods for tumor classification.

2D arrays (channels). Fully connected layers accumulate all the high-level combinations of the features learned by earlier layers. Since the working mechanism of DL is unclear, it is assumed to be a black box. However, the learned features can be visualized using the MATLAB function 'deepDreamImage'. In our experiment the features learned via the last fully connected layer (layer FC-3 in figure 4.6) during the training for all the above tumor classifications are depicted in figure 4.14. These images differentiate visually between high- and low-grade tumors and their combinations. It is also seen from the visual inspection of these images that each tumor type and its grade is different from each other. However, this can be justified by the feature analysis of these images. Thus, in order to understand what the network learns and validate the CNN model for tumor classification between high and low grades, we proposed a unique statistical test for scientific validation using the tumor separation index (TSI). The TSI defines the degree of separation (in percentage) between two classes. In order to calculate the TSI, the mean and standard deviation of the above-mentioned features were calculated for each grade separately. The TSI between two classes is the mathematical expression of equation (4.1), where $a$ and $b$ are the mean of all the features learned by the CNN during the training via the $K$-fold cross-validation protocol (here, $K = 5, 10$).

$$\text{TSI} = \frac{a - b}{a} \times 100. \tag{4.1}$$

The TSIs of the K5 and K10 CV protocol of AST (g2 versus g3), OLI (g2 versus g3), and LGG versus HGG classifications are shown in table 4.4. The features learned by the CNN in the form of images (one round of K10) are shown in figure 4.14. In the

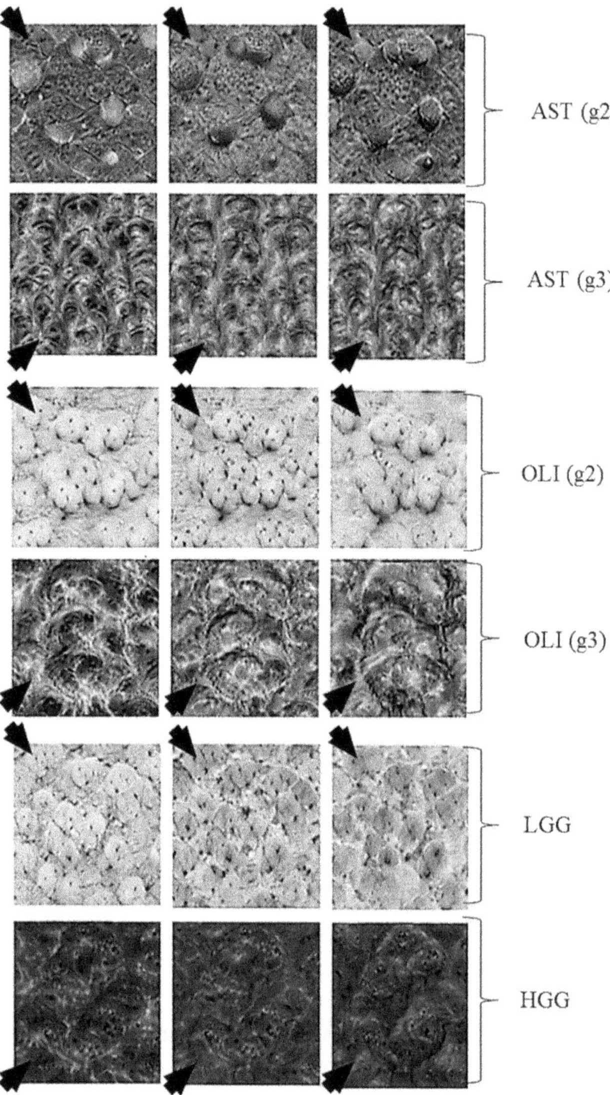

**Figure 4.14.** Features learned by the last fully connected layer of the CNN from training samples of different tumor classification.

**Table 4.4.** Tumor separation index: a statistical analysis test of deep learning features.

|  | AST (g2 versus g3) | OLI (g2 versus g3) | LGG versus HGG |
|---|---|---|---|
| TSI, K5 CV | 13.56% | 26% | 17.54% |
| TSI, K10 CV | 14.23% | 27.32% | 21.42% |

experiment we obtained a higher classification accuracy in the K10 CV protocol than in the K5 protocol and, interestingly, we note that the highest TSI difference is also found in the K10 CV protocol. The mean K10 TSIs of AST (g2 versus g3), OLI (g2 versus g3), and LGG versus HGG are 14.23%, 27.32%, and 21.42% respectively.

### 4.6.2 Software verification

In order to test the legitimacy of the selected classifier (AlexNet), we have chosen well-known known biometric data (Face95) [85] which already produced high classification accuracy. The dataset has 72 subjects, including 58 males and 14 females. It further consists of 20 images per subject. Male versus female image classification was performed using the same model (AlexNet) and under the same circumstances as tumor classification. A total of 1440 image samples, further consisting of 1160 for males and 280 for the females class, we used. Although there is a huge difference in the size of male and female class samples, our model attained overwhelming average performance: ACC (99.86% ± 0.31%), SE (100% ± 0.00%), SP (99.91% ± 0.00%), PPV (99.43% ± 2.00%), NPV (100.00% ± 0.00%), and AUC (100.00% ± 0.00%) for $p \leqslant 0.0001$.

## 4.7 Discussion

Due to the deep locations of tumors in the brain, biopsies can sometimes be life-threatening. However, they are an important gold standard and are always needed for validating the tumor grading processes. MRI is most popularly adapted for brain imaging due to its ability to capture the soft tissue characteristics in high-resolution displays. However, due to the large number of slices, manual delineation and classification (the grading process) is tedious and error-prone. Further, the manual process introduces inter- and intra-operator variabilities. The previous methods of segmentation and grading were based on classical tools of machine learning, which were not robust enough or required heavy number crunching with ad hoc local parameterization. Here, we adapted an automated feature selection combined with grading in a deep learning paradigm. This study used a special form of CNN, AlexNet, for tumor grading, which offered a more robust and accurate solution. To the best of our knowledge, our study is the first to design a grading solution where the tumor types AST and OLI were categorized into their subgrades g2 and g3. Further, our system categorized the tumors into low-grade and high-grade glioma classes. To better understand the grading solution, our study attempted to create an innovative solution for the understanding of the tumor grade separations during the training of the deep learning paradigm. We call it the tumor separation index (TSI). This helps us to validate our scientific hypothesis that showed that DL solutions are better than ML solutions.

*A special note on CNN/AlexNet*
Designing a suitable CNN from scratch is a challenging task. This process takes a lot of iterations of trial and error, and has high computational cost for suitable layer selection [58]. Therefore, utilizing some of the existing recognized models (such as LetNet, VGGNet, AlexNet, GoogLeNet, etc) for the grading mechanism is a wise

idea. Similarly, limited medical data is another serious issue and data are often highly insufficient for classification tasks via DL networks. Thus, we resolved these challenges via a pre-trained CNN network with the transfer learning technique. Transfer learning has been proven to be constructive in transferring feature knowledge learned from a large number of natural databases to medical tasks [78]. Moreover, it has been found in the literature [29] that a pre-trained network performs better compared to a network which is designed from scratch and optimized. Therefore we decided to go with a pre-trained CNN network with transfer learning to enhance the accuracy of tumor classification. As stated above, LetNet, VGGNet, AlexNet, GoogLeNet, etc, have shown remarkable performance in image classification and are able to classify more than 1000 objects. It has been proven that a deeper network such as GoogLeNet performs well compared to shallower networks [29]. However, it is also expensive in terms of computational cost. In the presence of limited computational resources (a CPU), we have chosen AlexNet (an eight-layer CNN) for brain tumor classification, which is a moderately deep network that produced an almost comparable performance compared to GoogLeNet.

*Challenges of deep learning and their solutions*

Although DL performs better for image classification, there are also many challenges associated with this approach. The choice of a suitable layer selection with respect to suitable data is a trade-off between time and accuracy in the DL design. Overfitting is another serious issue if the DL model is trained on limited data. In overfitting, the model shows maximum accuracy on the available data but fails to identify unseen data and the model is unable to generalize the problem. Consequently, the model shows poor performance on unseen data. In our method overfitting is diminished by using transfer learning, a dropout parameter, and the K5/K10 CV protocol. Transfer learning and the K5/K10 CV protocol are already discussed above. The dropout parameter needs to be fixed before the training during network design, and the dropout rate is a degree of randomly disconnecting the nodes between the top layers of the network to avoid overfitting. AlexNet has two dropout layers with a 50% dropout rate, which means that 50% of random nodes are disconnected in fully connected layers.

*A special note on Big-O complexity*

The question of the time complexity of deep learning architectures is not new. Deep learning models are complex architectures involving many components and sub-components and deriving the overall Big-$O$ complexity is very difficult, as the number of feature sizes, components, and sub-components change layer by layer. It has also been seen that the hardware (H/W) components such as GPUs with multiple cores and parallel computing ability have a significant impact on the time complexity when compared with the single-core CPUs on which Big-$O$ time complexity algorithms are estimated. There is also the additional overhead of the distribution of jobs on multiple core GPUs, therefore the desired scalability is also not achieved. However, based on certain empirical benchmarks [86] some theories have been put forward at both the micro- and macro-levels. The total number of attributes on which the time complexity of the AlexNet used for our purpose is dependent on are as follows: (i) the activation function used: ReLU; (ii) the loss function used: cross-entropy loss function; (iii) the batch size used: number of inputs ($B$); (iv) the CNN features such as (a) the matrix size—the size of input data trained on, (b) the kernel size—the size of filter data ($K_1$), (c) the number of kernels—($P$), and (d) strides—the number of movements a filter must make ($s$); and (v) pooling features such as kernel size: the size of pooling kernel ($K_2$). In a convolution operation, if $I$ is the input image and $O_c$ be the output of the operation [87], the operation being

$$O_c = B \times (s \times P \times (K_1 \times I)), \tag{4.2}$$

where $P$, $B$, and $s$ are constants and the size of $O_c$ is dependent on the size of the input image, the kernel size, and the number of strides. In this case for a single-core CPU, the Big-$O$ complexity is dependent on the number of input images (denoted as $n$) as well as the size of each input image. Gray images have a depth of two while color images have a depth of three (RGB). In this case, the Big-$O$ time complexity is $n \times R$, where $R$ represents the size of the image. Depending on the size, the number of operations can vary from $O(nxy)$ to $O(nxyz)$, where $x$, $y$, and $z$ represents the length, width, and height of the image. Similarly, the output of the pooling operation can be denoted as $O_p$, which is given by

$$O_p = K_2 \times (O_c). \tag{4.3}$$

The Big-$O$ complexity is the same as in the previous operation where the Big-$O$ complexity ranges from $O(nxy)$ to $O(nxyz)$. If the number of deep layers is considered as variable $l$, then the complexity scales from $O(lnxy)$ to $O(lnxyz)$ for the total network. The complexity of the loss function and backward propagation is $O(n^2)$ for a two-dimensional gray image but requires $O(n^3)$ for colored images [88]. Overall, time complexity by the overall estimate should not be greater than $O(nxyz)$. However, there are other theories as well. One theory uses Betti numbers [89] to estimate the time complexity of deep networks as $O(2^h)$, where $h$ is the number of hidden neurons. Another theory determines the time complexity of deep networks using the VC dimension [90] at $O(w \log w)$, where $w$ represents the number of weights in a deep network. In conclusion, several theories expound the theory of time complexity of deep networks but no literature clearly propagates the exactness of it, since the empirical time varies from hardware to hardware, and software to software.

*Strengths, weaknesses, and extensions*

Our study considered a relatively moderate deep network that wins the trade-off between computational cost and accuracy. As discussed above pre-training with transfer learning has many advantages that inherently add value in our state-of-the-art method. The network can better fine-tune the features with a slow learning rate and high epoch rate and, thus, we can proudly claim that we retained a slow learning rate (0.0001) and high epochs (1000), ensuring that the model is robust in character-izing the brain tumor classes. Even though the proposed system is stable, robust, and has high accuracy, it has a few limitations.

(i) Due to inadequate labeled data of brain tumor MRIs, our study was limited to single-institute brain tumor classification. This can be further extended to multicenter data with similar labels. There could be a website that would be the host for data collection. People could test their methods and leave their data there. This way, more datasets could be collected over time. Comparative papers could thus be developed using multicenter datasets.

(ii) Another concern involves sample selection. In this experiment all the tumorous 2D MRI slices were selected manually from MRI voxels, however, due to unpredictable tumor size, there exists variation in tumor area in individual neighboring slices. This may lead to a false label prediction. This problem may be solved by selecting approximately similar sized tumor ROI slices via (CS $\pm$ $n$), where CS is the center of the MRI slice, and $n$ is the number of slices before and after the central slice. The value of $n$ will be carefully chosen via manual observation. This technique may improve the classification accuracy. Another solution could be 3D MRI voxel classification of the tumor via 3D CNN.

The performance of the system can be improved in many respects, and this will be considered as future work. Here, we have used only T2W MRI in the axial view. Multi-view (axial, sagittal, and coronal) and multi-protocol MRI (T1W, T2W, and FLAIR) can provide more systematic tumor features. Hence, the performance of the system can be improved by incorporating multi-view and multi-protocol MRI. Moreover, features such as metabolic and genetic data and the anatomical attributes of the brain could also be incorporated into the system for better tumors grade classification.

*Benchmarking*

This subsection compares and contrasts our proposed method against the previously developed techniques. In our literature survey, we found that a very limited number of methods were proposed for brain tumors grading. This may be due mainly to the challenges in brain tumor classification, limited access to appropriate medical datasets, and, finally, a lack of ground truth (gold standard) information. We thus compare our method against the limited number of existing focused studies as shown in table 4.5.

## 4.8 Conclusion

In this study, a novel method was proposed to develop an automatic brain tumor grading tool using a deep learning model (CNN). The transfer learning method was used to attain superior performance. The pre-trained model enabled us to reuse an already learned feature map for a new problem in which the data sample was comparatively small and assist us to attain higher accuracy. Additionally, transfer learning makes the model more generalized via knowledge gained during earlier tasks (weights). A pre-trained CNN model (AlexNet) was selected for tumor classification. The brain MRI dataset of 79 patients, which consisted of three brain tumor type, astrocytoma (AST), oligodendroglioma (OLI), and glioblastoma-multiform (GBM). AST and OLI are available in grade II (g2) and grade III (g3), whereas GBM is only available in grade IV (g4). In this chapter, three studies of tumor classification were performed, including (i) AST (g2 versus g3), (ii) OLI (g2 versus g3), and (iii) low-grade glioma (LGG) versus high-grade glioma HGG classification.

**Table 4.5.** Benchmarking: a comparison between existing state-of-the-art techniques.

| C1 RF | C2 DS | C3 C | C4 P | C5 PP | C6 CT | C7 TPar | C8 ES | C9 HP |
|---|---|---|---|---|---|---|---|---|
| Pan *et al* (2015) [54] | BRATS 2014 | 2 | NA | Augmentation | CNN ANN | NA | H/W: NA<br>TP: 5-NA<br>P&V: NA | SE = NA,<br>SP = NA |
| Mohsen *et al* (2018) [90] | Harvard Medical School website | 4 | T2W | Segmentation (fuzzy C-means) feature reduction (PCA) | DNN (seven-layer) | NA | H/W: NA<br>TP: K7 CV<br>P&V: NA | ACC = 96.97%<br>Recall = 97%<br>Precision = 97%<br>F-measure = 97%<br>AUC = 98.4% |
| Ari and Hanbay (2018) [16] | Kwan R K S *et al* [4] | 2 | NA | Image enhanced | ELM CNN GW SF | NA | H/W: CPU<br>TP: NA<br>P&V: NA | ACC = 97.18%,<br>SE = 96.80%,<br>SP = 97.12% |
| Ge *et al* (2018) [57] | BraTS 2017 Mayo Clinic in the USA | 3 | T1W T2W FLAIR | Augmentation image enhanced | CNN (seven-layer) | Epochs = 50<br>LR = 0.0001<br>BS = NA | H/W: CPU, GPU<br>TP: NA<br>P&V: NA | ACC = 90.87% |
| Anaraki *et al* (2018) [58] | REMBRAND TTCGA | 2 | T1W | Raw image | CNN-1 (seven-layer)<br>CNN-2 (eight-layer) | Epochs = 100<br>LR = 0.0001<br>BS = NA | | Study 1 ACC = 90.9%<br>Study 2 ACC = 94.2% |

| | | | | | | | | |
|---|---|---|---|---|---|---|---|---|
| Yang *et al* (2018) [28] | REMBRANDT | 3 | T1W | ROI patching | Alxnet (eight-layer) GoogLeNet (22-layer) | Epochs = 30 LR = 0.1 BS = 50 | H/W: CPU TP: K5 CV | V. ACC = 86.7% T. ACC = 90.9% AUC = 93.9% |
| Proposed method | REMBRANDT | 2 | T2W | Skull stripping | CNN AlexNet (eight-layer) | Epochs = 1000 LR = 0.0001 BS = 128 | H/W: CPU TP: K5/ K10 CV P&V: Yes | ACC = 99.71% ± 0.93%, SE = 98.75% ± 3.95%, SP = 100.00% ± 0.00%, PPV = 100.00% ± 0.00%, NPV = 99.63% ± 1.17% and AUC = 99.38% ± 1.98% |

RN: row no; RF: reference; DS: data source; C: classes; P: MRI protocol; N: no. of patients; PP: pre-processing; CT: classification technique; ES: experimental set-up; HP: highest performance; R#: row no.; C#: column no.; T1W: T1 weighted; T2W: T2-weighted; FLAIR: fluid-attenuated inversion recovery; ACC: accuracy; SE: sensitivity; SP: specificity; AUC: area under curve; DSC: Dice coefficient; ANN: artificial neural network; DWT: discrete wavelet transform; DNN: deep neural network; GW: Gabor wavelets; SF: statistical, feature-based method; NA: not available; H/W: hardware; TP: training protocol; P&V: performance and validation; TPar: training parameters.

The performance of the classification was tested on a K5 and K10 cross-validation protocol. The model achieved the highest performance in OLI (g2 versus g3) categorization on the K10 protocol as follows: ACC (99.71% ± 0.93%), SE (98.75% ± 3.95%), SP (100.00% ± 0.00%), PPV (100.00% ± 0.00%), NPV (99.63% ± 1.17%), and AUC (99.38% ± 1.98%) for $p < 0.0001$. Additionally, the performance of the CNN was compared to different ML classifiers using first- and second-order statistical features of similar data. The CNN classifier gained 13.86% improvement in accuracy over the ML classifier (SVM). Moreover, the model was validated and verified using the tumor separation index (TSI) and face biometric data, respectively.

**Funding:** This research received no external funding.

**Conflicts of interest:** The authors declare no conflict of interest.

# Appendix A

**Table A1.** AlexNet architecture.

| Layer # | Layer initial | Layer | Remarks |
|---|---|---|---|
| 1 | Data | Image input | 227 × 227 × 3 images with 'zero center' normalization |
| 2 | 'conv1' | Convolution-1 | 96 11 × 11 × 3 convolutions with stride [4 4] and padding [0 0 0 0] |
| 3 | Relu1 | ReLu-1 | |
| 4 | Normal1 | Normal-1 | Cross-channel normalization with five channels per element |
| 5 | Pool1 | Max-pooling-1 | 3 × 3 max-pooling with stride [2 2] and padding [0 0 0 0] |
| 6 | Conv2 | Convolution | 256 5 × 5 × 48 convolutions with stride [1 1] and padding [2 2 2 2] |
| 7 | Relu2 | ReLu | |
| 8 | Normal2 | | Cross-channel normalization with five channels per element |
| 9 | Pool2 | Max-pooling | 3 × 3 max-pooling with stride [2 2] and padding [0 0 0 0] |
| 10 | Conv3 | Convolution | 384 3 × 3 × 256 convolutions with stride [1 1] and padding [1 1 1 1] |
| 11 | Relu3 | ReLu | |
| 12 | Conv4 | Convolution | 384 3 × 3 × 192 convolutions with stride [1 1] and padding [1 1 1 1] |
| 13 | Relu4 | ReLu | |
| 14 | Conv5 | Convolution | 256 3 × 3 × 192 convolutions with stride [1 1] and padding [1 1 1 1] |
| 15 | Relu5 | ReLu | |
| 16 | Pool5 | Max-pooling | 3 × 3 max-pooling with stride [2 2] and padding [0 0 0 0] |

| 17 | fc6 | Fully connected | 4096 fully connected layers |
| 18 | Relu6 | ReLu | |
| 19 | drop6′ | Dropout | 50% dropout |
| 20 | fc7 | Fully connected | 4096 fully connected layers |
| 21 | Relu7 | ReLu | |
| 22 | drop7′ | Dropout | 50% dropout |
| 23 | Modified1 | Fully connected | 2 fully connected layers |
| 24 | Modified2 | Softmax | |
| 25 | Modified3 | Classification output | Cross entropy |

**Table A2.** Texture feature expressions.

| Feature | Expression | Feature | Expression |
| --- | --- | --- | --- |
| Mean (m) | $m = \dfrac{1}{N}\sum_{i=1}^{n}\sum_{j=1}^{n} I(i,j)$ | Contrast ($C$) | $C = \sum_{i,j} \lvert i - j \rvert^2 \times p(i,j)$ |
| Variance (var) | $\mathrm{var} = \dfrac{\sum i \sum j(I(i,j) - m)^2}{N}$ | Energy (E) | $E = \sum_{i,j} P(i,j)^2$ |
| Standard deviation (std) | $\mathrm{std} = \sqrt{\dfrac{\sum i \sum j(I(i,j) - m)^2}{N}}$ | Entropy (etp) | $\mathrm{etp} = -\sum_{i,j} P(i,j) \times \log_2(P(i,j))$ |
| Skewness (skw) | $\mathrm{skw} = \dfrac{1}{N}\dfrac{\sum i \sum j(I(i,j) - m)^3}{\mathrm{std}^3}$ | Correlation (cor) | $\mathrm{cor} = \sum_{i,j} \dfrac{(i - m)(j - m) \times P(i,j)}{\mathrm{std}_i \times \mathrm{std}_j}$ |
| Kurtosis (kts) | $\mathrm{kts} = \dfrac{1}{N}\dfrac{\sum i \sum j(I(i,j) - m)^4}{\mathrm{std}^4}$ | Homogeneity (H) | $H = \sum_{i,j} \dfrac{P(i,j)}{1 + \lvert i - j \rvert}$ |

**Table A3.** Performance of different machine learning classifiers using texture features.

| Classifier | Kernel | Accuracy |
| --- | --- | --- |
| Tree | Fine | 68.6% |
| Tree | Medium | 67.1% |
| Tree | Curse | 64.5% |
| Linear discriminant | | 81.5% |
| Logistic regression | | 83.3% |
| SVM | Linear | 70.7% |
| SVM | Quadratic | 81.9% |
| **SVM** | **Cubic** | **84.7%** |
| KNN | Fine | 80.1% |
| KNN | Medium | 71.5% |

*(Continued)*

**Table A3.** (*Continued*)

| Classifier | Kernel | Accuracy |
| --- | --- | --- |
| KNN | Corse | 64.6% |
| KNN | Cosine | 75.9% |
| KNN | Cubic | 71.3% |
| KNN | Weighted | 77.6% |
| SVM | Fine Gaussian | 77.2% |
| SVM | Medium Gaussian | 69.3% |
| SVM | Coarse Gaussian | 63.9% |
| Ensemble | Boosted tree | 72.7% |
| Ensemble | Bagged tree | 77.7% |
| Ensemble | Subspace discriminant | 80.6% |
| Ensemble | Subspace KNN | 59.8% |
| Ensemble | RUS boosted trees | 73.8% |

# References


[1] *Latest Global Cancer Data* https://www.who.int/news-room/fact-sheets/detail/cancer

[2] Ngulde S I *et al* 2015 Improving brain tumor research in resource-limited countries: a review of the literature focusing on West Africa *Cureus* **7** 1–7

[3] American Society of Clinical Oncology 2019 *Brain Tumor: Statistics* https://cancer.net/cancer-types/brain-tumor/statistics

[4] The Brain Tumor Charity *Brain Tumor Basics* https://thebraintumourcharity.org/

[5] Louis D N *et al* 2016 The 2016 World Health Organization classification of tumors of the central nervous system: a summary *Acta Neuropathol.* **131** 803–20

[6] Louis D N *et al* 2014 International Society of Neuropathology-Haarlem consensus guidelines for nervous system tumor classification and grading *Brain Pathol.* **24** 429–35

[7] Gurcan M N, Boucheron L E, Can A, Madabhushi A, Rajpoot N M and Yener B 2009 Histopathological image analysis: a review *IEEE Rev. Biomed. Eng.* **2** 147–71

[8] Collins V P 2004 Brain tumours: classification and genes *J. Neurol. Neurosurg. Psych.* **75** ii2–11

[9] Behin A, Hoang-xuan K, Carpentier A F and Delattre J 2003 Primary brain tumours in adults *Lancet* **361** 323–31

[10] Ainsworth N L *et al* 2016 Quantitative and textural analysis of magnetization transfer and diffusion images in the early detection of brain metastases *Magn. Reson. Med.* **77** 1987–95

[11] McClure P *et al* 2014 *In-vitro* and *in-vivo* diagnostic techniques for prostate cancer: a review *J. Biomed. Nanotechnol.* **10** 2747–77

[12] Acharya U R, Faust O, Sree S V, Molinari F, Garberoglio R and Suri J S 2011 Cost-effective and non-invasive automated benign and malignant thyroid lesion classification in 3D contrast-enhanced ultrasound using combination of wavelets and textures: a class of ThyroScan™ algorithms *Technol. Cancer Res. Treat.* **10** 371–80

[13] Acharya U R, Faust O, Sree S V, Molinari F and Suri J S 2012 ThyroScreen system: high resolution ultrasound thyroid image characterization into benign and malignant classes using

novel combination of texture and discrete wavelet transform *Comput. Methods Programs Biomed.* **107** 233–41

[14] Pareek G *et al* 2013 Prostate tissue characterization/classification in 144 patient population using wavelet and higher order spectra features from transrectal ultrasound images *Technol. Cancer Res. Treat.* **12** 545–57

[15] Acharya U R *et al* 2013 Effect of complex wavelet transform filter on thyroid tumor classification in three-dimensional ultrasound *Proc. Inst. Mech. Eng.* H **227** 284–92

[16] Ari A and Hanbay D 2018 Deep learning based brain tumor classification and detection system *Turkish J. Electr. Eng. Comput. Sci.* **26** 2275–86

[17] Tandel G S *et al* 2019 A review on a deep learning perspective in brain cancer classification *Cancers* **11** 111

[18] Sasikala M and Kumaravel N 2008 A wavelet-based optimal texture feature set for classification of brain tumours *J. Med. Eng. Technol.* **32** 198–205

[19] Prasad H, Martis R J, Acharya U R, Min L C and Suri J S 2013 Application of higher order spectra for accurate delineation of atrial arrhythmia *2013 35th Annual Int. Conf. of the IEEE Engineering in Medicine and Biology Society* (Piscataway, NJ: IEEE) pp 57–60

[20] Araki T *et al* 2016 PCA-based polling strategy in machine learning framework for coronary artery disease risk assessment in intravascular ultrasound: a link between carotid and coronary grayscale plaque morphology *Comput. Methods Programs Biomed.* **128** 137–58

[21] Acharya U *et al* 2012 Evolutionary algorithm-based classifier parameter tuning for automatic ovarian cancer tissue characterization and classification *Ultraschall Med.—Eur. J. Ultrasound* **35** 237–45

[22] Shrivastava V K, Londhe N D, Sonawane R S and Suri J S 2016 Computer-aided diagnosis of psoriasis skin images with HOS, texture and color features: a first comparative study of its kind *Comput. Methods Programs Biomed.* **126** 98–109

[23] Saba L *et al* 2017 Plaque tissue morphology-based stroke risk stratification using carotid ultrasound: a polling-based PCA learning paradigm *J. Med. Syst.* **41** 98

[24] Araki T *et al* 2017 Stroke risk stratification and its validation using ultrasonic echolucent carotid wall plaque morphology: a machine learning paradigm *Comput. Biol. Med.* **80** 77–96

[25] Ayman El-Baz J S S 2019 *Big Data in Multimodal Medical Imaging* 1st edn (Boca Raton, FL: CRC Press)

[26] Tajbakhsh N *et al* 2016 Convolutional neural networks for medical image analysis: full training or fine tuning? *IEEE Trans. Med. Imaging* **35** 1299–312

[27] Shin H C *et al* 2016 Deep convolutional neural networks for computer-aided detection: CNN architectures, dataset characteristics and transfer learning *IEEE Trans. Med. Imaging* **35** 1285–98

[28] Yang Y *et al* 2018 Glioma grading on conventional MR images: a deep learning study with transfer learning *Front. Neurosci.* **12** 1–10

[29] Pan Y *et al* 2015 Brain tumor grading based on neural networks and convolutional neural networks *2015 37th Annual Int. Conf. of the IEEE Engineering in Medicine and Biology Society (EMBC)* (Piscataway, NJ: IEEE) pp 699–702

[30] Khawaldeh S, Pervaiz U, Rafiq A and Alkhawaldeh R 2017 Noninvasive grading of glioma tumor using magnetic resonance imaging with convolutional neural networks *Appl. Sci.* **8** 27

[31] Chang P *et al* 2018 Deep-learning convolutional neural networks accurately classify genetic mutations in gliomas *Am. J. Neuroradiol.* **39** 1201–7

[32] Yonekura A, Kawanaka H, Prasath V B S, Aronow B J and Takase H 2017 Glioblastoma multiforme tissue histopathology images based disease stage classification with deep CNN *6th Int. Conf. on Informatics, Electronics and Vision and 2017 7th Int. Symp. in Computational Medical and Health Technology* vol 2018 *(January)* pp 1–5

[33] Xu Y *et al* 2017 Large scale tissue histopathology image classification, segmentation, and visualization via deep convolutional activation features *BCM Bioinform.* **18** 281

[34] Salvado O, Hillenbrand C, Suri J and Wilson D L 2004 MR coil sensitivity inhomogeneity correction for plaque characterization in carotid arteries *Medical Imaging 2004: Image Processing Proc. SPIE* vol 5370 https://doi.org/10.1117/12.535958

[35] Suri J and Laxminarayan S 2005 *Handbook of Biomedical Image Analysis* (New York: Springer)

[36] Abiwinanda N, Hanif M, Hesaputra S T, Handayani A and Mengko T R 2019 Brain tumor classification using convolutional neural network *IFMBE Proc.* **68** 183–9

[37] Joshi D M, Rana N K and Misra V M 2010 Classification of brain cancer using artificial neural network *ICECT 2010—Proc. 2010 2nd Int. Conf. Electron. Comput. Technol.* pp 112–6

[38] Akkus Z, Galimzianova A, Hoogi A, Rubin D L and Erickson B J 2017 Deep learning for brain MRI segmentation: state of the art and future directions *J. Digit. Imaging* **30** 449–59

[39] Ryu Y J, Choi S H, Park S J, Yun T J, Kim J H and Sohn C H 2014 Glioma: application of whole-tumor texture analysis of diffusion-weighted imaging for the evaluation of tumor heterogeneity *PLoS One* **9** e108335

[40] Abdel-Maksoud E, Elmogy M and Al-awadi R 2015 Brain tumor segmentation based on a hybrid clustering technique *Egypt Inform. J.* **16** 71–81

[41] Veta M, Van Diest P J, Kornegoor R, Huisman A, Viergever M A and Josien P W 2013 Automatic nuclei segmentation in H & E stained breast cancer histopathology images *PLoS ONE* **8** e70221

[42] Balafar M A, Ramli A R, Saripan M I and Mashohor S 2010 Review of brain MRI image segmentation methods *Artif. Intell. Rev.* **33** 261–74

[43] Lavanyadevi R, MacHakowsalya M, Nivethitha J and Niranjil Kumar A 2017 Brain tumor classification and segmentation in MRI images using PNN *Proc. 2017 IEEE Int. Conf. Electr. Instrum. Commun. Eng.* (Piscataway, NJ: IEEE) pp 1–6

[44] Greenspan H, Ruf A and Goldberger J 2006 Constrained Gaussian mixture model framework for automatic segmentation of MR brain images *IEEE Trans. Med. Imaging* **25** 1233–45

[45] Zacharaki E I *et al* 2009 Classification of brain tumor type and grade using MRI texture and shape in a machine learning scheme *Magn. Reson. Med.* **62** 1609–18

[46] Abdullah N, Chuen L W, Ngah U K and Ahmad K A 2011 Improvement of MRI brain classification using principal component analysis *Proc. 2011 IEEE Int. Conf. on Control System, Computing and Engineering* (Piscataway, NJ: IEEE) pp 557–61

[47] Zacharaki E I *et al* 2009 Classification of brain tumor type and grade using MRI texture and shape in a machine learning scheme *Magn. Reson. Med.* **62** 1609–18

[48] Verma R *et al* 2008 Multiparametric tissue characterization of brain neoplasms and their recurrence using pattern classification of MR images *Acad. Radiol.* **15** 966–77

[49] Liberman G *et al* 2013 Automatic multi-modal MR tissue classification for the assessment of response to bevacizumab in patients with glioblastoma *Eur. J. Radiol.* **82** E87–94

[50] Stoyanov D, Taylor Z and Hutchison D 2018 *Understanding and Interpreting Machine Learning in Medical Image Computing Applications* vol 11038 (Cham: Springer)

[51] Biswas M *et al* 2019 State-of-the-art review on deep learning in medical imaging *Front. Biosci.* **24** 380–406

[52] Erickson B J, Korfiatis P, Akkus Z and Kline T L 2017 Machine learning for medical imaging *Radiographics* **37** 505–15

[53] Pan Y *et al* 2015 Brain tumor grading based on neural networks and convolutional neural networks *Proc. Annu. Int. Conf. IEEE Eng. Med. Biol. Soc.* (Piscataway, NJ: IEEE) pp 699–702

[54] Mohsen H, El-Dahshan E-S A, El-Horbaty E-S M and Salem A-B M 2018 Classification using deep learning neural networks for brain tumors *Futur. Comput. Inform. J.* **3** 68–71

[55] Kuppili V *et al* 2017 Extreme learning machine framework for risk stratification of fatty liver disease using ultrasound tissue characterization **41** 152

[56] Ge C, Gu I Y-H, Jakola A S and Yang J 2018 Deep learning and multi-sensor fusion for glioma classification using multistream 2D convolutional networks *2018 40th Annual Int. Conf. of the IEEE Engineering in Medicine and Biology Society* (Piscataway, NJ: IEEE) pp 5894–7

[57] Kabir Anaraki A, Ayati M and Kazemi F 2019 Magnetic resonance imaging-based brain tumor grades classification and grading via convolutional neural networks and genetic algorithms *Biocybern. Biomed. Eng.* **39** 63–74

[58] Clark K *et al* 2013 The Cancer Imaging Archive (TCIA): maintaining and operating a public information repository *J. Digit. Imaging* **26** 1045–57

[59] Scarpace L, Flanders A E, Jain R, Mikkelsen T and Andrews D W 2015 Public data (REMBRANDT). Data From REMBRANDT *The Cancer Imaging Archive* **10** k9

[60] NeuroImaging Tools and Resources Collaboratory https://nitrc.org/projects/mricron

[61] Shattuck D W and Leahy R M 2002 BrainSuite: an automated cortical surface identification tool *Med. Image Anal.* **6** 129–42 BrainSuite2018a

[62] Saba L *et al* 2019 The present and future of deep learning in radiology *Eur. J. Radiol.* **114** 14–24

[63] Biswas M *et al* 2019 Deep learning fully convolution network for lumen characterization in diabetic patients using carotid ultrasound: a tool for stroke risk *Med. Biol. Eng. Comput.* **57** 543–64

[64] Biswas M *et al* 2018 Symtosis: a liver ultrasound tissue characterization and risk stratification in optimized deep learning paradigm *Comput. Methods Programs Biomed.* **155** 165–77

[65] Pereira S, Pinto A, Alves V and Silva C A 2016 Brain tumor segmentation using convolutional neural networks in MRI images *IEEE Trans. Med. Imaging* **35** 1240–51

[66] Saba L *et al* 2015 Is there an association between cerebral microbleeds and leukoaraiosis? *J. Stroke Cerebrovasc. Dis.* **24** 284–9

[67] Ayman El-Baz J S S 2020 *Neurological Disorders and Imaging Physics: Application to Autism Spectrum Disorders and Alzheimer's* (Bristol: IOP Publishing)

[68] Manuel J S S, Casanova F and El-Baz A S 2013 *Imaging the Brain in Autism* (New York: Springer) p 387

[69] Krizhevsky A, Sutskever I and Hinton G E 2012 2012 AlexNet *Adv. Neural Inf. Process. Syst.* 1–9

[70] Szegedy C *et al* 2015 Going deeper with convolutions *2015 IEEE Conf. on Computer Vision and Pattern Recognition (7–12 June)* (Piscataway, NJ: IEEE) pp 1–9

[71] Schaetti N 2018 Character-based convolutional neural network and res: Net 18 for Twitter author profiling: notebook for PAN at CLEF 2018 *CEUR Workshop Proc.* **vol 2125**

[72] He K, Zhang X, Ren S and Sun J 2016 Deep residual learning for image recognition *2016 IEEE Conf. on Computer Vision and Pattern Recognition (December)* (Piscataway, NJ: IEEE) pp 770–8

[73] Ragab D A, Sharkas M, Marshall S and Ren J 2019 Breast cancer detection using deep convolutional neural networks and support vector machines *Peer J.* **2019** 1–23

[74] Dawud A M, Yurtkan K and Oztoprak H 2019 Application of deep learning in neuro-radiology: brain haemorrhage classification using transfer learning *Comput. Intell. Neurosci.* **2019** 4629859

[75] Mansour R F 2018 Deep-learning-based automatic computer-aided diagnosis system for diabetic retinopathy *Biomed. Eng. Lett.* **8** 41–57

[76] Taylor M E and Stone P 2009 Transfer learning for reinforcement learning domains: a survey *J. Mach. Learn. Res.* **10** 1633–85

[77] Pan S J and Yang Q 2010 A survey on transfer learning *IEEE Trans. Knowl. Data Eng.* **22** 1345–59

[78] Bottou L 1991 Stochastic gradient learning in neural networks *Proc. Neuro-Nımes* **91**

[79] El-Baz A S, Acharya U R and Mirmehdi M (ed) 2011 *Multi Modality State-of-the-Art Medical Image Segmentation and Registration Methodologies* vol 12 (New York: Springer)

[80] Saba L *et al* 2016 Automated stratification of liver disease in ultrasound: an online accurate feature classification paradigm *Comput. Methods Programs Biomed.* **130** 118–34

[81] Krizhevsky A, Sutskever I and Hinton G E ImageNet classification with deep convolutional neural networks *Advances in Neural Information Processing Systems 25 (NIPS 2012)*

[82] El-Baz A, Gimel'farb G and Suri J S 2015 *Stochastic Modeling for Medical Image Analysis* (Boca Raton, FL: CRC Press)

[83] Matlab https://in.mathworks.com/solutions/deep-learning.html

[84] Data: Face95 https://cmp.felk.cvut.cz/~spacelib/faces/faces95.html

[85] Justus D, Brennan J, Bonner S and McGough A S 2018 Predicting the computational cost of deep learning models *2018 IEEE Int. Conf. on Big Data* (Piscataway, NJ: IEEE) pp 3873–82

[86] Qian Y, Dong J, Wang W and Tan T 2015 Deep learning for steganalysis via convolutional neural networks *Proc. SPIE* **9409** 94090J

[87] Schmidhuber J 1992 A fixed size storage $O(n^3)$ time complexity learning algorithm for fully recurrent continually running networks *Neural Comput.* **4** 243–8

[88] Bianchini M and Scarselli F 2014 On the complexity of neural network classifiers: a comparison between shallow and deep architectures *IEEE Trans. Neural Networks Learn. Syst.* **25** 1553–65

[89] Sontag E D 1998 VC dimension of neural networks *NATO ASI Ser. F Comput. Syst. Sci.* **168** 69–96

[90] Mohsen H, El-Dahshan E-S A, El-Horbaty E-S M and Salem A-B M 2017 Classification using deep learning neural networks for brain tumors *Futur. Comput. Inform. J.* **3** 68–71

**IOP** Publishing

# Multimodality Imaging, Volume 1
### Deep learning applications
**Mainak Biswas and Jasjit S Suri**

# Chapter 5

# Magnetic resonance based Wilson's disease tissue characterization in an artificial intelligence framework using transfer learning

**Siva Skandha, Luca Saba, Suneet K Gupta, Vijaya K Kumar, Amer M Johri, Narendra N Khanna, Sophie Mavrogeni, John R Laird, Gyan Pareek, Petros P Sfikakis, Athanasios Protogerou, Monica Turk, Aditya M Sharma, Andrew Nicolaides, George D Kitas, Klaudija Viskovic, Tomaz Omerzu and Jasjit S Suri**

Wilson's disease (WD) is a rare disease which occurs in one in 30 000 people. The main cause of WD is excess copper accumulation in the liver and brain, and most of the diagnosis of this disease has been through non-imaging methods. However, magnetic resonance imaging (MRI) of WD brains shows white matter hyper-intensity (WMH), but suffers from the challenge that the intensity changes in WD are not much greater than in the control. We hypothesize that transfer learning (TL) will be able to classify and characterize WD versus control.

In the proposed study, five different kinds of optimized TL architectures are developed and validated for characterization and classification of WD.

We propose five kinds of optimized TL architectures, namely AlexNet, InceptionV3 (IV3), XceptionNet, DenseNet 169, and ResNet50 for characterization and classification. The performance of TL models is analyzed by varying the augmentation folds (from 1× to 6×) using the K10 cross-validation protocol. Moreover, the performance of the best-optimized TL architecture is compared to a novel architecture called SivaSuriNet. SivaSuriNet is built on top of UNet consisting of separable convolution layers with a varied number of filters. The hypothesis of characterization is validated with artificial intelligence (AI) based mean feature strength (MFS) and a signal processing based higher-order spectrum (HOS), known as bispecturm.

We achieve a best accuracy in the K10 combination of 97.82% ± 1.52% from IV3 and a mean best accuracy of 91.83% ± 2.81% with an area-under-the-curve (AUC) of 0.918 ($p < 0.0001$). For SivaSuriNet we achieve a best accuracy of 98.43% ± 1.45% and a mean best accuracy of 90.96% ± 4.63% with an AUC of 0.91 ($p < 0.0001$) using the K10 combination and it is 0.62% higher than IV3. We compare the performance of the proposed model with existing work on WD and our model has 2.45% superior performance. We characterize WD using MFS, and it shows that WD is higher than the control by 21.57%. We validate our hypothesis with the signal processing technique HOS, which shows that the result for WD is higher than the control by 9.24%.

## 5.1 Introduction

Wilson's disease (WD) is a rare disease and it occurs in one in 30 000 to 40 000 people worldwide, as per the report published by the National Organization for Rare Diseases (NORD). The main cause of WD is the deposition of excessive copper in the liver and brains [1]. As per a study published in 2005, there are 9000 people affected by WD in the United States [2]. There are many methods such as blood tests and histology to diagnose WD, but these approaches are very slow and unreliable [3, 4]. We can also diagnose WD by finding the ATP7B genes in genetic analysis [5]. Most of the methods used for WD identification are non-imaging and time-consuming processes [6]. However, the early diagnosis of WD reduces the severity of the disease and increases the mortality rate [7]. Therefore, there is a need to design an image-based WD diagnosis such as MRI analysis. MRI analysis shows promising results for diagnosing WD as it shows the white matter hyperintensity (WMH) [8, 9]. However, MRI as an imaging tool for WD suffers from inter-observer variability, and the subtle difference in WMH between WD and controls causes complications in diagnosis. Therefore, there is a need to develop an artificial intelligence-based computer-aided diagnosis (CAD) system to identify WD. Researchers have been applying advanced AI techniques to solve complex computer vision problems in the medical domain for the last two decades [10–20]. Machine learning (ML) is a subset of AI and shows efficiency in classification problems. However, it suffers from reliability issues because it require purpose-built features [21–26]. The performance of the ML model thus varies with the selected features. This limitation was addressed by deep learning (DL) techniques as features are extracted automatically during the training of a deep learning model. DL shows promising results in radiology [27], and stroke [28–37], diabetes [38–42], coronary [10, 15, 39, 43–47], liver [13, 48–50], lung [51–53], and COVID-19 [54–57] imaging.

Deep learning techniques show promising results in computer vision problems, but a huge amount of computation is required to implement DL models. The amount of training time required for DL models is also considerable. Thus, to avoid these issues in AI, another technique evolved, called transfer learning (TL). Transfer learning reduces the training time and the dependence on hardware resources [6, 58–60]. In the TL framework, we will use the pre-trained weights of the model (so-called pre-trained models) to retrain target variables. Due to weight transfer, the model training time is reduced and no sophisticated hardware resources are required.

We should consider three essential characteristics in AI models: speed, accuracy, and reliability. ML techniques are faster and accurate, DL techniques are accurate and reliable, and TL techniques are faster and reliable, and TL shows the same performance as DL based methods. When characterizing medical images, if any two characteristics are better and the third characteristic is comparable, that model can be used for characterization. Thus, we hypothesized that the training speed of TL is better than DL and that WMH is higher in the WD than the control images.

This paper aims to characterize and classify WD tissues into control and disease, and WMH intensity is higher in WD than in the control. Moreover, we have also validated the hypothesis using signal processing techniques. This paper uses five TL architectures and is optimized on augmentation folds starting from 1× to 6× for detecting the best performance. We implemented all the architectures using the K10 cross-validation protocol (i.e. 90% training 10% testing) and characterized the WD using novel means feature strength (MFS) and validated with bispectrum. We benchmarked using existing systems and a novel architecture called SivaSuriNet.

This paper is divided into the following sections. The background literature section deals with the current work on the classification and characterization of WD. The methodology section explains the patient demographics, pre-processing techniques, TL architectures, and SivaSuriNet. The results section deals with the optimization results for the TL and SivaSuriNet. The characterization section explains the characterization of the WD using MFS and higher-order spectra (HOS). Finally, the discussion section covers benchmarking, claims, strengths and weaknesses, and extensions.

## 5.2 Background literature

Most of the diagnostic methods for WD are non-imaging, and AI has not given impressive results in WD classification. In the literature survey we consider both AI and non-AI based techniques for WD classification. In the last two decades, different biomarkers such as those in serum and urine have been used to diagnose WD [1, 3, 61]. These procedures included 24 hour urine or serum laboratory testing to identify WD. Ferenci *et al* [61] proposed another diagnosis method using Kayser–Fleischer rings, neurological symptoms, and low ceruloplasmin for detection of WD. Other methods for the detection of WD are a laboratory-based test (blood test) and gene code detection test. Vrabelova *et al* [62] identified WD using the H1069Q mutation of ATP7B in people of European origin. Rosencrantz and Schilsky *et al* [4] detected WD using mutation analysis of ATP7B, neurological disorders, Kayser–Fleischer rings, elevated urine and hepatic copper, and histological changes in the liver.

All of the above mentioned WD detection techniques are non-imaging, and they are time-consuming. Due to the recent advancements in MRI it is now easy for doctors to identify the WD by visualizing the WMH in the brain [63]. Moreover, WMH can also be utilized to detect other diseases such as stroke [64], Willis configuration [65], and mood disorders [66]. Kim *et al* [67] used T2-weighted MRI images of pediatric patients and divided them into three groups based on initial MRI findings, and observed that the WMH is high in the caudate nucleus, globus pallidus, thalamus, midbrain, and pons. Recent advances in AI have introduced the detection of WD by applying AI architectures on fMRI. Hu *et al* [68] performed a study of WD patients with frequency-dependent channels in the amplitude of the low-frequency functions. Kaden *et al* [69] classified WD using a support vector machine and parametrized generalized learning vector quantization (PGLVQ) and achieved accuracies of 87.5% and 90.1%, respectively. Jing *et al* [70] studied the classification of WD using the MRI scans of 30 patients. They extracted the features using the independent component analysis method (ICA). After extracting the features, Jing *et al* applied an SVM and achieved an area-under-the-curve (AUC) of 0.94 with an accuracy of 89.4%. Agarwal *et al* [71] used MRI scans of WD patients for classification and characterization using 'improved convolutional neural networks', which were optimized on the augmentation folds. They achieved an accuracy of 98.28% with an AUC of 0.99. They also characterized WD using AI based FMS and bispectrum.

It can be seen that WD classification and characterization using AI has few research works in the literature. Our study is the first of its kind, using the transfer learning paradigm for characterization and classification of WD.

## 5.3 Methodology

### 5.3.1 Patient demographics

T2W-TSE MRI scans from a cohort of 46 patients (average age: 40.73 ± 11.3 years, balanced M/F ratio) collected between the years 2011 and 2015 were analyzed (approval was obtained from the Institutional Ethics Committee, Azienda Ospedaliero Universitaria (AOU), Cagliari, Italy). The imaging examinations were performed using a 1.5 Tesla superconducting magnet (Philips, Best, The Netherlands) with a head coil according to a standardized protocol. For each subject, the conventional diffusion-weighted imaging (DWI) was performed with a single-shot spin-echo with two diffusion-sensitivity values of 0 and 1000 s mm$^{-2}$ along the transverse axis. As part of our general brain protocol, axial and sagittal 2D FLAIR images (10000/140/2200 ms for TR/TE/TI; matrix: 512 × 512; FOV: 240 × 240 mm$^2$; section thickness: 5 mm) were acquired. In addition to the FLAIR and DWI sequences, axial spin-echo T1-weighted images (500–600/15/2 for TR/TE/excitations) and fast spin-echo T2-weighted images (2200–3200/80–120/1,2 for TR/TE/excitations; turbo factor, 2) were also obtained with the same section thickness.

### 5.3.2 Data augmentation

The initial MRI data were manually classified by our radiologist team, and were then prepared for further processing. Because the cohort consisted of 37 controls and 9 WD patients, we had an unequal number of images in both classes. As each patient MRI study had 12–13 slices, this resulted in 458 control images and 115 WD images. For optimal performance with an unbalanced dataset, the augmentation protocol using the Python 'augmentor' API was applied to the WD class and resulted in 343 more WD images. As deep CNN (DCNN) needs a large number of images for proper training and performance, we increased the number of images from 458 by two-, three-, four-, and five-fold in both classes, and the system was then trained and tested to find which augmented inputs yielded optimal results. To avoid unrealistic brain MRI scans during the augmentation protocol, we followed the acceptable protocol of rotating the image by −10 to 10 degrees randomly. This would prevent methods such as horizontal or vertical flipping, or rotating by larger angles.

### 5.3.3 Pre-processing: skull and background removal

Pre-processing is an essential component of the classification process which helps to extract the region-of-interest (ROI) from the MRI images. In pre-processing, there are two important steps: (i) removal of the skull region and (ii) removal of the black background to prepare for the segmented ROI. For the segmentation and removal of the background from images, the popular tools BrainSuite [35] and VolBrain [36] were used. The BrainSuite tool was used to read DICOM images that were converted to .nii files (the .nii file type is primarily associated with the NIfTI-1 Data Format by Neuroimaging Informatics Technology Initiative) and obtained grayscale images of the brain with the skull. VolBrain helps to create a mask of the brain which can be used to remove the skull from the original MRI grayscale images. After pre-processing the data using BrainSuite and VolBrain, the resultant images were passed to segmentation to remove the background. A sample pair of images from a patient with WD and a control is shown in figure 5.1. The WD segmented brain images had brighter regions (higher WMH) in the convoluted zones of the brain (as shown inside the yellow dashed rectangles figure 5.1) compared to the control images. Figures 5.2 and 5.3 present a sample the segmented MRI scans of six patients from the control and diseased cohorts.

## 5.4 Global architecture: transfer learning

We have developed a TL global architecture for the characterization and classification of WD. Figure 5.4 presents the global architecture of the TL framework. We have developed similar models for stroke [12, 35–37, 47, 72–74], intima-media thickness [75–77], liver [13, 48–50], and diabetes [39, 78, 79] using DL and ML architectures. This study is the first of its kind to use the TL paradigm for the

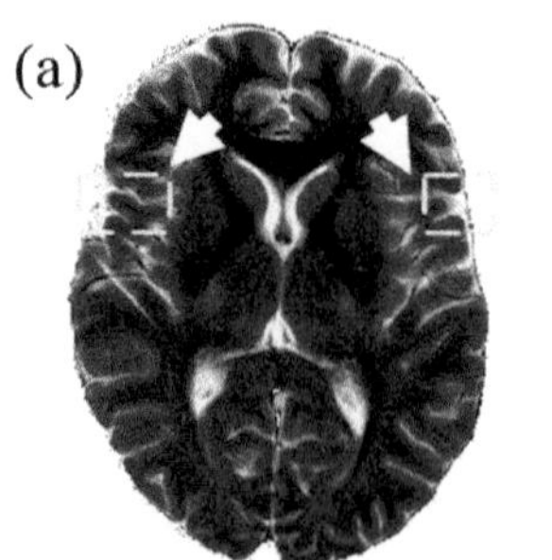
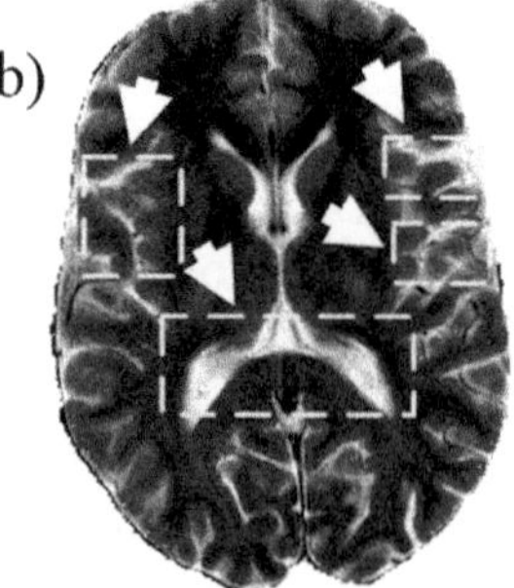

**Figure 5.1.** Control (left) and WD (right). Both images show the skull with the removal of background.

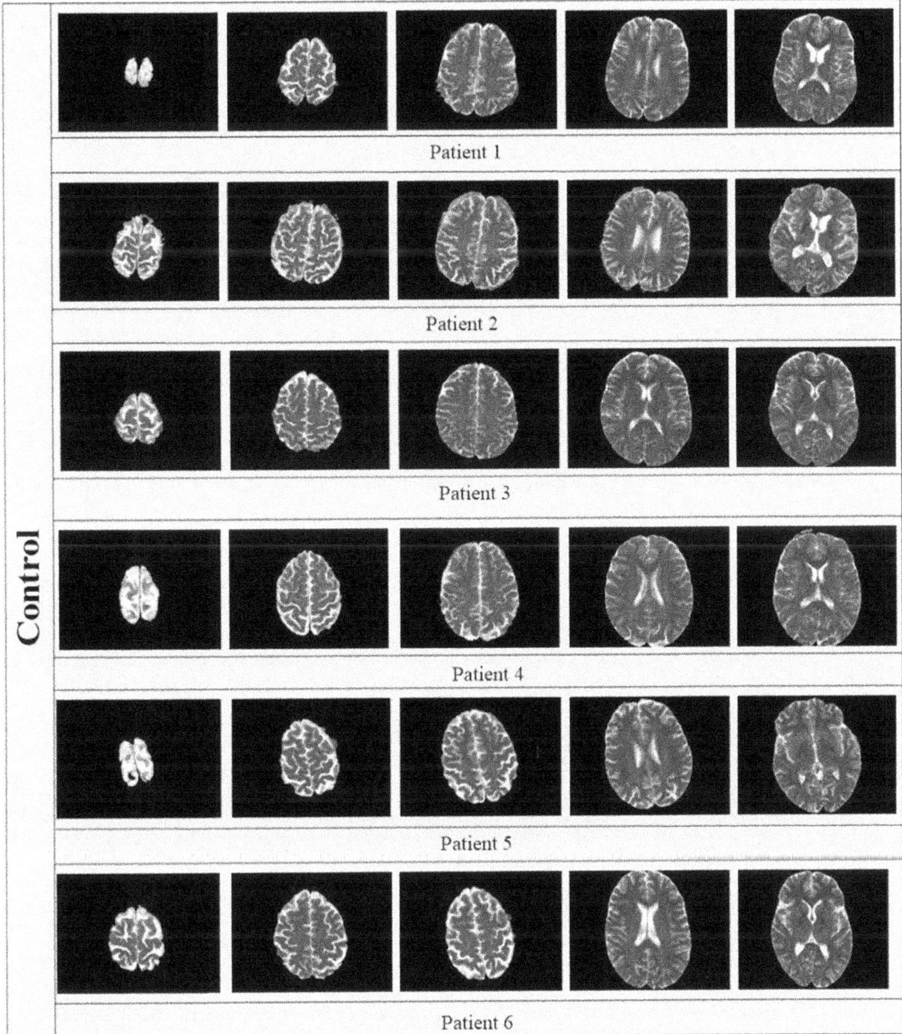


**Figure 5.2.** Samples of the segmented MRI scans of six patients from the control dataset.

classification of WD. In this study we have used four TL architectures, AlexNet [80], ResNet [81], InceptionV3 [82], and DenseNet121 [83] which achieved the best performance in the ImageNet challenge.

### 5.4.1 AlexNet

AlexNet was the first intelligent model designed for the classification of ImageNet using a deep convolution neural network. This architecture was proposed by Krizhevsky *et al* [80] in 2012 for classifying 1.2 million high-resolution images with 1000 classes in the ImageNet challenge. The AlexNet architecture won the first prize in the ImageNet challenge. This architecture uses a combination of

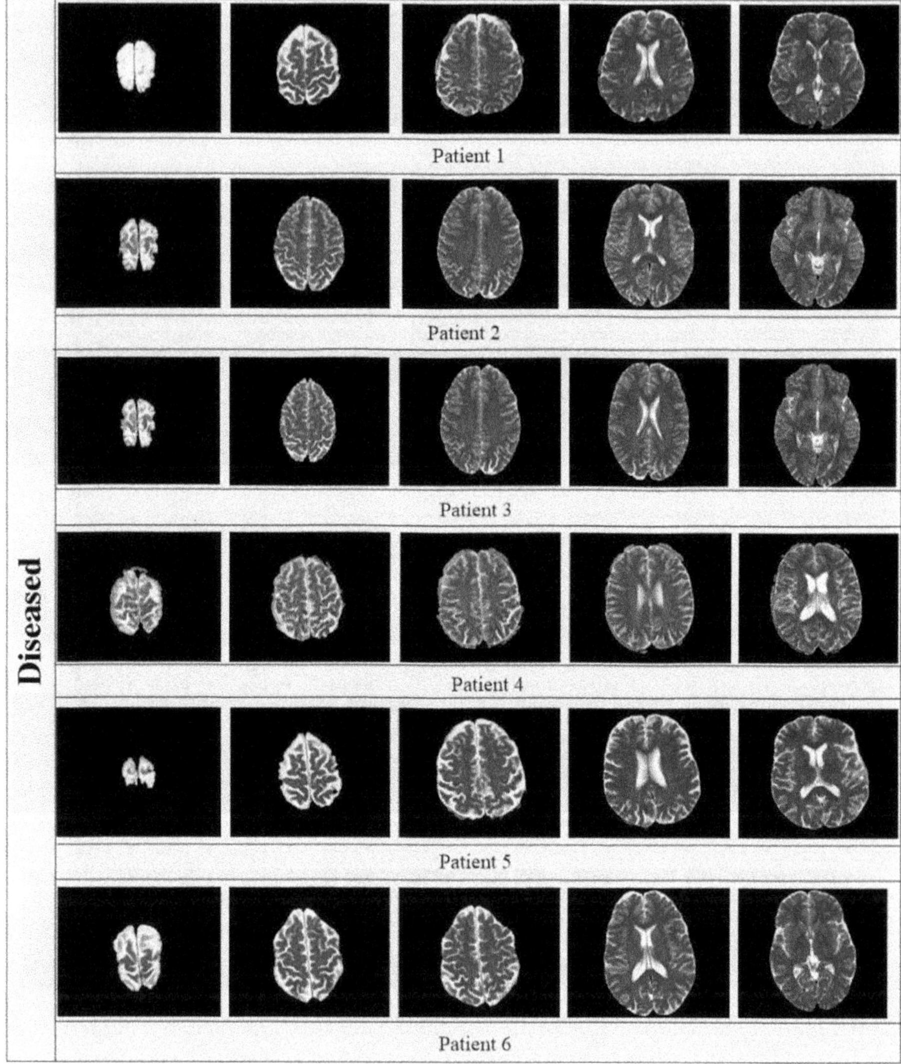


**Figure 5.3.** Samples of segmented MRI scans of six patients from the diseased dataset.

convolution and max-pooling with input image sizes of 224 × 224 and achieves a top-five test error rate of 15.3%. The architecture of AlexNet is shown schematically in figure 5.5.

In the TL paradigm, we took the pre-trained weights of the AlexNet model and used these weights at initial and intermediate layers. We fed the WD cohort in at the dense layer (FC6) for retraining AlexNet for WD classification. Moreover, an extra dropout layer was added between FC6 and FC7 with a 50% rate and the number of nodes in the output layer was changed from 1000 to 2 as our cohort has only two classes.

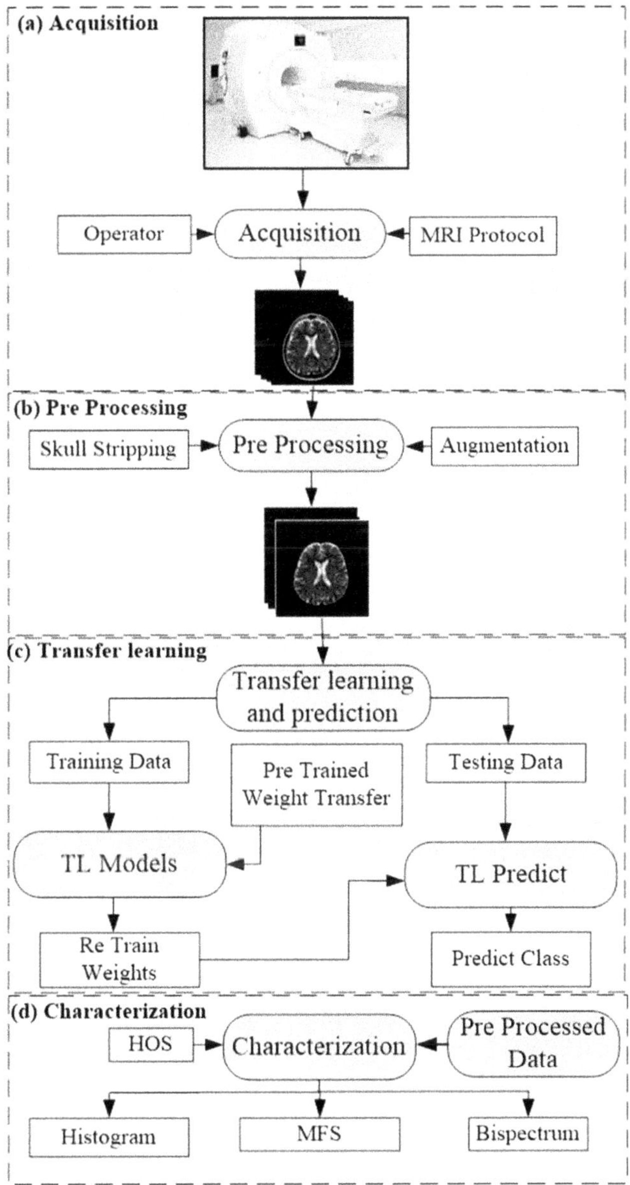


**Figure 5.4.** The global architecture of our study. (MRI: magnetic resonance imaging; TL: transfer learning; HOS: higher-order spectra; MFS: mean feature strength.)

## 5.4.2 ResNet50

He *et al* [81] built the ResNet architecture and achieved 3.57% error in the top 5%. This model also won the challenge and the authors proposed a wide range of different ResNet architectures containing from 20 layers to 1202 layers. In our study we selected ResNet50 for experimental purposes. In ResNet50 there are 50 layers

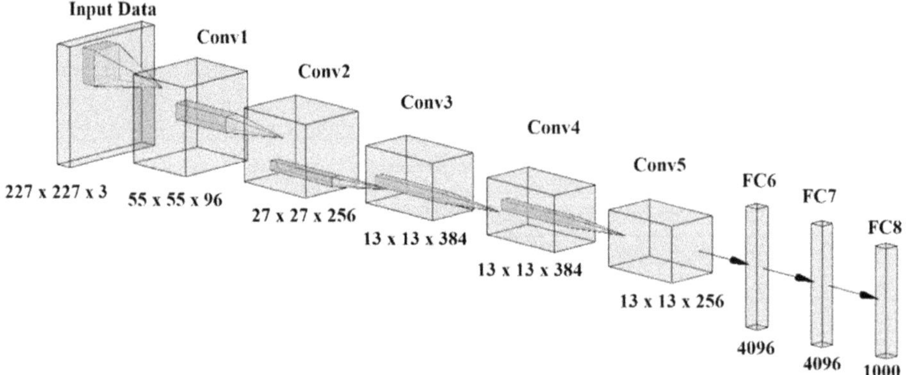


**Figure 5.5.** AlexNet architecture.

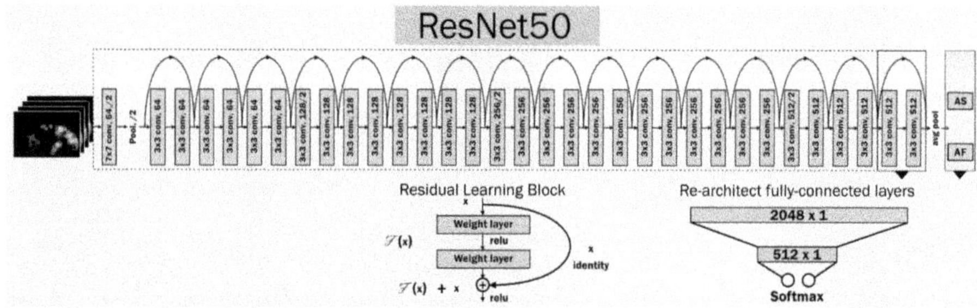


**Figure 5.6.** Resnet50 architecture.

with 25 residual blocks and one fully connected network (FCN) followed by a softmax layer. We implemented this model under the TL paradigm by freezing all the layer weights with the pre-trained weights of ResNet on ImageNet. We modified the end layers by adding dropout layers after the first FCN, and then added one more FCN followed by a softmax layer. We retrained the model from the first FCN using the WD cohort. Here we also changed the number of nodes at the end layer from 1000 to 2 as the WD cohort only has two classes. The architecture of ResNet is presented in figure 5.6.

### 5.4.3 DenseNet161

DenseNet is another popular architecture and it won the best paper award at CVPR in 2017. Huang *et al* [83] proposed densely connected convolutional neural networks for solving the vanishing gradient problem and better feature propagation with a smaller number of training parameters. This architecture contains dense blocks which concatenate the features from the current block and previous block. Unlike ResNet, in DenseNet all the features are fused, not summing with the previous blocks. They proposed a range of architectures with varying number of feature maps from 32 to 48 with varying depths from 121 to 161. This model achieved a 6.15% error rate at the top 5% in the ImageNet 2017 challenge.

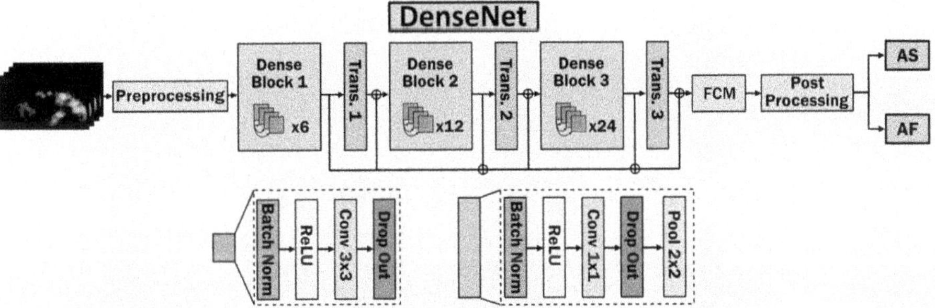


**Figure 5.7.** DenseNet161 architecture.

In our study, we choose DenseNet161 which consists 161 layers and implemented it in the TL framework. To make the DenseNet fit in the TL framework, we froze all the initial to intermediate layers with Imagenet 2017 weights. After that we altered the end layer architecture in DenseNet161 by adding a dropout and dense layer. The modified DenseNet architecture is presented in figure 5.7.

### 5.4.4 XceptionNet

Developed by Google, XceptionNet is another popular architecture for image classification and it is based on InceptionNet [84]. Moreover, XceptionNet introduced a modified depth-wise separable convolution technique for ImageNet classification and achieved an error rate of 0.790 in top-one accuracy and 0.945 in top-five accuracy.

In our study, we used XceptionNet in the TL paradigm by freezing the initial layers and modifying the end layers by retraining the model using the WD cohort for classification. The modified architecture of XceptionNet is depicted in figure 5.8.

### 5.4.5 InceptionV3

Szegedy *et al* [82] developed InceptionV3 for solving computer vision problems such as label smoothing and propagating label information deep down into the neural network. The authors proposed a series of architectures under the Inception family, among which InceptionV3 (IV3) achieved the lowest error rate pf 4.2% among the top 1%.

We considered IV3 under the TL paradigm by transferring the pre-trained weights of IV3 (initially trained on ImageNet) from the initial to intermediate layers. We modified the end layers, i.e. from the FCN to output layer, and added two dropout layers between the FCNs with a 50% drop rate with two nodes in the output layer. We fed the WD cohort at the FCN for retraining the IV3 model for target labels. The architecture of InceptionV3 is presented in figure 5.9.

### 5.4.6 SivaSuriNet

The performance of the proposed TL architectures was compared with a novel DL architecture called SivaSuriNet. SivaSuriNet uses a separable convolution neural network with batch and normalization combinations. After each combination of

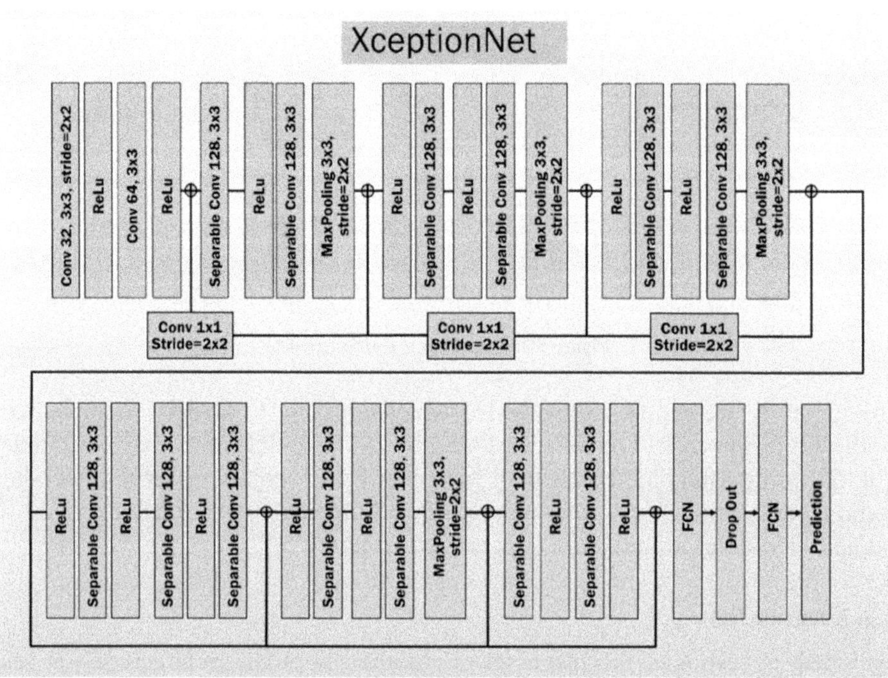



**Figure 5.8.** Proposed architecture of XceptionNet.

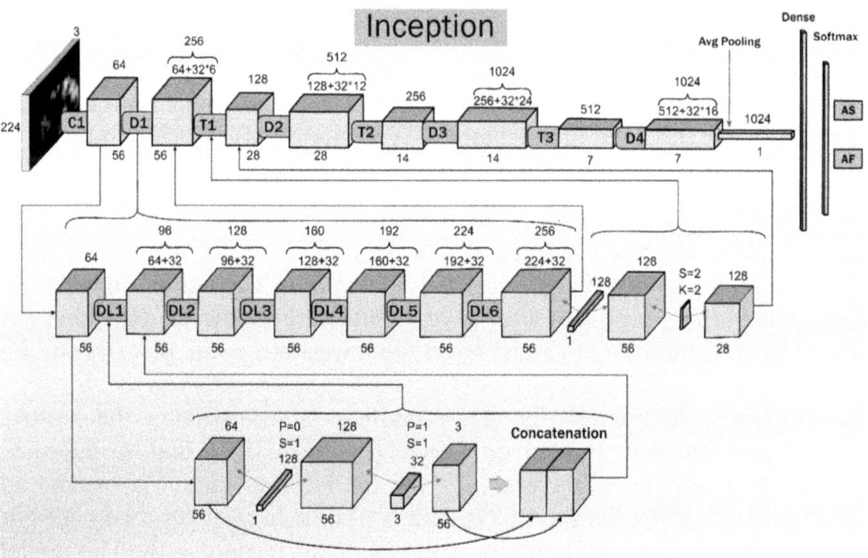



**Figure 5.9.** Modified architecture of InceptionV3.

these set of layers, we used the max-pooling layer. Initially, we used a small number of filters, i.e. 32, and then increased the number to 256. The shape of the architecture is like a funnel and is presented in figure 5.10. Details of the training parameters of SivaSuriNet are shown in table 5.1.

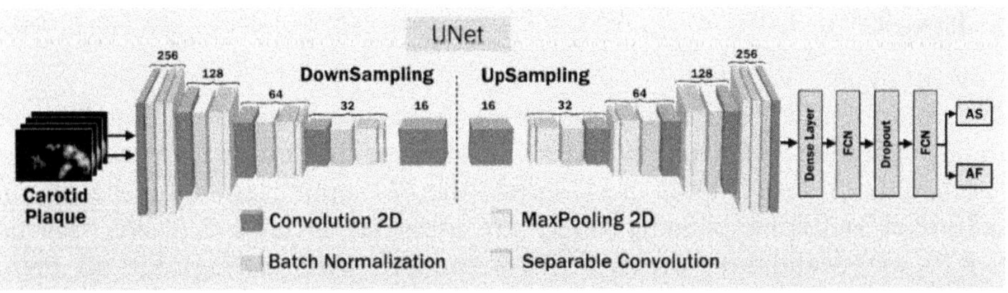


**Figure 5.10.** SivaSuriNet architecture.

**Table 5.1.** SivaSuriNet architecture parameters.

| Layer type | Shape | Parameter # |
| --- | --- | --- |
| Convolution 2D | $128 \times \times 128 \times 32$ | 896 |
| Batch normalization | $128 \times 128 \times 32$ | 128 |
| Separable convolution 2D | $128 \times 128 \times 64$ | 2400 |
| Batch normalization | $128 \times 128 \times 64$ | 256 |
| Max-pooling 2D | $64 \times 64 \times \times 64$ | 0 |
| Separable convolution 2D | $64 \times 64 \times 128$ | 8896 |
| Batch normalization | $64 \times 64 \times 128$ | 512 |
| Max-pooling 2D | $32 \times 32 \times 128$ | 0 |
| Separable convolution 2D | $32 \times 32 \times 256$ | 34176 |
| Batch normalization | $32 \times 32 \times 256$ | 1024 |
| Max-pooling 2D | $16 \times 16 \times 256$ | 0 |
| Separable convolution 2D | $16 \times 16 \times 64$ | 18752 |
| Batch normalization | $16 \times 16 \times 64$ | 256 |
| Max-pooling 2D | $8 \times 8 \times \times 64$ | 0 |
| Separable convolution 2D | $8 \times 8 \times 128$ | 8896 |
| Batch normalization | $8 \times 8 \times 128$ | 512 |
| Max-pooling 2D | $4 \times 4 \times 128$ | 0 |
| Separable convolution 2D | $4 \times 4 \times 256$ | 34176 |
| Batch normalization | $4 \times 4 \times 256$ | 1024 |
| Max-pooling 2D | $2 \times 2 \times 256$ | 0 |
| Flatten | 1024 | 0 |
| Dense | 1024 | 1049600 |
| Dropout | 0.5 | 0 |
| Dense | 512 | 524800 |
| Dropout | 0.5 | 0 |
| Dense (softmax) | 2 | 1026 |
| **Total trainable parameters** | | **1 687 330** |

## 5.5 Results

*TL optimization*

We optimized the performance of the TL framework using six TL architectures with augmentation folds from 1× to 6× and achieved the best performance (in terms of accuracy) in IV3. Figure 5.11 shows the 3D optimization of the TL and SivaSuriNet and corresponding values are presented in table 5.2. We have used the K10 cross-validation protocol (90% training, 10% testing) for all the AI architectures and achieved the best accuracy from IV3 with 97.82%± 1.52% while the mean accuracy of 91.83% ± 2.23%.

We compared the TL performance with SivaSuriNet using the K10 cross-validation protocol and SivaSuriNet achieved the best accuracy of 98.43% ± 1.45% with a mean accuracy of 90.96% ± 4.63%. SivaSuriNet shows an improvement of 0.62% in the best accuracy combination with a comparable mean accuracy. We also computed the area-under-the-curve (AUC) of all the AI architectures, which are shown in figure 5.12.

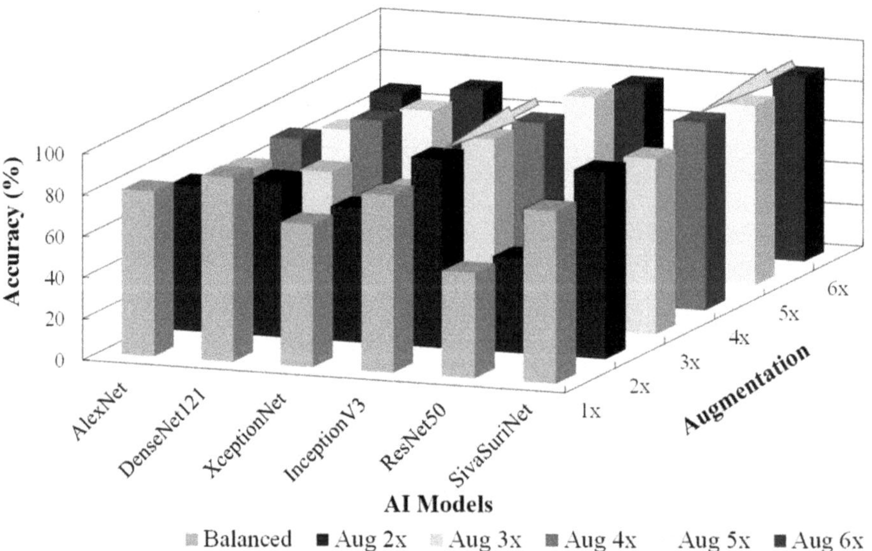


**Figure 5.11.** Optimization of TL and SivaSuriNet with augmentation.

**Table 5.2.** 3D optimization of TL architectures and SivaSuriNet (asterisks represent the highest and second highest accuracy values).

| Model | Balanced | Aug 2× | Aug 3× | Aug 4× | Aug 5× | Aug 6× | Mean ± SD |
|---|---|---|---|---|---|---|---|
| AlexNet | 79.98 ± 4.18 | 71.10 ± 32.60 | 67.21 ± 63.63 | 69.81 ± 47.52 | 62.49 ± 4.10 | 68.05 ± 3.25 | 71.94 ± 2.71 |
| DenseNet121 | 89.36 ± 17.76 | 75.766 ± 33.29 | 68.39 ± 46.16 | 81.41 ± 40.50 | 74.47 ± 15.50 | 72.39 ± 25.12 | 78.74 ± 4.70 |
| XceptionNet | 69.30 ± 17.78 | 65.33 ± 42.60 | 62.1 ± 35.01 | 61.87 ± 30.04 | 57.81 ± 15.56 | 52.72 ± 25.52 | 64.67 ± 3.05 |
| InceptionV3 | 85.98 ± 4.06 | **91.83 ± 2.23*** | 89.40 ± 2.53 | 85.41 ± 2.13 | 86.23 ± 1.13 | 79.51 ± 3.25 | 88.16 ± 1.74 |
| ResNet50 | 50.86 ± 4.18 | 46.35 ± 2.70 | 48.24 ± 3.41 | 53.90 ± 2.48 | 50.34 ± 5.35 | 51.55 ± 5.45 | 49.84 ± 1.79 |
| SivaSuriNet | 83.57 ± 4.03 | 90.79 ± 2.76 | 85.23 ± 6.35 | **90.96 ± 4.63*** | 87.49 ± 5.35 | 89.58 ± 2.58 | 87.64 ± 2.28 |
| Mean+SD | 76.45 ± 7.71 | 73.53 ± 8.90 | 70.10 ± 8.11 | 73.89 ± 8.51 | 67.81 ± 8.56 | 67.52 ± 6.25 | |

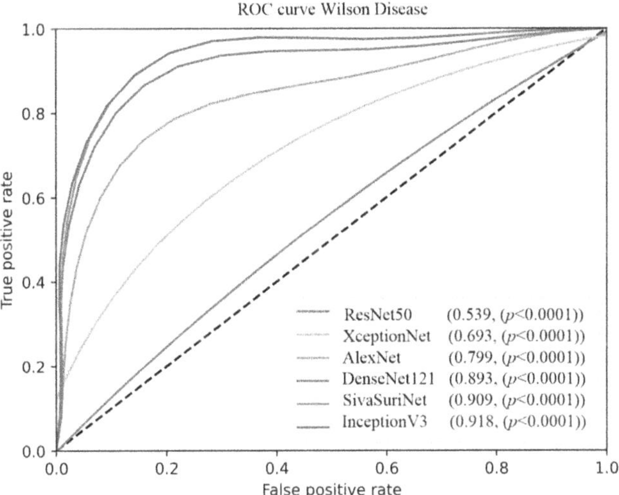


**Figure 5.12.** ROC analysis of six AI models.

## 5.6 Characterization

We hypothesized that the white matter hyperintensity (WMH) of WD is higher than that of a control and to validate the proposed hypothesis we have used two strategies: (i) AI based mean feature strength (MFS) and (ii) signal processing based on higher-order spectra (HOS) called bispectrum.

### 5.6.1 Mean feature strength

AI based MFS [37] is a novel technique for computing the strength of the features learned by the DL models at every layer. The MFS will be calculated from DL models only because TL models take ImageNet features at the initial and intermediate layers, so it is not a good option to compute the strength of the features. Therefore, the trained weight of SivaSuriNet was taken for computation of MFS. Here we passed all the training images of the best accuracy combination and computed the feature strength for each image at every layer. After computation of the feature strength, the mean of all the feature strengths of all the training images at every layer was computed. Figure 5.13 shows the MFS of SivaSuriNet and also shows that the MFS of WD is higher than that of the control by 21.57%, thus we conclude that MFS (WD) > MFS (control).

### 5.6.2 Higher-order spectrum

For validating the MFS we choose a non-AI based technique, i.e. signal processing based HOS by computing the bispectrum ($B$) of the same combination used for computing the MFS. Here also we computed the mean $B$ value of every image by rotating at a random angle from $0°$ to $180°$. Figure 5.14(a) represents the mean $B$ value of the training images at every angle from $0°$ to $180°$ with a step size of $15°$. Figure 5.14(b) shows the HOS of the WD at the best possible angle of 60 degrees.

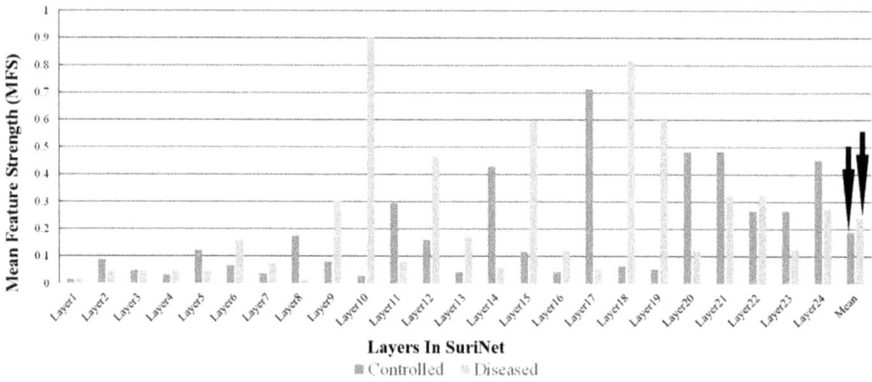

**Figure 5.13.** Mean feature strength of SivaSuriNet.

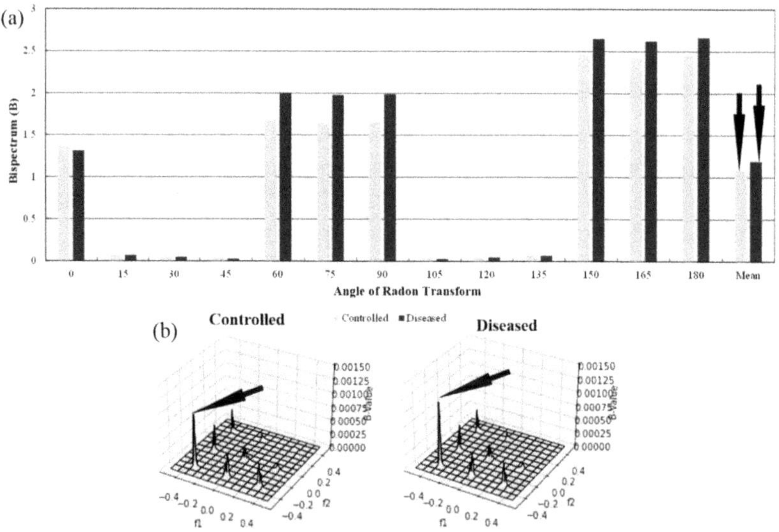

**Figure 5.14.** (a) Mean bispectrum analysis. (b) Higher-order spectra of WD.

HOS also exhibits similar results to MFS and the bispectrum of WD is higher than that of the control by 9.24%.

## 5.7 Discussion

In this section we discuss the benchmarking of our work with existing works. First a discussion of the characterization of WD is presented and then we discuss our claims, and the strengths and weaknesses of the proposed work, followed by our conclusion and future work.

Our study is the first of its kind on WD classification using an optimized TL paradigm with augmentation. We compare its performance with the novel architecture SivaSuriNet. Both architectures exhibit similar performance but IV3 shows 0.94% higher performance than SivaSuriNet.

*Benchmarking*

We benchmarked our current TL paradigm against existing works, as shown in table 5.3. We found that apart from our team, the other existing works have not performed any classification. The current study is the first of its kind on TL optimization in comparison with SivaSuriNet for classification of WD. The benchmarking table contains the first author's name with the year, the application and the modality, the techniques used for the application, and the accuracy with area-under-the-curve (AUC) shown in table 5.3.

*A short note on WD characterization*

We performed the characterization of WD using an AI method, i.e. MFS, and a non-AI method, i.e. bispectrum. The characterization based on the AI method depends on the input data. However, we have a cohort which is initially unbalanced, and it has a larger control class. However, bispectrum is based on the pixel intensity. Both characterization techniques are valid for our datasets and a large amount of data is required to strengthen the validation under these paradigms.

**Table 5.3.** Benchmarking with existing systems.

| Author, year, reference | Application | Modality | Technique | AI | ACC (%) (AUC) |
|---|---|---|---|---|---|
| Kaden *et al* (2015) [69] | WD | MRI | SVM and PGLVQ | ML | **90.1** |
| Suk *et al* (2017) [85] | ALZ | MRI (ADNI) | DBN + NN | DL | **85.91 (0.91)** |
| Abiwinanda *et al* (2018) [86] | BT | MRI | CNN | DL | **84.19** |
| Zhang *et al* (2019) [87] | MCI | fMRI | SSGSR | ML | **88.50 (0.965)** |
| Abrol *et al* (2018) [88] | MCI versus ALZ | MRI | ResNet | DL | **82.7 (0.89)** |
| Jing *et al* (2019) [70] | WD | fMRI | SVM and ICA | ML | **89.4 (0.94)** |
| Richhariya *et al* (2020) [89] | MCI | MRI | SVM | ML | **90** |
| Liu *et al* (2020) [90] | MCI | MRI | SVM | ML | **88.5 (0.90)** |
| Saba *et al* (2020) [6] | CNN | MRI | VGG-19 | DL | 95.46 (0.954) |
| Agarwal *et al* (2020) [71] | iDCNN | MRI | CNN | DL | 98.28 (0.990) |
| **Proposed work** | **CNN** | **MRI** | **Optimized InceptionV3** | **TL** | **Optimized Comb: 97.82, 91.83 (0.918)** |
| **Proposed work** | **CNN** | **MRI** | **SivaSuriNet** | **DL** | **Optimized Comb: 98.43, 90.96 (0.91)** |

ALZ: Alzheimer's disease; BT: brain tumor; MCI: mild cognitive impairment.

*Power analysis*

We also performed a power analysis to find the sample size required for this study and used the following equation for the calculation of sample size:

$$\text{sample size} = \frac{2 \times \left(Z_\alpha + Z_{1-\beta}\right)^2 \times \sigma^2}{\Delta^2}. \tag{5.1}$$

Here we choose $Z_\alpha = 3.2905$ for type 1 error having a value of 1%, and $Z_{1-\beta} = 1.6449$ for type 2 error having a value of 1%. Here $\sigma$ (standard deviation) $= 2.53$ and $\Delta$ (mean difference) $= 0.627$. By substituting these values in equation (5.1), 793 samples are required for the proposed study. Moreover, the number of samples in the cohort was 916 under the balanced category, which is 13.42% higher than required.

*Strengths, weaknesses, and extensions*

This study is the first of its kind that uses optimized TL architectures for the classification and characterization of WD. All the TLs were optimized by varying the augmentations and architectures. Moreover, the performance of all optimized TL models was compared with a novel architecture called SivaSuriNet. The hypothesis was validated using AI based MFS and signal processing based HOS, and both presented similar characterization of WD.

Though the current study yielded impressive results with a greater number of samples which support the use of the model to study more complex patterns, since our original cohort contains more control and fewer WD samples, to balance the number of samples in each class augmentation was performed. This system can be extended to the fusion of two or more architectures, known as hybrid architecture, for studying WD. Further, we can also extend this work to multicenter data for a larger training set.

## 5.8 Conclusion

This study is the first of its kind to use the TL framework for the classification and characterization of WD. We achieved the best-optimized combination in IV3 with 97.82% $\pm$ 1.52% accuracy. Moreover, we have also compared the performance of TL with a novel DL architecture called SivaSuriNet. SivaSuriNet achieved the best accuracy of 98.43% $\pm$ 1.45% from augmentation 4 with K10-fold. After classification, we are further able to carry out the characterization of WD using AI based MFS and signal-based bispectrum and concluded that the value for WD was higher than the control by 21.57%. Using bispectrum we obtained that the value for WD is higher than the control by 9.24%. We validated our systems using widely accepted heart and lung data.

Sections 5.3.1, 5.3.2, and 5.3.3 of this chapter contain excerpts reproduced with permission from [71]. Copyright (2021) Springer.

# References


[1] Dusek P, Litwin T and Czlonkowska A 2015 Wilson disease and other neurodegenerations with metal accumulations *Neurol. Clin.* **33** 175–204

[2] Brewer G J and Askari F K 2005 Wilson's disease: clinical management and therapy *J. Hepatol.* **42** S13–21

[3] Medici V, Rossaro L and Sturniolo G 2007 Wilson disease—a practical approach to diagnosis, treatment, and follow-up *Dig. Liver Dis.* **39** 601–9

[4] Rosencrantz R and Schilsky M 2011 Wilson disease: pathogenesis and clinical considerations in diagnosis and treatment *Seminars in Liver Disease* (Stuttgart: Thieme Medical)

[5] Saba L *et al* 2019 Wilson's disease: a new perspective review on its genetics, diagnosis and treatment *Front. Biosci.* **11** 166–85

[6] Saba L *et al* 2020 Brain MRI-based Wilson disease tissue classification: an optimised deep transfer learning approach *Electron. Lett.* **56** 1395–8

[7] Roberts E A and Schilsky M L 2003 A practice guideline on Wilson disease *Hepatology* **37** 1475–92

[8] Parekh J R and Agrawal P R 2014 Wilson's disease: 'face of giant panda' and 'trident' signs together *Oxford Med. Case Rep.* **2014** 16–7

[9] Yousaf M *et al* 2009 Atypical MRI features involving the brain in Wilson's disease *Radiol. Case Rep.* **4** 312

[10] Banchhor S K *et al* 2017 Well-balanced system for coronary calcium detection and volume measurement in a low resolution intravascular ultrasound videos *Comput. Biol. Med.* **84** 168–81

[11] Radeva P and Suri J S 2019 *Vascular and Intravascular Imaging Trends, Analysis, and Challenges* vol 2 (Bristol: IOP Publishing)

[12] Jamthikar A *et al* 2020 Ultrasound-based stroke/cardiovascular risk stratification using Framingham Risk Score and ASCVD Risk Score based on 'Integrated Vascular Age' instead of 'Chronological Age': a multi-ethnic study of Asian Indian, Caucasian, and Japanese cohorts *Cardiovasc. Diag. Ther.* **10** 939

[13] Biswas M *et al* 2017 Symtosis: a liver ultrasound tissue characterization and risk stratification in optimized deep learning paradigm *Comput. Methods Programs Biomed.* **155** 165–77

[14] Maniruzzaman M *et al* 2019 Statistical characterization and classification of colon micro-array gene expression data using multiple machine learning paradigms *Comput. Methods Programs Biomed.* **176** 173–93

[15] Araki T *et al* 2015 Shape-based approach for coronary calcium lesion volume measurement on intravascular ultrasound imaging and its association with carotid intima-media thickness *J. Ultrasound Med.* **34** 469–82

[16] Jamthikar A D *et al* 2019 Rheumatoid arthritis: its link to atherosclerosis imaging and cardiovascular risk assessment using machine-learning-based tissue characterization *Vascular and Intravascular Imaging Trends, Analysis, and Challenges* vol 2 (Bristol: IOP Publishing) pp 4-1–29

[17] Acharya U R *et al* 2013 Plaque tissue characterization and classification in ultrasound carotid scans: a paradigm for vascular feature amalgamation *IEEE Trans. Instrum. Meas.* **62** 392–400

[18] Acharya U R *et al* 2012 Ovarian tumor characterization and classification: a class of GyneScan systems *2012 Annual Int. Conf. of the IEEE Engineering in Medicine and Biology Society* (Piscataway, NJ: IEEE)

[19] Saba L *et al* 2013 Inter- and intra-observer variability analysis of completely automated cIMT measurement software (AtheroEdge) and its benchmarking against commercial ultrasound scanner and expert Readers *Comput. Biol. Med.* **43** 1261–72

[20] Molinari F *et al* 2018 An efficient data mining framework for the characterization of symptomatic and asymptomatic carotid plaque using bidimensional empirical mode decomposition technique *Med. Biol. Eng. Comput.* **56** 1579–93

[21] Banchhor S K *et al* 2017 Wall-based measurement features provides an improved IVUS coronary artery risk assessment when fused with plaque texture-based features during machine learning paradigm *Comput. Biol. Med.* **91** 198–212

[22] Singh B K *et al* 2017 Risk stratification of 2D ultrasound-based breast lesions using hybrid feature selection in machine learning paradigm *Measurement* **105** 146–57

[23] Sharma A M *et al* 2015 A review on carotid ultrasound atherosclerotic tissue characterization and stroke risk stratification in machine learning framework *Curr. Atheroscler. Rep.* **17** 55

[24] Shrivastava V K *et al* 2015 Reliable and accurate psoriasis disease classification in dermatology images using comprehensive feature space in machine learning paradigm *Expert Syst. Appl.* **42** 6184–95

[25] Schurischuster S *et al* 2018 A preliminary study of image analysis for parasite detection on honey bees *Image Analysis and Recognition* (Cham: Springer)

[26] Saba L *et al* 2017 Plaque tissue morphology-based stroke risk stratification using carotid ultrasound: a polling-based PCA learning paradigm *J. Med. Syst.* **41** 98

[27] Saba L *et al* 2019 The present and future of deep learning in radiology *Eur. J. Radiol.* **114** 14–24

[28] Acharya U R *et al* 2013 Understanding symptomatology of atherosclerotic plaque by image-based tissue characterization *Comput. Methods Programs Biomed.* **110** 66–75

[29] Acharya U R *et al* 2013 Computed tomography carotid wall plaque characterization using a combination of discrete wavelet transform and texture features: a pilot study *Proc. Inst. Mech. Eng.* H **227** 643–54

[30] Acharya U R *et al* 2013 Automated classification of patients with coronary artery disease using grayscale features from left ventricle echocardiographic images *Comput. Methods Programs Biomed.* **112** 624–32

[31] Acharya U R *et al* 2012 Atherosclerotic risk stratification strategy for carotid arteries using texture-based features *Ultrasound Med. Biol.* **38** 899–915

[32] Acharya U R *et al* 2013 Atherosclerotic plaque tissue characterization in 2D ultrasound longitudinal carotid scans for automated classification: a paradigm for stroke risk assessment *Med. Biol. Eng. Comput.* **51** 513–23

[33] Acharya U R *et al* 2011 Atheromatic™: symptomatic vs asymptomatic classification of carotid ultrasound plaque using a combination of HOS, DWT and texture *2011 Annual Int. Conf. of the IEEE Engineering in Medicine and Biology Society* (Piscataway, NJ: IEEE)

[34] Acharya U R *et al* 2012 An accurate and generalized approach to plaque characterization in 346 carotid ultrasound scans *IEEE Trans. Instrum. Meas.* **61** 1045–53

[35] Saba L *et al* Ultrasound-based internal carotid artery plaque characterization using deep learning paradigm on a supercomputer: a cardiovascular disease/stroke risk assessment system *Int. J. Cardiovasc. Imaging* **35** 1–18

[36] Saba L *et al* 2021 A multicenter study on carotid ultrasound plaque tissue characterization and classification using six deep artificial intelligence models: a stroke application *IEEE Trans. Instrum. Meas.* **70** 1–12

[37] Skandha S S *et al* 2020 3-D optimized classification and characterization artificial intelligence paradigm for cardiovascular/stroke risk stratification using carotid ultrasound-based delineated plaque: Atheromatic™ 2.0 *Comput. Biol. Med.* **125** 103958

[38] Saba L *et al* 2019 Ultrasound-based carotid stenosis measurement and risk stratification in diabetic cohort: a deep learning paradigm *Cardiovasc. Diagn. Ther.* **9** 439–61

[39] Banchhor S K *et al* 2017 Relationship between automated coronary calcium volumes and a set of manual coronary lumen volume, vessel volume and atheroma volume in Japanese Diabetic Cohort *J. Clin. Diagn. Res.: JCDR* **11** TC09

[40] Khanna N N *et al* 2019 Performance evaluation of 10-year ultrasound image-based stroke/cardiovascular (CV) risk calculator by comparing against ten conventional CV risk calculators: a diabetic study *Comput. Biol. Med.* **105** 125–43

[41] Kabir C S, Haq M M and Cader F A 2017 Disparity in coronary artery diameter in diabetic and nondiabetic subjects undergoing percutaneous coronary intervention in Bangladesh: a 2-year retrospective analysis *Bangladesh Heart J.* **32** 23–8

[42] Biswas M *et al* 2018 Deep learning strategy for accurate carotid intima-media thickness measurement: an ultrasound study on Japanese diabetic cohort *Comput. Biol. Med.* **98** 100–17

[43] Jamthikar A D *et al* 2019 Wall quantification and tissue characterization of the coronary artery *Vascular and Intravascular Imaging Trends, Analysis, and Challenges* vol 2 (Bristol: IOP Publishing) pp 3-1–36

[44] Jamthikar A D *et al* 2019 Risk of coronary artery disease: genetics and external factors *Vascular and Intravascular Imaging Trends, Analysis, and Challenges* vol 2 (Bristol: IOP Publishing) pp 2-1–38

[45] Araki T *et al* 2016 PCA-based polling strategy in machine learning framework for coronary artery disease risk assessment in intravascular ultrasound: a link between carotid and coronary grayscale plaque morphology *Comput. Methods Programs Biomed.* **128** 137–58

[46] Araki T *et al* 2016 A new method for IVUS-based coronary artery disease risk stratification: a link between coronary and carotid ultrasound plaque burdens *Comput. Methods Programs Biomed.* **124** 161–79

[47] Jamthikar A D *et al* 2020 Multiclass machine learning vs conventional calculators for stroke/CVD risk assessment using carotid plaque predictors with coronary angiography scores as gold standard: a 500 participants study *Int. J. Cardiovasc. Imaging* **37** 1–17

[48] Kuppili V *et al* 2017 Extreme learning machine framework for risk stratification of fatty liver disease using ultrasound tissue characterization *J. Med. Syst.* **41** 152

[49] Acharya U R *et al* 2012 Data mining framework for fatty liver disease classification in ultrasound: a hybrid feature extraction paradigm *Med. Phys.* **39** 4255–64

[50] Saba L *et al* 2016 Automated stratification of liver disease in ultrasound: an online accurate feature classification paradigm *Comput. Methods Programs Biomed.* **130** 118–34

[51] Sanagala S S *et al* 2019 A fast and light weight deep convolution neural network model for cancer disease identification in human lung(s) *18th IEEE Int. Conf. on Machine Learning And Applications* (Piscataway, NJ: IEEE)

[52] Alakwaa W, Nassef M and Badr A 2017 Lung cancer detection and classification with 3D convolutional neural network (3D-CNN) *Lung Cancer* **8** 409

[53] Skourt B A, El Hassani A and Majda A 2018 Lung CT image segmentation using deep neural networks *Procedia Comput. Sci.* **127** 109–13

[54] Agarwal M *et al* 2021 A novel block imaging technique using nine artificial intelligence models for COVID-19 disease classification, characterization and severity measurement in lung computed tomography scans on an Italian cohort *J. Med. Syst.* **45** 1–30

[55] Suri J S *et al* 2021 A narrative review on characterization of acute respiratory distress syndrome in COVID-19-infected lungs using artificial intelligence *Comput. Biol. Med.* **130** 104210

[56] Mallio C A *et al* 2021 Deep learning algorithm trained with COVID-19 pneumonia also identifies immune checkpoint inhibitor therapy-related pneumonitis *Cancers* **13** 652

[57] Saba L *et al* 2021 Six artificial intelligence paradigms for tissue characterisation and classification of non-COVID-19 pneumonia against COVID-19 pneumonia in computed tomography lungs *Int. J. Comput. Assist. Radiol. Surg.* **16** 423–34

[58] Singh M *et al* 2021 Transfer learning based ensemble support vector machine model for automated COVID-19 detection using lung computerized tomography scan data *Med. Biol. Eng. Comput.* **59** 825–39

[59] Apostolopoulos I D and Mpesiana T A 2020 Covid-19: automatic detection from x-ray images utilizing transfer learning with convolutional neural networks *Phys. Eng. Sci. Med.* **43** 635–40

[60] Lam C *et al* 2018 Automated detection of diabetic retinopathy using deep learning *AMIA Jt. Summits Transl. Sci. Proc.* **2017** 147–55

[61] Ferenci P 2006 Regional distribution of mutations of the ATP7B gene in patients with Wilson disease: impact on genetic testing *Human Genet.* **120** 151–9

[62] Vrabelova S *et al* 2005 Mutation analysis of the ATP7B gene and genotype/phenotype correlation in 227 patients with Wilson disease *Mol. Genet. Metab.* **86** 277–85

[63] Brickman A M *et al* 2011 White matter hyperintensities and cognition: testing the reserve hypothesis *Neurobiol. Aging.* **32** 1588–98

[64] Saba L *et al* 2018 Volumetric distribution of the white matter hyper-intensities in subject with mild to severe carotid artery stenosis: does the side play a role? *J. Stroke Cerebrovasc. Dis.* **27** 2059–66

[65] Saba L *et al* 2017 Relationship between white matter hyperintensities volume and the circle of Willis configurations in patients with carotid artery pathology *Eur. J. Radiol.* **89** 111–6

[66] Porcu M *et al* 2018 Clinical neuroimaging markers of response to treatment in mood disorders *Neurosci. Lett.* **669** 43–54

[67] Kim T *et al* 2006 MR imaging of the brain in Wilson disease of childhood: findings before and after treatment with clinical correlation *Am. J. Neuroradiol.* **27** 1373–8

[68] Hu X *et al* 2017 Frequency-dependent changes in the amplitude of low-frequency fluctuations in patients with Wilson's disease: a resting-state fMRI study *Metab. Brain Dis.* **32** 685–92

[69] Kaden M *et al* 2015 Border-sensitive learning in generalized learning vector quantization: an alternative to support vector machines *Soft Comput.* **19** 2423–34

[70] Jing R *et al* 2020 Altered large-scale functional brain networks in neurological Wilson's disease *Brain Imaging Behav.* **14** 1445–55

[71] Agarwal M *et al* 2021 Wilson disease tissue classification and characterization using seven artificial intelligence models embedded with 3D optimization paradigm on a weak training brain magnetic resonance imaging datasets: a supercomputer application *Med. Biol. Eng. Comput.* **59** 511–33

[72] Jamthikar A *et al* 2020 Cardiovascular/stroke risk prevention: a new machine learning framework integrating carotid ultrasound image-based phenotypes and its harmonics with conventional risk factors *Indian Heart J.* **72** 258–264

[73] Jamthikar A *et al* 2019 A low-cost machine learning-based cardiovascular/stroke risk assessment system: integration of conventional factors with image phenotypes *Cardiovasc. Diag. Ther.* **9** 420

[74] Jamthikar A *et al* 2020 Ultrasound-based stroke/cardiovascular risk stratification using Framingham Risk Score and ASCVD Risk Score based on "Integrated Vascular Age" instead of "Chronological Age": a multi-ethnic study of Asian Indian, Caucasian, and Japanese cohorts *Cardiovasc. Diag. Ther.* **10** 939

[75] Suri J S 2014 System and method for creating and using intelligent databases for assisting in intima-media thickness (IMT) *US patent* 8805043

[76] Molinari F *et al* 2012 Ultrasound IMT measurement on a multi-ethnic and multi-institutional database: our review and experience using four fully automated and one semi-automated methods *Comput. Methods Programs Biomed.* **108** 946–60

[77] Jamthikar A *et al* 2019 A low-cost machine learning-based cardiovascular/stroke risk assessment system: integration of conventional factors with image phenotypes *Cardiovasc. Diagn. Ther.* **9** 420–30

[78] Biswas M *et al* 2019 Deep learning fully convolution network for lumen characterization in diabetic patients using carotid ultrasound: a tool for stroke risk *Med. Biol. Eng. Comput.* **57** 543–64

[79] Beswick A D *et al* 2008 *A Systematic Review of Risk Scoring Methods and Clinical Decision Aids Used in the Primary Prevention of Coronary Heart Disease (Supplement)* (London: Royal College of General Practitioners)

[80] Krizhevsky A, Sutskever I and Hinton G E 2017 ImageNet classification with deep convolutional neural networks *Commun. ACM* **60** 84–90

[81] He K *et al* 2016 Deep residual learning for image recognition *Proc. of the IEEE Conf. on Computer Vision and Pattern Recognition* (Piscataway, NJ: IEEE)

[82] Szegedy C *et al* 2016 Rethinking the inception architecture for computer vision *Proc. of the IEEE Conf. on Computer Vision and Pattern Recognition* (Piscataway, NJ: IEEE)

[83] Huang G *et al* 2017 Densely connected convolutional networks *Proc. of the IEEE Conf. on Computer Vision and Pattern Recognition* (Piscataway, NJ: IEEE)

[84] Chollet F 2017 Xception: deep learning with depthwise separable convolutions *Proc. of the IEEE Conf. on Computer Vision and Pattern Recognition* (Piscataway, NJ: IEEE)

[85] Suk H-I *et al* 2017 Deep ensemble learning of sparse regression models for brain disease diagnosis *Med. Image Anal.* **37** 101–13

[86] Abiwinanda N *et al* 2019 Brain tumor classification using convolutional neural network *World Congress on Medical Physics and Biomedical Engineering 2018* (Berlin: Springer)

[87] Zhang Y *et al* 2019 Strength and similarity guided group-level brain functional network construction for MCI diagnosis *Pattern Recognit.* **88** 421–30

[88] Abrol A *et al* 2020 Deep residual learning for neuroimaging: an application to predict progression to Alzheimer's disease *J. Neurosci. Methods* **339** 108701

[89] Richhariya B *et al* 2020 Diagnosis of Alzheimer's disease using universum support vector machine based recursive feature elimination (USVM-RFE) *Biomed. Signal Process. Control.* **59** 101903

[90] Liu J *et al* 2020 Enhancing the feature representation of multi-modal MRI data by combining multi-view information for MCI classification *Neurocomputing.* **400** 322–32

# Part III

Deep learning in cardiovascular imaging

**IOP** Publishing

## Multimodality Imaging, Volume 1
### Deep learning applications
**Mainak Biswas and Jasjit S Suri**

# Chapter 6

# Artificial intelligence based carotid plaque tissue characterisation and classification from ultrasound images using a deep learning paradigm

**Luca Saba, Sanagala S Skandha, Suneet K Gupta, Anudeep Puvvula, Vijaya K Koppula, Amer M Johri, Narendra N Khanna, Sophie Mavrogeni, John R Laird, Gyan Pareek, Martin Miner, Petros P Sfikakis, Athanasios Protogerou, Durga P Misra, Vikas Agarwal, Aditya M Sharma, Vijay Viswanathan, Vijay S Rathore, Monika Turk, Raghu Kolluri, Klaudija Viskovic, Elisa Cuadrado-Godia, George D Kitas, Vijay Nambi, Deepak L Bhatt, Andrew Nicolaides and Jasjit S Suri**

Characterising and classifying atherosclerotic plaque lesions manually is difficult, error-prone, and time-consuming. This study uses a deep learning (DL) framework and a supercomputer to automatically classify carotid artery plaque into sympto-matic and asymptomatic forms.

On ultrasound images, symptomatic carotid plaques exhibit (a) a low greyscale median due to a histologically large lipid core and relatively little collagen and calcium, and (b) a higher chaotic (heterogeneous) greyscale distribution due to their composition. The methodology consisted of building a DL model of artificial intelligence (called Atheromatic) that uses a classic convolution neural network consisting of 13 layers implemented on a supercomputer. The DL model used a cross-validation protocol for estimating the classification accuracy (ACC) and area-under-the-curve (AUC).

A total of 346 carotid ultrasound based defined plaques (196 symptomatic and 150 asymptomatic, mean age 69.9 + 7.8 years, 39% females) were employed in the study. The geometric transformation was used, resulting in 2312 plaques (1191 symptomatic and 1120 asymptomatic plaques). The K10 (90% training and 10% testing) cross-validation DL technique was used. The validation of the two hypotheses: (a) mean feature strength (MFS) and (b) Mandelbrot's fractal dimension (FD) for evaluating chaotic behaviour comprised the DL characterisation system. We found that the MFS and FD were 64.15% + 0.73% ($p$-value < 0.0001) and 6% + 0.13% ($p$-value < 0.0001) greater in symptomatic plaques than in asymptomatic plaques, respectively. The benchmarking findings demonstrate that DL with augmentation (ACC: 89.7%; AUC: 0.91 ($p$-value < 0.0001)) is 6.0% better than a previously published machine learning method (ACC: 83.7%). Atheromatic runs the test patient in less than 2 s.

## 6.1 Introduction

Every year 17.9 million people die from cardiovascular disease (CVD), with 647 000 deaths in the United States [1, 2], or one death every 37 s [3]. The development of atherosclerosis with plaque formation in the vasculature, such as the coronary and carotid arteries [4], is the primary cause of CVD. When a plaque ruptures or ulcerates, it blocks the blood flow and causes myocardial infarction [5] or stroke [5]. This study focuses solely on the characterisation and classification of atherosclerotic plaques in the carotid artery. Plaque is visualised using a variety of medical imaging modalities, the most frequently used of which are magnetic resonance imaging (MRI) [6], computed tomography (CT) [7], and ultrasound (US) [8]. US has established itself as a standard not only as a first-line diagnostic modality in symptomatic patients but also as a potent screening tool in asymptomatic persons [9] during the last decade. It is a low-cost [10] safe test that is also simple to perform, has a small footprint, and is radiation-free [11]. Furthermore, carotid ultrasound imaging with a resolution of less than 0.2 mm can be used to examine plaque texture and identify if it is stable or unstable [12–14]. Due to substantial inter-observer variability, the carotid plaque texture as observed on ultrasonography is varied and difficult to classify with the naked eye [15]. When compared with asymptomatic plaques [16–18], symptomatic plaques create substantial stenosis, which is more hypoechoic. They have a large juxtaluminal black area near the lumen without a discernible echogenic cap, and have discrete white regions in hyperechoic locations. These results were discovered in cross-sectional examinations of symptomatic and asymptomatic patients, as well as in a large prospective study of asymptomatic individuals, and then confirmed by comparing histology to *in vivo* ultrasound imaging [18]. As the pixel data obtained from spatial ultrasound images is big and fuzzy, and computers have a stronger learning ability to handle linear and nonlinear fluctuations in plaque distribution, artificial intelligence (AI) has been utilised to define and categorise [19] plaques using machine learning (ML). This machine learning technique necessitates manually calculating greyscale features [20–25], which are subsequently trained using a training classifier to generate offline signatures and patterns. These are then applied to the test pattern in order to forecast its class risk [26, 27]. Such ML based solutions are ad hoc, sluggish, and non-generalised [27], not to mention unreliable and unstable.

Deep learning (DL) technology has swept the globe, in particular in the field of radiological imaging [28–30]. DL technology offers a better alternative to ML techniques. It includes the capacity to automatically build a down-sampled representation (a so-called feature map) of the original pattern and dynamically modify greyscale contrast variations via the DL architecture's [31] neural network layers. Lekadir *et al* [32] constructed a CNN model for categorisation of a total of 90 000 patches from 50 *in vivo* ultrasound image plaque components with a 0.90 correlation coefficient. It is seen that that the symptomatic plaque has variable tissue characteristics such as (a) hypoechoic regions with a low greyscale median (GSM) due to a large lipid core, low calcium, or intraplaque bleeding, and (b) a more chaotic (heterogeneous) representation in ultrasound scans [13, 14, 25]. It occurs due to the frequent presence of neovascularisation alternating with collagen or lipid areas. Asymptomatic plaques, on the other hand, are generally hyperechoic with increased GSM due to a large and diffused collagen content, high calcification, and a small lipid core. We created a new DL based carotid plaque tissue characterisation and classification system. The design addresses several ML flaws such as (i) feature extraction by hand and (ii) classification within a single layer. The K10 cross-validation methodology (90% training and 10% testing) is used to determine the classification system's accuracy. The plaque is characterised by (a) determining the mean feature strength (MFS) at distinct DL layers [33] and (b) using the fractal dimension (FD) to quantify the randomness in these plaque images [34]. We then evaluate the DL system against a previously created ML system (on the same cohort) and, finally, we use a supercomputer framework to optimise the system speed. The following is the layout of this chapter. The experimental technique and new DL architectural design are presented in section 6.2. In section 6.3 the classification findings and plaque characterisation utilising a DL framework, as well as the hypothesis validation, are reported. The benchmarking of the DL method versus the ML paradigm is presented in section 6.4. The study's findings are presented in section 6.5.

## 6.2 Methodology

### 6.2.1 Patient demographics

The experts detected an internal carotid artery (ICA) stenosis of 50%–99% in 346 referred consecutive patients (mean age 69.9 + 7.8 years and 39% female) (permission was acquired from the Institutional Ethics Committee, St Mary's Hospital, Imperial College, London, UK). 150 of the 346 patients were asymptomatic, meaning they had no neurological problems. The remaining 196 developed carotid artery atherosclerosis-related ipsilateral cerebral hemisphere symptoms (amaurosis fugax (AF), transient ischemic episodes, or stroke with excellent recovery). There were a total of 196 distributions, with 88 strokes, 70 TIA, and 38 AF. Patients' medical histories and physical examinations [17] were recorded by a neurologist. This patient data were previously used in our machine learning research [20, 35].

### 6.2.2 Exclusion criteria

Patients with cardioembolic symptoms or long-term symptoms (six months) were ruled out of the trial. The European Carotid Surgery Trialists' Collaborative Group (ECST) revealed that surgery reduces overall stroke risk in patients with 70% to 99% stenosis. The surgery is deleterious in patients with mild stenosis (0%–29%)

(European Carotid Surgery Trialists' Collaborative Group 1991), and had little effect in patients with 30%–49% or 50%–69% stenosis (European Carotid Surgery Trialists' Collaborative Group 1996). Carotid endarterectomy was recommended for most patients with 80% stenosis (European Carotid Surgery Trialists' Collaborative Group 1998). Surgery was found to be highly advantageous for 70% to 99% stenosis and somewhat favourable for 50% to 69% stenosis in our earlier study [20, 35]. Strokes are rarely linked to plaques in the carotid artery that cause less than 50% stenosis. Noise and prejudice would have been introduced if such plaques had been included. We excluded plaques with less than 50% stenosis based on these data. The ultrasonographers were not blinded to the plaque categorisation or characterisation because they did routine testing for the presence of stenosis and rated the plaques. However, the persons who processed the images were blinded to classification and, hence, they did not know whether the plaques were symptomatic or asymptomatic.

### 6.2.3 Ultrasound data acquisition and preprocessing

The ultrasound scans were taken at Saint Mary's Hospital in the UK's Irvine Laboratory for Cardiovascular Investigation and Research. The ultrasound equipment used was an ATL HDI 3000 (Advanced Technology Laboratories, Seattle, WA, USA) with a duplex scanner and a linear broadband width 4–7 MHz (multifrequency) transducer with a resolution of 20 pixels/mm.

The following conditions were met in order to achieve satisfactory image normalisation (as outlined in our prior study): (i) as the ultrasonic beam was not attenuated while it travelled through blood, the dynamic range was used; (ii) frames were averaged; and (iii) the time gain compensation curve that was sloping across the tissues was positioned vertically across the lumen of the artery. This was done to guarantee that the adventitia on the anterior and posterior walls were both bright. (iv) The gain was adjusted, (v) a linear transfer curve was employed for postprocessing, (vi) the ultrasonic beam was held at 90 degrees to the artery wall, (vii) the lowest depth was chosen so that the plaque took up a considerable portion of the image, and (viii) the probe was calibrated to identify adventitia next to the plaque as a hyperechoic region that could be used to normalise the data. The cohort was taken ten years ago and the machine's resolution has improved since then. However, if the probe's orientation is maintained in both circumstances, ML is more likely to compound better. As the features will be more widely separated, the separation index will improve (APSI). In the case of DL, it is more likely that the architecture will improve. This is because achieving the same performance would necessitate a smaller number of layers.

### 6.2.4 Plaque delineation

As in earlier investigations [20, 35], the Plaque Texture Analysis software (Iconsoft International Ltd, Greenford, London, UK) is also used for normalisation and plaque ROI selection. The median grey-level intensity of blood after normalisation was in the range of 0–5, while that of the adventitia layer was in the

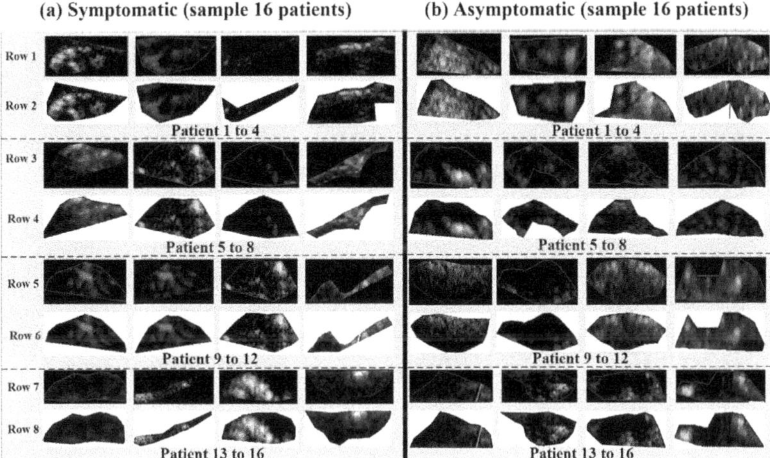


**Figure 6.1.** Left: Symptomatic. Right: Asymptomatic. Rows 1, 3, 5, and 7 are original carotid plaque scans. Rows 2, 4, 6, and 8 are pre-processed plaque delimited cut sections by the vascular surgeon (AN%). (Reproduced with permission from [49]. Copyright 2021 Springer.)

range of 180–190, as was the case in our earlier study [20, 35]. The plaque in the normalised image was outlined with using a mouse and saved as a new file in the case of ROI selection. The software allowed users to choose between the anterior and posterior walls of a plaque. The acoustic shadow was not taken into consideration when selecting calcified plaques. Figure 6.1 depicts these chopped portions.

### 6.2.5 Ultrasound plaque data augmentation

We employed a standardised approach called 'augmentation' since the DL systems often require larger databases (tens of thousands to millions). This necessitates a random geometric alteration of the demarcated plaque, such as (i) skewing and (ii) rotation. The cohort's original size was 346 plaques, with 196 symptomatic and 150 asymptomatic plaques. Using the augmentation process described above, a new cohort of 2311 plaques was created, with 1191 symptomatic plaques and 1120 asymptomatic plaques. A total of 196 symptomatic plaques were transformed to 1191 (804 were newly unique and boosted to 1000, and then the original 196 were added for a total of 1196). Five images were eliminated because they were distorted and inappropriate, resulting in a total of 1191. For the amplification of asymptomatic plaques, a similar method was used. Thus 150 plaques were added to make a total of 1000, then 150 original plaques were added to make a total of 1150. A total of 30 skewed plaques were deemed unsatisfactory and they were removed from the final tally, leaving 1120 asymptomatic plaques.

### 6.2.6 Supercomputer specifications

Our DL model was implemented on the NVIDIA-DGX v100-1 Tesla, as illustrated in figure 6.2(a), which is a ground-breaking system. It has unrivalled performance

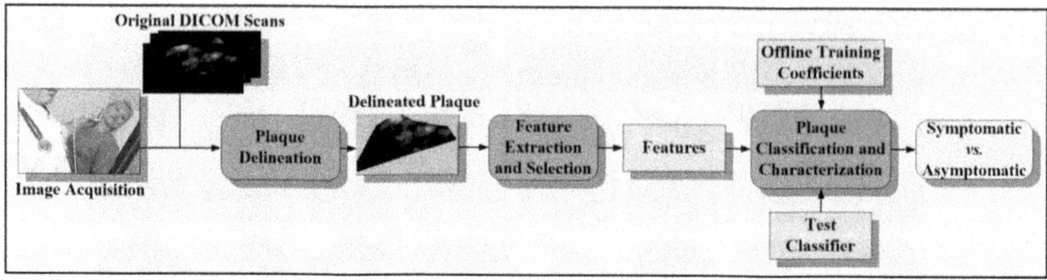


**Figure 6.2.** Global ML architecture. (Reproduced with permission from [49]. Copyright 2021 Springer.)

when it comes to deep learning training. It has eight NVIDIA Tesla V100 graphics processing unit (GPU) accelerators that are connected via an NVIDIA NVLink. As illustrated in figure 6.2(b), all of the GPUs are connected to the cube-mesh network. This facilitates the GPU load-sharing procedure. Deep learning training [36] is more scalable using the high-performance NVLink GPU interface.

### 6.2.7 Deep learning architecture

The creation of the ideal DL architecture for tissue characterisation and plaque categorisation is the fundamental uniqueness of our proposed system. The number of convolution neural network (CNN) layers and the types of the layers [37] in a DL architecture are heavily influenced by the dataset's complexity and the hyper-parameters when combined. Our lab has previously built DL architectures with up to 22 layers, which are typically designed to accept large sample sizes (in the hundreds) and larger image sizes (W × H) [38]. We used a 13-layer CNN architecture with five convolution layers (CLs), five average pooling layers (APLs), two dense layers (DenL-1, DenL-2), and one dropout layer, as shown figure 6.2(c), because our cohort size and image size are both reasonable (i.e. 346 images with sizes ranging from 55 × 43 pixel$^2$ to 593 × 107 pixel$^2$ without augmentation or a cohort size of 2311 with augmentation). By adjusting the number of layers, dropout rate, momentum rate, and learning rate, we were able to fine-tune the hyper-parameters. The softmax layer, which computes the categorical cross-entropy loss function ($E$) between the two classes of symptomatic and asymptomatic and is mathematically as follows:

$$E = -(y^*\log(p)) + [(1 - y)^*\log(1 - p)], \tag{6.1}$$

where $p$ is the DL-estimated predicted probability of the plaque belonging to a specific class, $y$ is the binary indication for the observed class, and * is the product. Equation (6.2) gives the number of output features from the convolution process [33] and equation (6.3) gives the number of output features from average pooling feature mappings (APFMs):

$$n_{\text{out}} = \left[ \frac{n_{\text{in}} + (2^*P) - M}{S} \right] + 1 \tag{6.2}$$

**Table 6.1.** CNN with 13 layers (dropout layer not included) with trainable parameters. (Reproduced with permission from [49]. Copyright 2021 Springer.)

| C1<br>Name of the layer | C2<br>Feature map size | C3<br>Trainable parameter |
| --- | --- | --- |
| 2D convolution layer 1 | (240, 240, 256) | 7168 |
| 2D average pooling layer 1 | (120, 120, 256) | 0 |
| 2D convolution layer 2 | (120, 120, 128) | 295 040 |
| 2D average pooling layer 2 | (60, 60, 128) | 0 |
| 2D convolution layer 3 | (60, 60, 64) | 73 792 |
| 2D average pooling layer 3 | (30, 30, 64) | 0 |
| 2D convolution layer 4 | (30, 30, 32) | 18 464 |
| 2D average pooling layer 4 | (15, 15, 32) | 0 |
| 2D convolution layer 5 | (15, 15, 16) | 4624 |
| 2D average pooling layer 5 | (7, 7, 16) | 0 |
| 1D flatten | 784 | 0 |
| 1D dense 1 | 128 | 100 480 |
| Dropout layer | 128 | 0 |
| 1D dense 2 | 2 | 258 |
| Total trainable parameters | | 499 826 |

$$a_{\mathrm{out}} = \left(\frac{w - f}{S}\right) + 1, \tag{6.3}$$

where $n_{\mathrm{in}}$ and $n_{\mathrm{out}}$ denote the number of input and output characteristics that each CL represents. $M$ is the size of the convolution kernel, $P$ is the size of the convolution padding, $S$ is the stride size (representing the kernel movement), $a_{\mathrm{out}}$ is the number of APL output features, $w$ is the width of the input feature map, and $f$ is the kernel size. Table 6.1 provides the names of the layers, the sizes of the feature maps, and the training parameters in three columns (per epoch). The input size must be the same for the DL system to work. As a result, we employed padding zeros to convert the cut portions images to same-sized images, but we made sure they did not get used throughout the DL process. Moreover, we utilise a 'mask image' the same size as the chopped part to only consider the greyscale pixels of the chopped parts. As a result this masked image is employed to ensure that DL only uses the chopped sections' greyscale pixels.

### 6.2.8 Experimental protocol

Our experimental technique entails employing the K10 technique to determine classification accuracy using a cross-validation (CV) paradigm (90% training and 10% testing). We conducted the CV technique on both datasets to demonstrate the potential of the augmentation (default 346 plaques and augmented 2311 plaques). During the training and testing of the DL system, the following factors were taken into account: 10 000 total epochs, 0.001 learning rate, 32 batch size, 0.001 regularisation (L2), and 0.5 dropout, Adam as the optimiser. According to our

DL laboratory experience [39], a single combination can have somewhere between 10 000 and 20 000 iterations during training. We chose 10 000 epochs as a reliable value because the image size was moderate. The learning rate and regularisation values are typically 0.001 and 0.001, respectively, as they are in the industry. As the overall data size for ten combinations of the K10 protocol was 311 (90% of 346) or 2080 (90% of 2311), the batch size for training was 32. MEDCALC 17.0 was used to evaluate the final predicted class value and compute the received operating characteristics (ROC) and areas-under-the-curve (AUCs).

### 6.2.9 Machine learning for benchmarking deep learning

We compare textural characteristics, AUC, and accuracy to see how well the DL model compares to ML models. Figure 6.2 depicts the overall architecture of the ML models. The effectiveness of ML models is determined by the features retrieved and chosen. We collected the histogram, Haralick, and Hu-moment features, which we then entered into linear discriminant analysis (LDA), $k$-nearest neighbours ($k$-NN), and the K10 cross-validation protocol [40, 41].

### 6.2.10 Performance parameters using the DL and ML methods

If $\eta_{K10}^{DL}(c)$ represents the accuracy of the DL method using the K10 protocol for the combination, $\bar{\eta}_{K10}^{DL}$ represents the mean of the $C$ combinations, and $\sigma_{K10}^{DL}$ represents the corresponding standard deviation, then these may be stated mathematically as

$$\bar{\eta}_{K10}^{DL} = \frac{\displaystyle\sum_{c=1}^{C=10} \eta_{K10}^{DL}(c)}{C} \tag{6.4}$$

$$\sigma_{K10}^{DL} = \sqrt{\frac{\displaystyle\sum_{c=1}^{C=10}\left[\eta_{K10}^{DL}(c) - \bar{\eta}_{K10}^{DL}\right]^2}{C-1}}. \tag{6.5}$$

Following a similar notation for the ML based strategy, we can compute the mean and standard deviation as follows:

$$\bar{\eta}_{K10}^{ML} = \frac{\displaystyle\sum_{c=1}^{C=10} \eta_{K10}^{ML}(c)}{C} \tag{6.6}$$

$$\sigma_{K10}^{ML} = \sqrt{\frac{\displaystyle\sum_{c=1}^{C=10}\left[\eta_{K10}^{ML}(c) - \bar{\eta}_{K10}^{ML}\right]^2}{C-1}}. \tag{6.7}$$

This process was repeated for two different kinds of validation datasets and performance evaluated.

# 6.3 Results

The first step is to evaluate our DL system's performance (PE) using a cross-validation technique, followed by plaque characterisation based on the two hypotheses. The PE is shown in section 6.3.1, while the characterisation is discussed in section 6.3.2. Finally, we employed the most extensively used facial biometric dataset to evaluate our DL architecture.

## 6.3.1 Deep learning data analysis and benchmarking against machine learning

### 6.3.1.1 DL classification accuracy with and without augmentation

As illustrated in figure 6.3, we used a 13-layer CNN to build the DL architecture, with the last layer serving as the softmax layer. The likelihood that the risk belongs to either symptomatic or asymptomatic groups is categorically (binary) estimated in the output layer. Without augmentation, the K10 protocol has the best accuracy and an AUC (calculated using MEDCALC 17.0) of 86.17% and 0.86 ($p$-value $< 0.0001$), respectively, whereas with augmentation it has 89.7% and 0.91 ($p$-value $< 0.0001$). We compute the real-time accuracy of each combination at the end of the 500th step value to have a better picture of the model's performance.

### 6.3.1.2 Benchmarking of deep learning against machine learning

The benchmarking protocol involves comparing a DL system to an ML system on the same cohort. Furthermore, we compare our innovative DL system to a previously reported ML system that used identical plaque data [20, 35]. Figure 6.4 shows the results of using the K10 cross-validation technique for both

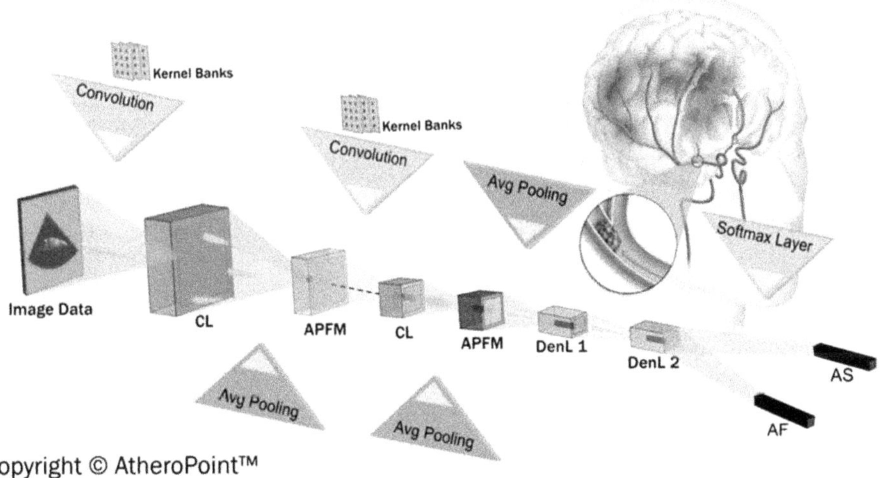




**Figure 6.3.** The conceptual view of the deep learning architecture consisting of five CLs, five APLs, two DenLs, and one flattened layer. The dotted line in the middle shows the missing three CLs and three APLs. (Reproduced with permission from [49]. Copyright 2021 Springer.)

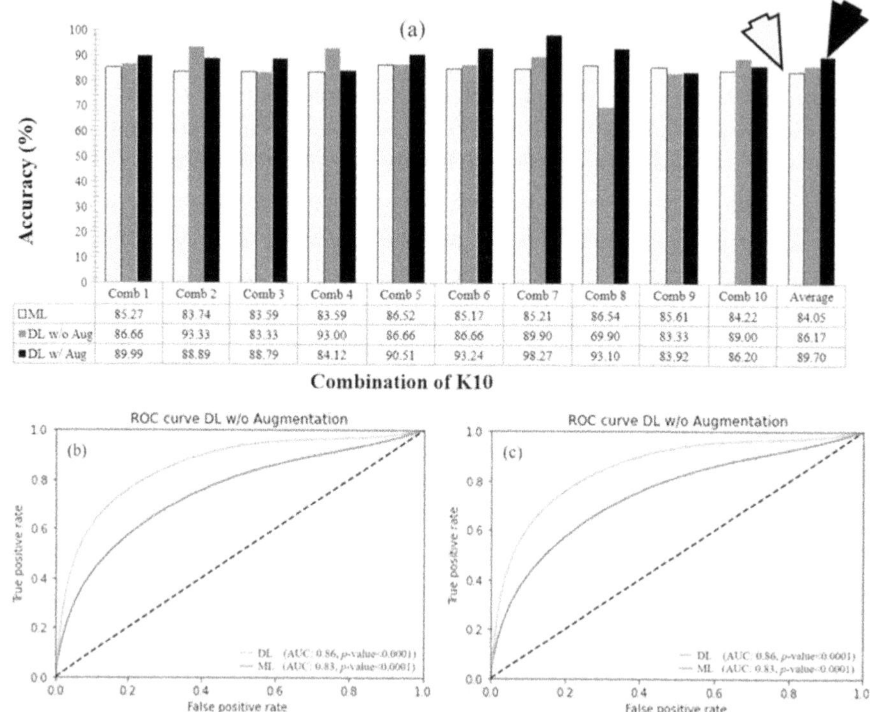



| | Comb 1 | Comb 2 | Comb 3 | Comb 4 | Comb 5 | Comb 6 | Comb 7 | Comb 8 | Comb 9 | Comb 10 | Average |
|---|---|---|---|---|---|---|---|---|---|---|---|
| □ML | 85.27 | 83.74 | 83.59 | 83.59 | 86.52 | 85.17 | 85.21 | 86.54 | 85.61 | 84.22 | 84.05 |
| ▨DL w/o Aug | 86.66 | 93.33 | 83.33 | 93.00 | 86.66 | 86.66 | 89.90 | 69.90 | 83.33 | 89.00 | 86.17 |
| ■DL w/ Aug | 89.99 | 88.89 | 88.79 | 84.12 | 90.51 | 93.24 | 98.27 | 93.10 | 83.92 | 86.20 | 89.70 |

**Figure 6.4.** (a) Bar charts showing the accuracy comparison between (i) machine learning (light grey), (ii) deep learning without augmentation (DL w/o Aug) (dark grey), and (iii) deep learning with augmentation (DL w/ Aug) (black). (b) ROC curves showing the AUC comparison: ML (0.83, $p$-value < 0.0001) versus DL w/o Aug (0.86, $p$-value < 0.0001) systems showing an improvement of 3.61%. (c) ROC curves showing the AUC comparison: ML (AUC: 0.83, $p$-value < 0.0001) versus DL w/ Aug (AUC: 0.91, $p$-value < 0.0001) showing an improvement of 8%. (Reproduced with permission from [49]. Copyright 2021 Springer.)

AI algorithms. While the ML system had an accuracy of 84.05%, the DL system had an accuracy of 86.17% (without augmentation), 89.7% (with augmentation) (as shown in figure 6.4(a)). The AUCs for the ML and DL systems were 0.83 ($p$-value < 0.0001) and 0.86 ($p$-value < 0.0001) (without augmentation), and 0.91 ($p$-value < 0.0001) (with augmentation), respectively. The crucial element to remember is that, unlike in DL systems, the design and implementation approach for ML systems involves precise stages. The features had to be manually computed, and the offline training system had to be monitored for optimisation at all times. In the discussion section, we will cover the significant distinctions in more detail.

### 6.3.1.3 Power analysis

Power analysis is used to check the validity of the recruited sample size. With a 95% confidence interval (CI) and a 5% margin of error, the sample size was 183 samples. As a result, our study's recruited sample size (346 scans) was around 47% larger than the needed. The power analysis is calculated using the formula

$$n = \left[ (z^*)^2 \times \left( \frac{\hat{p}(1 - \hat{p})}{\mathrm{MoE}^2} \right) \right], \tag{6.8}$$

where $z^*$ is the $z$-score, MoE is the margin of error, and $\hat{p}$ is the data proportion.

### 6.3.1.4 Validation of the DL and ML systems using a biometric facial data (BFD) set

The gold standard for plaque photos is reliant on the clinician's experience and the lighting circumstances under which the images are described manually. As a result, we employed the most widely used 'biometric facial data' [42] with a strong categorical gold standard to further test the DL architecture. This database contained 1440 photos divided into 72 classes, each with 20 samples. Supplemental material A contains sample photos of the dataset. We adjusted the DL model output layer to 72 nodes for the BFD experiment mainly because the DL model output layer had two classes. Our accuracy with the K10 technique was 99.84%, with an AUC of 0.99 ($p$-value < 0.0001). The ML system was used to benchmark this, and it produced an accuracy of 97.9% and an AUC of 0.95 ($p$-value < 0.0001), which was almost identical to the DL method. We used a diagnostics odds ratio (DOR) to validate the suggested model's performance; the DOR of DL was found to be greater than that of ML.

### 6.3.2 Plaque characterisation in a deep learning framework

Plaque characterisation necessitates determining the plaque area's (a) intensity distribution and (b) roughness (or chaotic behaviour). When compared to asympto-matic plaques [13, 14, 43, 44], pictures of symptomatic and susceptible plaques are (i) more hypoechoic (darker) and (ii) exhibit a more patchy (chaotic) greyscale representation. As part of the characterisation process, we must observe and justify these two properties based on this premise. We also need to see if these can distinguish between symptomatic and asymptomatic plaques. Using our DL architecture, this separation must be confirmed in terms of classification accuracy. Our calculations suggest that the GSM for symptomatic (25.67 ± 26.27) is 86% higher than the GSM for asymptomatic plaque (3.53 ± 10.38; CI: 24.33–29.91, $p$-value < 0.0001). It has a greater standard deviation, necessitating the use of an automated DL based approach to characterise the plaque. This section focuses on establishing the two most important components of the hypothesis.

### 6.3.2.1 Hypothesis 1: intensity distribution

The CNN is made up of 13 layers, the first eight of which are low-level character-istics (CL-1, APL-1, CL-2, APL-2 CL-3, APL-3, CL-4, and APL-4). As the asymptomatic plaque has a larger collagen/calcium (hyperechoic) composition, the first eight layers must catch the textured plaque's bright surface. The low-level features (LLFs) will, on the other hand, miss symptomatic plaque's high lipids and low calcium. The CNN should be able to detect the high lipid/low calcium (hypoechoic) aspects of symptomatic plaque as a part of the high-level features.

The intensity distribution, which is a realistic representation of the plaque type, is characterised using the feature map's strength (symptomatic versus asymptomatic). These strengths are simply maps of the plaque features; hence they must be referred to as 'feature maps'. We compute these feature maps for both sets of data (symptomatic and asymptomatic) to represent this in a deep learning framework. This is performed by determining the strengths at all the output points of layers using the traditional CNN approach for both the symptomatic and asymptomatic feature maps.

### 6.3.2.2 Quantification of feature maps at different DL stages

The DL-2 layer, we believed, would provide the model's feature map strength. Figure 6.5(a) depicts the strength of our model's feature maps in the K10 protocol. For the remaining two DL layers, the 2D FM is transformed into a 1D FM. For each class, such vectors are produced for all of the photos (symptomatic and asymptomatic). The quantification of the vector length corresponding to each of the classes is used to compare the strength of both classes. The difference between the classes can be observed for the DL system in figure 6.5(b).

### 6.3.2.3 Justification of the higher MFS of symptomatic plaques against the MFS of asymptomatic plaques

The understanding of the intensity distribution of the lobes of the histograms of these plaques is another way to justify why MFS (symptomatic plaques) > MFS (asymptomatic plaques). The symptomatic and asymptomatic histograms are depicted in figure 6.6(a) (b), with lobes A1 and A2 for symptomatic and lobes B1 and B2 for asymptomatic. It is worth noting that when determining the area of the major lobes, the side lobes of A1 and A2 are also taken into account. This means that lobe A1 covers the greyscale range of 0 to 200, while lobe A2 covers the range of 200 to 255. The lobes B1 and B2 are shown using the same logic. Table 6.2 shows the areas of these symptomatic lobes A1, A2, and A1+A2, and asymptomatic lobes B1, B2, and B1+B2. It is observed that the symptomatic lobe regions are (32.86%, 31.05%, and 32.73%) larger than the asymptomatic lobe areas. This explains why the

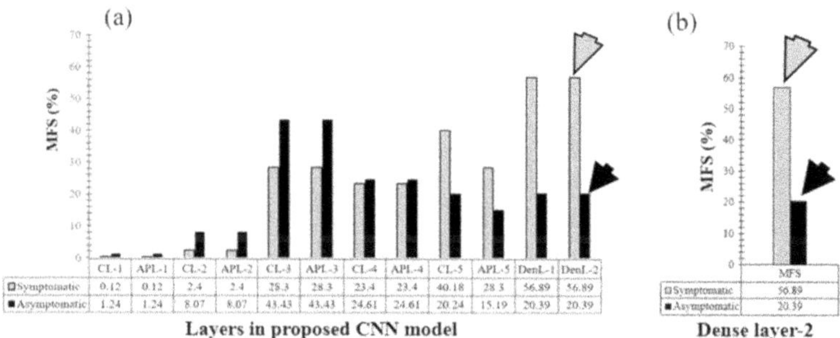



| | CL-1 | APL-1 | CL-2 | APL-2 | CL-3 | APL-3 | CL-4 | APL-4 | CL-5 | APL-5 | DenL-1 | DenL-2 |
|---|---|---|---|---|---|---|---|---|---|---|---|---|
| Symptomatic | 0.12 | 0.12 | 2.4 | 2.4 | 28.3 | 28.3 | 23.4 | 23.4 | 40.18 | 28.3 | 56.89 | 56.89 |
| Asymptomatic | 1.24 | 1.24 | 8.07 | 8.07 | 43.43 | 43.43 | 24.61 | 24.61 | 20.24 | 15.19 | 20.39 | 20.39 |

**Layers in proposed CNN model**

| | MFS |
|---|---|
| Symptomatic | 56.89 |
| Asymptomatic | 20.39 |

**Dense layer-2**

**Figure 6.5.** (a) Mean feature strength for every output layer of the 13-layer CNN model. (b) MFS at the final output of the DenL-2 layer of the CNN model. Note that the dropout layer is not included. (Reproduced with permission from [49]. Copyright 2021 Springer.)

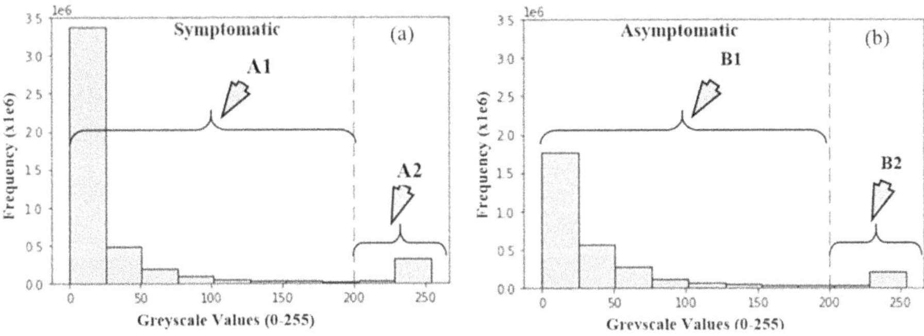


**Figure 6.6.** Histogram distribution of (a) symptomatic and (b) asymptomatic classes. (Reproduced with permission from [49]. Copyright 2021 Springer.)

**Table 6.2.** Strength of the histogram lobes for symptomatic versus asymptomatic in the form of area. (Reproduced with permission from [49]. Copyright 2021 Springer.)

| Symptomatic plaque area | | | Asymptomatic plaque area | | | Quantify |
|---|---|---|---|---|---|---|
| Lobes | Bin range | Area (pix$^2$) | Lobes | Bin range | Area (pix$^2$) | Difference |
| A1 | 0–200 | 4 256 297 | B1 | 0–200 | 2 857 284 | 32.86% |
| A2 | 200–255 | 341 801 | B2 | 200–255 | 235 670 | 31.05% |
| A1+A2 | All bins | **4 598 098** | B1+B2 | All bins | **3 092 954** | 32.73% |
| Area/plaque | All bins | **23 459** | Area/plaque | All bins | **20 619** | 32.73% |

DL model can indicate that the MFS of the symptomatic class is higher than the MFS of the asymptomatic class. In terms of the number of CNN layers, this further confirms the deep learning architectural design. It is worth noting that the ratio of symptomatic to asymptomatic plaques is 196:150. When calculating the lobe area per plaque (as shown in row R5), symptomatic outnumbers asymptomatic by 32.73%, confirming our prediction.

### 6.3.2.4 Visual representation of the visual feature maps using non-augmented data as an example

After loading of the trained weight file and training the model, the feature map for each layer is computed. The average output of three channels was gathered from all of the filters (RGB). An example feature map in layer 3 with 64 filters is shown in figures 6.7(b) and (c). A grid of filter output with four rows and 16 columns (16 × 4 = 64) is shown in figures 6.7(b) and (c). We use all of the images from the training set to compute mean values for the symptomatic (176 images) and asymptomatic (135 images) classes. The mean feature map view of the symptomatic and asymptomatic classes was calculated. The purple block in figures 6.7(b) and (c) is a filtered image, and the turquoise colour represents the features learned by the model at layer three, which are texture features of greyscale images (figure 6.7).

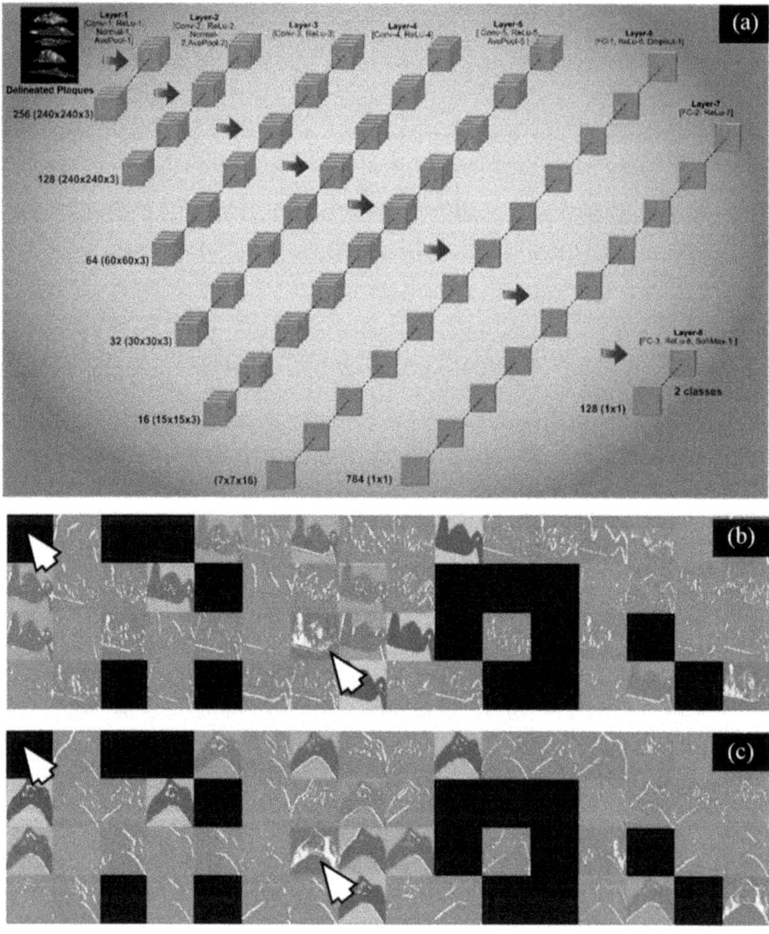

**Figure 6.7.** (a) Layers of the DL architecture, (b) sample feature maps for the symptomatic class, and (c) sample feature maps for the asymptomatic class. (Reproduced with permission from [49]. Copyright 2021 Springer.)

### 6.3.2.5 Hypothesis 2: chaotic distribution

As stated in the introduction, the second hypothesis is that the plaque assumes more unpredictability (chaotic or heterogeneous) in symptomatic individuals compared to asymptomatic individuals. One straightforward technique to quantify plaque randomness is to portray it as a chaotic behaviour using Mandelbrot's fundamental equation of chaotic measurement, the fractal dimension (FD), which is represented by $D$. We compute the FD for the symptomatic and asymptomatic class pools based on this assumption. If the FD of symptomatic is higher than the FD of asymptomatic, our hypothesis holds true for the plaque's characterisation assumption, implying that DL based categorisation is valid. Using the Mandelbrot equation, we compute $D$ for both pools using the standardised equation and technique, as shown in figure 6.8:

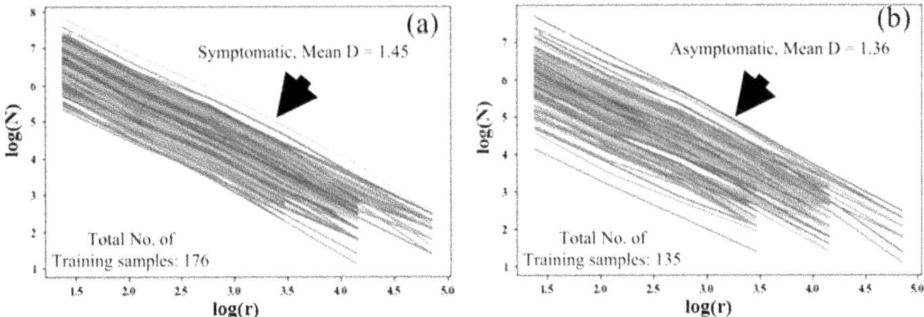


**Figure 6.8.** Fractal dimension ($D$) analysis of (a) the symptomatic class and (b) the asymptomatic class. (Reproduced with permission from [49]. Copyright 2021 Springer.)

$$D = \frac{\log(N)}{\log(r)} \tag{6.9}$$

where $D$ is the dimensionless number of self-similar objects, $N$ denotes the number of boxes that cover the pattern, and $r$ denotes the magnification. It is worth noting that the log is treated as a Napier log to the base 'e'. The symptomatic and asymptomatic $D$ values are 1.45 (derived from 176 photos) and 1.36 (derived from 135 images), respectively, with (CI: 0.05 to 0.09), verifying that $D$ (symptomatic) > $D$ (asymptomatic).

### 6.3.2.6 Atheromatic plaque segregation index using deep learning

The goal of this indicator is to examine how DL views the distinction between symptomatic and asymptomatic patients. As a result, the training datasets are sent into the DL system, which computes the mean feature strength (MFS) as described previously. The atheromatic plaque segregation index (APSI) [20] is calculated based on the percentage difference between symptomatic and asymptomatic datasets. The categorisation accuracy of 90% is further justified by the separation. The APSI is calculated using ML and DL techniques, as illustrated in table 6.3, and is mathematically given by

$$\mathrm{APSI}_{\mathrm{K10}}^{\mathrm{AI\,model}}(\%) = \left(\frac{|\,\mathrm{MFS}_{\mathrm{AF}} - \mathrm{MFS}_{\mathrm{AS}}\,|}{\mathrm{MFS}_{\mathrm{AF}}}\right)*100. \tag{6.10}$$

Here the AI model represents ML or DL, K10 is the CV protocol, and $\mathrm{MFS}_{\mathrm{AF}}$ and $\mathrm{MFS}_{\mathrm{AS}}$ represent the MFS for symptomatic and asymptomatic, respectively.

Table 6.3 shows that the DL model has a greater APSI than the ML model. Feature learning is balanced between two classes in the deep learning paradigm, but not in machine learning.

## 6.4 Discussion

Our research is the first of its kind, using deep learning architecture to classify symptomatic versus asymptomatic plaques. When compared to machine learning, our model had higher accuracy and AUC, with a 6.0% ($p$-value < 0.0001) increase in

**Table 6.3.** Atheromatic plaque separation index (APSI) using ML and DL systems. (Reproduced with permission from [49]. Copyright 2021 Springer.)

| Types of artificial intelligence | Mean feature strength for symptomatic (AF) | Mean feature strength for asymptomatic (AS) | Atheromatic plaque segregation index (APSI) |
| --- | --- | --- | --- |
| Deep learning | 56.89 ± 25.80 | 20.39 ± 25.80 | $\text{APSI}_{\text{K10}}^{\text{DL}}$ = **64.15%** ($p$-value < 0.0001; CI: 0.48 to 0.23) |
| Machine learning | 72.38 ± 25.93 | 35.7 ± 25.93 | $\text{APSI}_{\text{K10}}^{\text{ML}}$ = **50.67%** ($p$-value < 0.0001; CI:0.67 to 0.38) |

accuracy over previously published work [20, 35]. We tested the DL system by running the algorithm on a set of facial biometric data [42], which yielded a 99% accuracy. We next used the widely acknowledged animal dataset (ASSIRA) [45] to further validate the DL system, which resulted in an accuracy of 97.56%.

### 6.4.1 A note on the unbalanced datasets for symptomatic and asymptomatic plaques

Our dataset comprised two types of plaques: symptomatic (196 images) and asymptomatic (150 images), with a little imbalance in the class sizes. This was due to the fact that they were gathered from individuals who had been referred for ultrasound examination on a regular basis. In order to avoid overfitting, we employed two alternative strategies: (i) L2 regularisation in the DL system's dense layer and (ii) the probability of a dropout layer in the control. To avoid overfitting, the dropout layer's probability was fixed at 0.5. This made it easier to keep the weights in each tier of the DL system balanced.

### 6.4.2 Benchmarking against techniques available in the literature

Table 6.4 summarises the comparison of our DL model against previous works [21, 35, 46–48].

FDR: Fisher discrimination ratio; DWT: discrete wavelet transformation; #: chronological order; NG: not given.

### 6.4.3 A special note on the comparison of supercomputer hardware to a local machine

The impact of hardware resources on the in-depth learning approach was investigated. The speed of training on a supercomputer was compared to the speed of training on a local computer. The DL training on the supercomputer (with eight GPUs) takes 2 min for each epoch, whereas it took 28 min on the local PC.

**Table 6.4.** Benchmarking table showing previous classification paradigms. (Reproduced with permission from [49]. Copyright 2021 Springer.)

| Author, year[#] | Plaque | Data size | Feature extraction | Type of classifier | ML versus DL | ACC (%) | AUC ($p$-value) |
| --- | --- | --- | --- | --- | --- | --- | --- |
| Christodoulou *et al* (2003) | Plaque only | 230 (-) | Texture | SOM $k$-NN | ML | 73.1 68.8 | 0.753 0.738 NG |
| Acharya *et al* (2011) | Plaque only | 346 (Cyprus) | Texture | SVM | ML | 83 | NG ($p =$ 0.0000) |
| Acharya *et al* (2012) | Plaque only | 346 (Cyprus) | DWT | SVM | ML | 83.7 | NG ($p <$ 0.0001) |
| Acharya *et al* (2013) | Plaque only | 146 (Portugal) | Trace transform and fuzzy texture | Fuzzy classifier | ML | 93.1 | NG ($p <$ 0.0001) |
| | | 346 (UK) | | SVM with RBF | ML | 85.3 | NG ($p <$ 0.0001) |
| Gastounioti *et al* (2014) | Plaque without cut section | 56 US images | FDR | SVM | ML | 88.0 | 0.90 NG |
| Proposed DL without augmentation | Plaque only | 346 (Cyprus) | DL automated features | CNN (13 layers) | DL | **86.17** | **0.86 ($p <$ 0.0001)** |
| Proposed DL with augmentation | Plaque only | 2318 (Cyprus augmented) | DL automated features | CNN (13 layers) | DL | **89.7** | **0.91 ($p <$ 0.0001)** |

### 6.4.4 Strengths, weaknesses, and extensions

Using 13 layers of CNN architecture, we were able to successfully classify symptomatic and asymptomatic plaque. This is the first study of its kind to use a supercomputer to differentiate between symptomatic and asymptomatic plaque pictures. Two well-known datasets were used to test the system. We were able to employ six of the eight GPUs available while running our procedures because our university has numerous investigators accessing the supercomputer. Despite the system's virtues, there are also drawbacks. Due to the enormous number of iterations (or epochs) required, the training mechanism for a DL framework is always difficult. It was difficult to arrange these activities on a shared supercomputer because there were ten different combinations and three different types of datasets. The outcomes of the pilot study seem promising. More types of AI models can be exploited and tested in this framework because the AI field is always changing. Because the dataset was not particularly large, it was possible to enhance accuracy even more by augmenting it. Furthermore, AI systems based on transfer learning could be modified to eliminate the need for repetitive training.

## 6.5 Conclusion

This is the first study of its kind to use a deep learning model and a supercomputer to define and classify carotid plaques into symptomatic and asymptomatic groups. The accuracy and AUC of the deep learning system were 89.7% and 0.91 ($p$-value < 0.0001), respectively, indicating a 6.0% improvement over earlier methods. We exhibited tissue characterisation utilising (a) the mean strengths of the feature map and (b) Mandelbrot's fractal dimension, based on the notion that symptomatic plaque is heterogeneous and has a more chaotic representation in ultrasound. To demonstrate the class separation, we employed the atheromatic separation index. During the online test patient prediction, the system took less than two seconds per image and was constructed on a supercomputer with eight GPU configurations. The DL architecture was validated against two well acknowledged datasets, yielding consistent results.

**Disclosure/Conflict of interest**

Dr Jasjit S Suri is a stroke/cardiovascular imaging expert at AtheroPoint™.

Dr Vijay Nambi has the following relationships: (i) site PI research sponsored by Merck (completed), Amgen (to begin); (ii) a provisional patent with Roche and the Baylor College of Medicine for the use of biomarkers in heart failure prediction; and (iii) he is supported by a VA MERIT grant.

Dr Deepak L Bhatt discloses the following relationships—Advisory Board: Cardax, Cereno Scientific, Elsevier Practice Update Cardiology, Level Ex, Medscape Cardiology, PhaseBio, PLx Pharma, Regado Biosciences; Board of Directors: Boston VA Research Institute, Society of Cardiovascular Patient Care, TobeSoft; Chair: American Heart Association Quality Oversight Committee; Data Monitoring Committees: Baim Institute for Clinical Research (formerly Harvard Clinical Research Institute, for the PORTICO trial, funded by St Jude Medical, now

Abbott), Cleveland Clinic (including for the ExCEED trial, funded by Edwards), Duke Clinical Research Institute, Mayo Clinic, Mount Sinai School of Medicine (for the ENVISAGE trial, funded by Daiichi Sankyo), Population Health Research Institute; Honoraria: American College of Cardiology (Senior Associate Editor, Clinical Trials and News, ACC.org; Vice-Chair, ACC Accreditation Committee), Baim Institute for Clinical Research (formerly Harvard Clinical Research Institute; RE-DUAL PCI clinical trial steering committee funded by Boehringer Ingelheim; AEGIS-II executive committee funded by CSL Behring), Belvoir Publications (Editor in Chief, Harvard Heart Letter), Duke Clinical Research Institute (clinical trial steering committees, including for the PRONOUNCE trial, funded by Ferring Pharmaceuticals), HMP Global (Editor in Chief, Journal of Invasive Cardiology), Journal of the American College of Cardiology (Guest Editor; Associate Editor), Medtelligence/ReachMD (CME steering committees), Level Ex, MJH Life Sciences, Population Health Research Institute (for the COMPASS operations committee, publications committee, steering committee, and USA national co-leader, funded by Bayer), Slack Publications (Chief Medical Editor, Cardiology Today's Intervention), Society of Cardiovascular Patient Care (Secretary/Treasurer), WebMD (CME steering committees); Other: Clinical Cardiology (Deputy Editor), NCDR-ACTION Registry Steering Committee (Chair), VA CART Research and Publications Committee (Chair); Research Funding: Abbott, Afimmune, Amarin, Amgen, AstraZeneca, Bayer, Boehringer Ingelheim, Bristol-Myers Squibb, Cardax, Chiesi, CSL Behring, Eisai, Ethicon, Ferring Pharmaceuticals, Forest Laboratories, Fractyl, Idorsia, Ironwood, Ischemix, Lexicon, Lilly, Medtronic, Pfizer, PhaseBio, PLx Pharma, Regeneron, Roche, Sanofi Aventis, Synaptic, The Medicines Company; Royalties: Elsevier (Editor, Cardiovascular Intervention: A Companion to Braunwald's Heart Disease); Site Co-Investigator: Biotronik, Boston Scientific, CSI, St Jude Medical (now Abbott), Svelte; Trustee: American College of Cardiology; Unfunded Research: FlowCo, Merck, Novo Nordisk, Takeda.

# Appendix

**Table A1.** Symbol table.

| Symbol | Description |
| --- | --- |
| $E$ | Categorical cross-entropy |
| $p$ | Predicted probability of a class in DL |
| $y$ | Binary indicator |
| $n_{in}$ | Number of input features to CL |
| $n_{out}$ | Number of output features from CL |
| $K$ | Convolution kernel size |
| $P$ | Convolution padding size |

(*Continued*)

**Table A1.** (*Continued*)

| Symbol | Description |
| --- | --- |
| $S$ | Stride size |
| $a_{\text{out}}$ | Number of output features from APL |
| $W$ | Width of the input feature map to APL |
| $F$ | Size of the kernel |
| $\eta_{\text{K10}}^{\text{DL}}(c)$ | Accuracy of DL at $c$ |
| $C$ | Number of the combinations in K10 protocol |
| $\bar{\eta}_{K10}^{\text{DL}}$ | Mean accuracy of the DL |
| $\eta_{\text{K10}}^{\text{ML}}(c)$ | Accuracy of ML at $c$ |
| $\bar{\eta}_{K10}^{\text{ML}}$ | Mean accuracy of ML |
| $\sigma_{\text{K10}}^{\text{DL}}$ | Standard deviation of DL |
| $\sigma_{\text{K10}}^{\text{ML}}$ | Standard deviation of ML |
| $D$ | Fractal dimension |
| $N$ | Number of boxes that covering the pattern |
| $r$ | Magnification |
| $\text{APSI}_{\text{K10}}^{\text{AImodel}}$ | Atheromatic plaque separation index |
| $\text{MFS}_{\text{AF}}$ | Mean feature strength for AF class |
| $\text{MFS}_{\text{AS}}$ | Mean feature strength for AS class |

**Table A2.** Abbreviation table.

| Abbreviation | Description |
| --- | --- |
| DL | Deep learning |
| ML | Machine learning |
| AI | Artificial intelligence |
| AUC | Area-under-the-curve |
| ROC | Receiver operating characteristic |
| CNN | Convolution neural network |
| AF | Symptomatic plaque |
| AS | Asymptomatic plaque |
| MFS | Mean feature strength |
| CVD | Cardiovascular disease |
| USA | United States of America |
| MRI | Magnetic resonance imaging |
| CT | Computer tomography |
| US | Ultrasound |
| ICA | Internal carotid artery |
| TIA | Transient ischemic attack |
| GPU | Graphics processing unit |
| CL | Convolution layer |

| | |
|---|---|
| CLFM | Convolution layer feature map |
| APL | Average pooling layer |
| APFM | Average pooling feature map |
| DenL | Dense layer |
| LDA | Linear discriminant analysis |
| BFD | Biometric facial data |
| 2D FM | Two dimensional feature map |
| 1D FM | One dimensional feature map |
| RGB | Red, green, blue |
| FD | Fractal dimension |
| APSI | Atheromatic plaque separation index |
| $k$-NN | $k$-nearest neighbour |
| CV | Cross-validation |
| SOM | Self-organising map |
| SVM | Support vector machine |
| FDR | Fisher discrimination ratio |
| PCA | Principal component analysis |

# References


[1] Benjamin E J, Muntner P and Bittencourt M S 2019 Heart disease and stroke statistics—2019 update: a report from the American Heart Association *Circulation* **139** e56–528

[2] Fryar C D, Chen T-C and Li X 2012 *Prevalence of Uncontrolled Risk Factors for Cardiovascular Disease: United States, 1999–2010* (Washington, DC: US Department of Health and Human Services, Centers for Disease Control and Prevention, National Center for Health Statistics)

[3] Heron M P 2017 Deaths: leading causes for 2015, National Center for Health Statistics (U.S.). Division of Vital Statistics *National vital statistics reports vol 66 no. 5 DHHS publication no. (PHS) 2018–1120* URL: https://stacks.cdc.gov/view/cdc/50010

[4] Suri J S, Kathuria C and Molinari F 2010 *Atherosclerosis Disease Management* (Berlin: Springer)

[5] Sirimarco G *et al* 2013 Carotid atherosclerosis and risk of subsequent coronary event in outpatients with atherothrombosis *Stroke* **44** 373–9

[6] Liu Y *et al* 2019 Size of carotid artery intraplaque hemorrhage and acute ischemic stroke: a cardiovascular magnetic resonance Chinese atherosclerosis risk evaluation study *J. Cardiovasc. Magn. Reson.* **21** 36

[7] Chien J D *et al* 2013 Demographics of carotid atherosclerotic plaque features imaged by computed tomography *J. Neuroradiol.* **40** 1–10

[8] Seabra J and Sanches J 2012 *Ultrasound Imaging: Advances and Applications* (New York: Springer)

[9] Aichner F *et al* 2009 High cardiovascular event rates in patients with asymptomatic carotid stenosis: the REACH Registry *Eur. J. Neurol.* **16** 902–8

[10] Viswanathan V *et al* 2020 Low-cost preventive screening using carotid ultrasound in patients with diabetes *Front. Biosci.* **25** 1132–71

[11] Saba L *et al* 2014 *Multi-modality Atherosclerosis Imaging and Diagnosis* (Berlin: Springer)

[12] Kotsis V *et al* 2018 Echolucency-based phenotype in carotid atherosclerosis disease for risk stratification of diabetes patients *Diabetes Res. Clin. Pract.* **143** 322–31

[13] Nicolaides A N *et al* 2005 Effect of image normalization on carotid plaque classification and the risk of ipsilateral hemispheric ischemic events: results from the asymptomatic carotid stenosis and risk of stroke study *Vascular* **13** 211–21

[14] Nicolaides A N *et al* 2002 Ultrasound plaque characterisation, genetic markers and risks *Pathophysiol. Haemos. Thromb.* **32** 371

[15] Hussain M A *et al* 2018 Association between statin use and cardiovascular events after carotid artery revascularization *J. Am. Heart Assoc.* **7** e009745

[16] Nicolaides A N *et al* 2010 Asymptomatic internal carotid artery stenosis and cerebrovascular risk stratification *J. Vasc. Surg.* **52** 1486–96

[17] Kakkos S K *et al* 2013 The size of juxtaluminal hypoechoic area in ultrasound images of asymptomatic carotid plaques predicts the occurrence of stroke *J. Vasc. Surg.* **57** 609–18

[18] Paraskevas K I, Nicolaides A N and Kakkos S K 2020 Asymptomatic Carotid Stenosis and Risk of Stroke (ACSRS) study: what have we learned from it? *Ann. Trans. Med.* **8** 1271

[19] Sharma A M *et al* 2015 A review on carotid ultrasound atherosclerotic tissue characterization and stroke risk stratification in machine learning framework *Curr. Atheroscler. Rep.* **17** 55

[20] Acharya U R *et al* 2012 Atherosclerotic risk stratification strategy for carotid arteries using texture-based features *Ultrasound Med. Biol.* **38** 899–915

[21] Acharya U R *et al* 2012 Carotid far wall characterization using LBP, Laws' texture energy and wall variability: a novel class of atheromatic systems *2012 Annual Int. Conf. of the IEEE Engineering in Medicine and Biology Society* (Piscataway, NJ: IEEE)

[22] Acharya U R *et al* 2012 Carotid ultrasound symptomatology using atherosclerotic plaque characterization: a class of Atheromatic systems *2012 Annual Int. Conf. of the IEEE Engineering in Medicine and Biology Society* (Piscataway, NJ: IEEE)

[23] Acharya U *et al* 2013 Computed tomography carotid wall plaque characterization using a combination of discrete wavelet transform and texture features: a pilot study *Proc. Instit. Mech. Eng.* H **227** 643–54

[24] Acharya U R *et al* 2013 Understanding symptomatology of atherosclerotic plaque by image-based tissue characterization *Comp. Methods Prog. Biomed.* **110** 66–75

[25] Araki T *et al* 2017 Stroke risk stratification and its validation using ultrasonic echolucent carotid wall plaque morphology: a machine learning paradigm *Comp. Biol. Med.* **80** 77–96

[26] Araki T *et al* 2016 PCA-based polling strategy in machine learning framework for coronary artery disease risk assessment in intravascular ultrasound: a link between carotid and coronary grayscale plaque morphology *Comp. Methods Prog. Biomed.* **128** 137–58

[27] Saba L *et al* 2017 plaque tissue morphology-based stroke risk stratification using carotid ultrasound: a polling-based PCA learning paradigm *J. Med. Syst.* **41** 98

[28] Saba L *et al* 2019 The present and future of deep learning in radiology *Eur. J. Radiol.* **114** 14–24

[29] Khanna N N *et al* 2019 Rheumatoid arthritis: atherosclerosis imaging and cardiovascular risk assessment using machine and deep learning-based tissue characterization *Curr. Atheroscler. Rep.* **21** 7

[30] Biswas M *et al* 2019 State-of-the-art review on deep learning in medical imaging *Front. Biosci.* **24** 392–426

[31] Huang X *et al* 2017 Evaluation of carotid plaque echogenicity based on the integral of the cumulative probability distribution using grayscale ultrasound images *PloS One* **12** e0185261

[32] Lekadir K *et al* 2016 A convolutional neural network for automatic characterization of plaque composition in carotid ultrasound *IEEE J. Biomed. Health Inform.* **21** 48–55

[33] Liu B *et al* 2019 Feature generation by convolutional neural network for click-through rate prediction *The World Wide Web Conf.* 1119–29

[34] Yasar F and Akgunlu F 2005 Fractal dimension and lacunarity analysis of dental radiographs *Dentomaxillofac. Radiol.* **34** 261–7

[35] Acharya U R *et al* 2011 An accurate and generalized approach to plaque characterization in 346 carotid ultrasound scans *IEEE Trans. Instrum. Measur.* **61** 1045–53

[36] Nvidia 2017 NVIDIA DGX-1 with Tesla V100 system architecture the fastest platform for deep learning *Nvidia Whitepaper* WP-08437-002_v01 p 43

[37] Monkam P *et al* 2018 CNN models discriminating between pulmonary micro-nodules and non-nodules from CT images *Biomed. Eng.* **17** 96

[38] Biswas M *et al* 2018 Symtosis: a liver ultrasound tissue characterization and risk stratification in optimized deep learning paradigm *Comp. Methods Prog. Biomed.* **155** 165–77

[39] Sanagala S S *et al* 2019 A fast and light weight deep convolution neural network model for cancer disease identification in human lung(s) *18th IEEE Int. Conf. on Machine Learning and Applications (ICMLA)* (Piscataway, NJ: IEEE)

[40] Acharya U R *et al* 2011 Atheromatic™: symptomatic vs asymptomatic classification of carotid ultrasound plaque using a combination of HOS, DWT and texture *2011 Annual Int. Conf. of the IEEE Engineering in Medicine and Biology Society* (Piscataway, NJ: IEEE)

[41] Than J C *et al* 2017 Lung disease stratification using amalgamation of Riesz and Gabor transforms in machine learning framework *Comp. Biol. Med.* **89** 197–211

[42] Rujirakul K, So-In C and Arnonkijpanich B 2014 PEM-PCA: a parallel expectation-maximization PCA face recognition architecture *Sci. World J.* **2014** 1–16

[43] Kurosaki Y *et al* 2017 Asymptomatic carotid T1-high-intense plaque as a risk factor for a subsequent cerebrovascular ischemic event *Cerebrovasc. Dis.* **43** 250–6

[44] Ho S S Y 2016 Current status of carotid ultrasound in atherosclerosis *Quan. Imaging Med. Surg.* **6** 285

[45] Golle P 2008 Machine learning attacks against the Asirra CAPTCHA *Proc. of the 15th ACM Conf. on Computer and Communications Security*

[46] Acharya U R *et al* 2013 Atherosclerotic plaque tissue characterization in 2D ultrasound longitudinal carotid scans for automated classification: a paradigm for stroke risk assessment *Med. Biol. Eng. Comput.* **51** 513–23

[47] Gastounioti A *et al* 2014 A novel computerized tool to stratify risk in carotid atherosclerosis using kinematic features of the arterial wall *IEEE J. Biomed. Health Inform.* **19** 1137–45

[48] Christodoulou C I *et al* 2003 Texture-based classification of atherosclerotic carotid plaques *IEEE Trans. Med. Imaging* **22** 902–12

[49] Saba L 2021 Ultrasound-based internal carotid artery plaque characterization using deep learning paradigm on a supercomputer: a cardiovascular disease/stroke risk assessment system *Int. J. Cardiovasc. Imaging* **37** 1511–28

# Chapter 7

# Quantification of plaque volume using a two-stage deep learning paradigm

**Mainak Biswas and Jasjit S Suri**

Throughout the world cardiovascular diseases (CVDs) kill more people than any other disease. Early diagnosis and treatment can save lives. CVD risk prognosis can be performed if atherosclerotic plaque volume quantification can be performed. Such quantification is currently manual and time-consuming, and is not inspired by knowledge-based systems, i.e. artificial intelligence (AI). In this chapter an AI based model, a class of AtheroEdge™ systems, for quantification of plaque volume is presented. The plaque volume is represented as carotid intima–media thickness (cIMT) and total plaque area (TPA). The model applied two independent AI based models in two stages to quantify plaque volume from ultrasound common carotid artery (CCA) images. Initially, the images are divided into rectangular small patches of equal sizes. The first stage AI based deep learning model characterizes the patches into wall and non-wall segments. The wall patches of every image form the region-of-interest (ROI), which is again fed into the second stage. In the next stage, another DL based model is tested on the mined ROI, to demarcate the far-wall segment. Finally, the media–adventitial (MA) and lumen–intima (LI) boundaries are delineated, and then the cIMT and TPA quantification is performed. The database consists of 250 CCA scans. The mean cIMT bias error is $0.0935 \pm 0.0637$ mm and the mean TPA bias error is $2.7939 \pm 2.3702$ mm$^2$. The correlation coefficient between the gold standard and AI based results for cIMT and TPA are 0.99 ($p < 0.0001$) and 0.89 ($p < 0.0001$), respectively. The strategy appears to be unique, in that it provided plaque volume measurement in the form of both cIMT and TPA measurement with improved performance.

## 7.1 Introduction

Cardiovascular diseases (CVDs) account for over 17.6 million deaths around the world each year. The majority, 85%, of which are caused by heart attacks [1, 2]. CVD is mainly caused by artery inflammation disease, i.e. atherosclerosis [3].

The atherogenesis process begins with endothelial dysfunction, which leads to effluents from the blood such as low-density lipoprotein (LDL) cholesterol and macrophages diffusing through the damaged walls of an artery [4–6]. These effluents accumulate over time within the arterial walls leading to narrowing and hardening of blood vessels. Further, other factors such as diabetes, smoking, alcohol, an homocysteine concentrations escalate the formation of complex lesions within the arterial walls. With increasing age, macrophages and lymphocytes from the blood enter the arterial walls, forming a necrotic core [7–10]. Continued pathophysiological changes lead to the formation of a lipid core and necrotic tissue underneath a thin fibrous cap (shown in figure 7.1). Over time the cap ruptures, leading to thrombosis, blood flow blockage, and ultimately stroke.

Generally, blood tests, i.e. C-reactive protein, and cardiac catheterization angiography are considered the gold standard for diagnosing atherosclerosis. However, being invasive, these tests carry the usual risks. Conversely, non-invasive tests such as imaging provide visualization and measurement of these lesions without the risks of invasive procedures. There are multiple imaging modalities, i.e. magnetic resonance (MR), computed tomography (CT), [11] and ultrasound [12]. Among these, the ultrasound is considered the safest. It is non-invasive, radiation free, portable, and less expensive than the other imaging modalities. There are two paradigms of carotid artery disease management, which are the quantification of

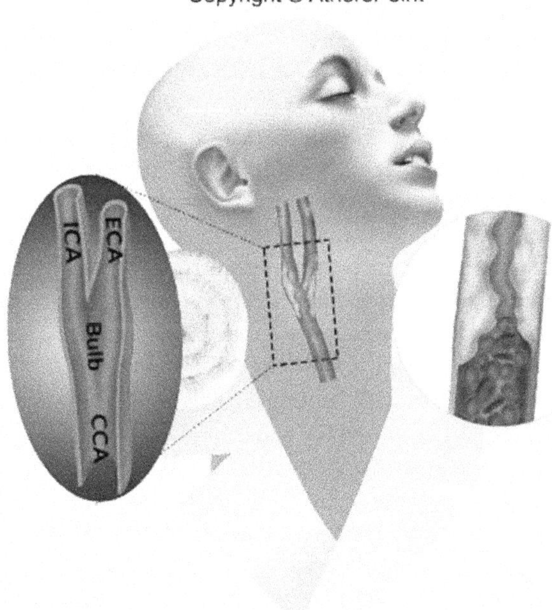



**Figure 7.1.** The development of plaque in the common carotid arteries' common and bulb regions. (Reproduced with permission of AtheroPoint™. Reproduced with permission from [84]. Copyright 2020 Elsevier.)

carotid plaque (CP) volume and the characterization of morphology [13–15]. The quantification of plaque burden is generally used since it captures the thickness (carotid intima–media thickness (cIMT)) in metric measurements and also the area of carotid plaque (in the form of total plaque area (TPA)) [4]. Therefore, both the cIMT and TPA measurements will from now on be referred to as the cumulative assessment of CP and carotid wall thickness (cWT). In this work, a two-stage system is used to segment the CP region and, thereafter, is used to compute both the cIMT and TPA. The computation of the plaque volume is performed on both sides of the neck. However, due to limited image data, the TPA is computed for the total area between the media-adventitia (MA) and lumen–intima (LI) wall (shown in figure 7.1).

In the past, approximating the plaque burden involved the low-level image processing methods of edge recognition and threshold estimation [16, 17]. Thereafter, the model based paradigm came into existence, where the plaque parts of common carotid artery (CCA) images were segmented using model based techniques [18, 19]. Further scale-space models were developed for segmenting the CP [20]. In recent times the rise of AI has significantly increased the performance and speed of automatic techniques for plaque segmentation. Machine learning (ML) and deep learning (DL) are the two paradigms of AI used extensively in categorization and segmentation. In a recent work on lumen measurement a 13 layer DL fully convolutional network (FCN) was applied for segmentation [21]. The 13 layers were used for feature map extraction, which in turn was up-sampled using three layers to the size of the ground truth (GT) for pixel-to-pixel characterization. There are two DL techniques for segmentation: full image and patching based segmentation. Here, we have used a patching based technique for quantification of the plaque burden (details given in the methodology). A two-stage based DL was used, where the ROI was first detected and in the next stage the CP borders were delineated. The first stage of the two-stage system envisages the detection of image patches comprising the plaque within the far-wall area. The far-wall is divided into patches. The AI module is used in the first stage DL convolutional neural network (CNN) [22] for characterizing the patches into wall and non-wall areas. The wall patches are then combined, as shown in the flow diagram in figure 7.2. Once again, we would like to remind the reader that stage one is for ROI detection [23], and the second stage is for the actual border delineation of the CP from the ROI. This stage uses an FCN [24] to delineate the LI and MA borders from the ROI images mined from the first stage [25, 26]. After the LI and MA borders are detected, the polyline distance measurement algorithm is applied to calculate the cIMT [27]. The area within the boundaries is estimated to find the TPA. It is to be noted that both stages work independently using two different DL models and paradigms of segmentation and characterization. The two-stage process is hypothesized on the premise that better localization will lead to better results in terms of accuracy and performance.

The literature survey, data acquisition, and methodology are provided in sections 7.2, 7.3, and 7.4, respectively. The experimental protocol, results, and statistical analysis are provided in sections 7.4, 7.5, and 7.6, respectively. Finally, section 7.7 contains the discussion and section 7.8 is the conclusion.

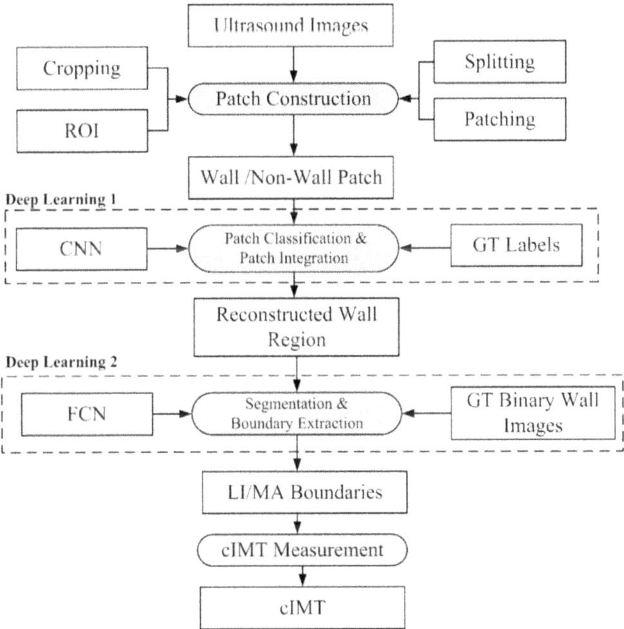


**Figure 7.2.** Two-stage DL design for quantifying carotid area (cWT+CP) estimation. (Reproduced with permission from [84]. Copyright 2020 Elsevier.)

## 7.2 Background

The previous techniques of plaque segmentation involved low-level image processing including edge detection/threshold based combination or both. These techniques were based on the human perception of images and could not be generalized. These techniques belong to first generation techniques. With better understanding of medical imaging and the advent of new-generation techniques, the segmentation technology advanced towards pixel characterization and deformable models. With these advances the segmentation of the carotid wall also improved in terms of accuracy and other performance parameters. These models belong to the second generation. A comparative literature survey for both generations was provided recently in [18, 19]. In the third-generation, scale-space technology was used for carotid wall detection [14, 28, 29]. The methodology uses higher-order Gaussian derivatives for edge detection, a Gaussian standard deviation function called the stretcher which captures the plaque, where the edges are captured at zero-crossing. This technology was a game changer in the realm of risk assessment using ultrasound [28] and was validated using image restoration principles [30]. Several models were constructed as variations of this technique, such as CALEX [31], CARES [32], CAMES [28], CLAUDLES [33], and FOAM [18, 34]. All the systems discussed pave the way for the development of the commercial AtheroEdgeTM system, which is used widely by the research community for plaque segmentation [35]. The generalized architecture of the models discussed can be divided into two stages: (i) ROI recognition and (ii) segmentation. The ROI recognition consists of the detection of the far-wall region while the segmentation consists of delineating the LI and MA borders. Plaque area (PA), as well as cIMT, has also been used increasingly as a novel biomarker for stroke risk estimation [36]. Although the first and second generation techniques did not use PA much, the third generation started emphasizing it as well as cIMT [37]. The methods discussed until now were created under the class of AtheroEdgeLink™ [38, 39]. There are many beneficial aspects to this class. One is knowing the relationship between the image based phenotypes and other contemporary risk parameters, (a) the SYNTAX score, ankle-brachial index, and risk characterization among leukoaraiosis patients [40–43]; (b) the investigative study of plaque [35]; (c) the correlation between wall variability and stroke risk [44–47]; and (d) the development of new risk estimators such as AECRS 2.0 [48, 49] and its comparative analysis with conventional stroke risk parameters [39, 50]. The studies conducted so far have also been validated for variability analysis and reproducibility [51–53]. Morphological studies such as echolucency [50] were also performed under the aegis of the Atheromatic™ [51, 52] and AtheroRisk™ applications [53]. With the advent of AI based systems, Suri and his team developed new intelligence based models which could learn dynamically and adjust its parameters as per the grayscale characteristics of the ROI [21, 54, 55]. AI models for carotid imaging can be categorized into two groups. The first group deals with the characterization of plaque based on the morphology, while the second group deals with segmenting the plaque region for estimating the cIMT and TPA. The morphological categorization of plaque until now has used ML based techniques [43–45]. The ML paradigm works in two phases. In the first phase various image parameters, such as texture and frequency parameters, are computed. Then, in the second phase, the learning model is trained on the parameters mined for characterization [56]. The most general ML algorithms are $K$-nearest neighbors (KNN) [57], support vector machines (SVM) [58], decision trees (DTs) [59], artificial neural networks (ANNs) [60], and extreme learning machines [61]. The working of DL methods consists of extracting the features and characterizing of the images within a single phase. The most general DL systems are CNN [22], FCN [24], deep belief networks (DBNs) [62], autoencoders [63], residual neural networks (RNNs) [64], etc. In the recent past there have been some implementations of DL based carotid volume computation. Autoencoders have been used by Rosa *et al* [65] for mining features. Shin *et al* [66] used CNN for cIMT measurement from motion video frames of the CCA. Recently, both Biswas *et al* [21] and Maria *et al* [67] used semantic segmentation by employing an FCN to measure the cIMT. However, all these models used full-scale images, which can lead to the inclusion of features with similar echolucency, such as the jugular vein, leading to bias errors in detection and measurement. FCNs and CNNs share similar configurations of layers such as convolution and pooling, however, they differ in output. CNNs employ fully connected networks, such as ANNs, for characterization. Also, FCNs use up-sampling layers to upscale the feature maps to characterize the images pixel-wise, i.e. segmentation. In the next few sections we discuss the implementation of a DL network for segmentation.

## 7.3 Data acquisition

A sonographer with fifteen years of experience used a sonographic scanner for the scans. A total of 408 scans of the left and right CCA were obtained from 204 individuals for the study. One of the CCA images on the left was missing. Due to the costs of manual tracing, 250 US CCA photos were used for this pilot investigation. The cohort's mean age was 69.11 years. There were 92 smokers among them. The IRB at Toho University in Japan provided ethical approval. The patient data are given in table 7.1.

## 7.4 Methodology

In this work a two-stage DL based model for estimation of plaque volume (cIMT + CP area) is studied (shown in figure 7.2). The first stage is the patch characterization and the subsequent stage is the segmentation based on which the LI–MA border delineation is performed. The patching process which is done prior to the two-stage characterization process is discussed next.

*Patch process*
The background information, such as text information, patient data, non-grayscale region, and radiologist remarks are removed using AtheroEdge™, an FDA-approved software. The cropping role puts the lumen region in the middle of the grayscale image (shown in figure 7.3(b)). The standard protocols are followed, which have been used and published in several works over the years [20, 68, 69]. After the cropping stage the CCA images are split horizontally into equal halves (figure 7.3(c)). The lower part consists of the lumen, far-wall, and the tissue region. The lower region

Table 7.1. Patients' baseline parameters. (Reproduced with permission from [84]. Copyright 2020 Elsevier.)

| Parameters | Values |
| --- | --- |
| Average hemoglobin(HbA1c) | $5.8 \pm 1.0\,\mathrm{dL}^{-1}$ |
| Average glucose | $108 \pm 31\,\mathrm{dL}^{-1}$ |
| Average low-density lipoprotein cholesterol | $99.80 \pm 31.30\,\mathrm{dL}^{-1}$ |
| Average high-density lipoprotein cholesterol | $50.40 \pm 15.40\,\mathrm{dL}^{-1}$ |
| Average total cholesterol | $174.6 \pm 37.7\,\mathrm{dL}^{-1}$ |

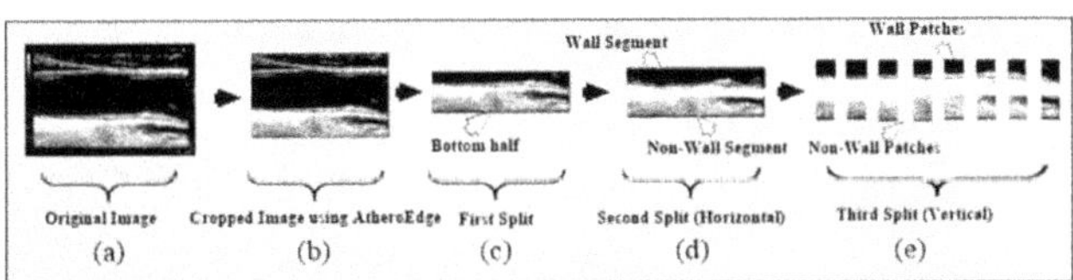


**Figure 7.3.** The patching process. (a) The scan; (b) the image cropped by employing AtheroEdge™; (c) the initial split; (d) the wall and non-wall regions shown in the second split; and (e) the third split showing vertical division of eight patches into the wall and non-wall regions. (Reproduced with permission from [84]. Copyright 2020 Elsevier.)

is further split horizontally into equal halves, where the upper half contains the lumen and far-wall, while the lower half contains the tissue region (figure 7.3(d)). The top half in this case represents the ROI which forms the ground truth in the next stage. The lower half consists of the tissue region. The patching process starts here where the ROI and tissue halves are vertically split into eight parts, each giving rise to 16 patches (figure 7.3(e)). A total of 4000 patches were generated from the 250 CCA images. The tissue patches from here on will be known as the non-ROI patches. The ROI and non-ROI patches were input into the two-stage DL model for segmentation and characterization. The online learning module is shown in figure 7.4. The characterization and segmentation module is discussed next.

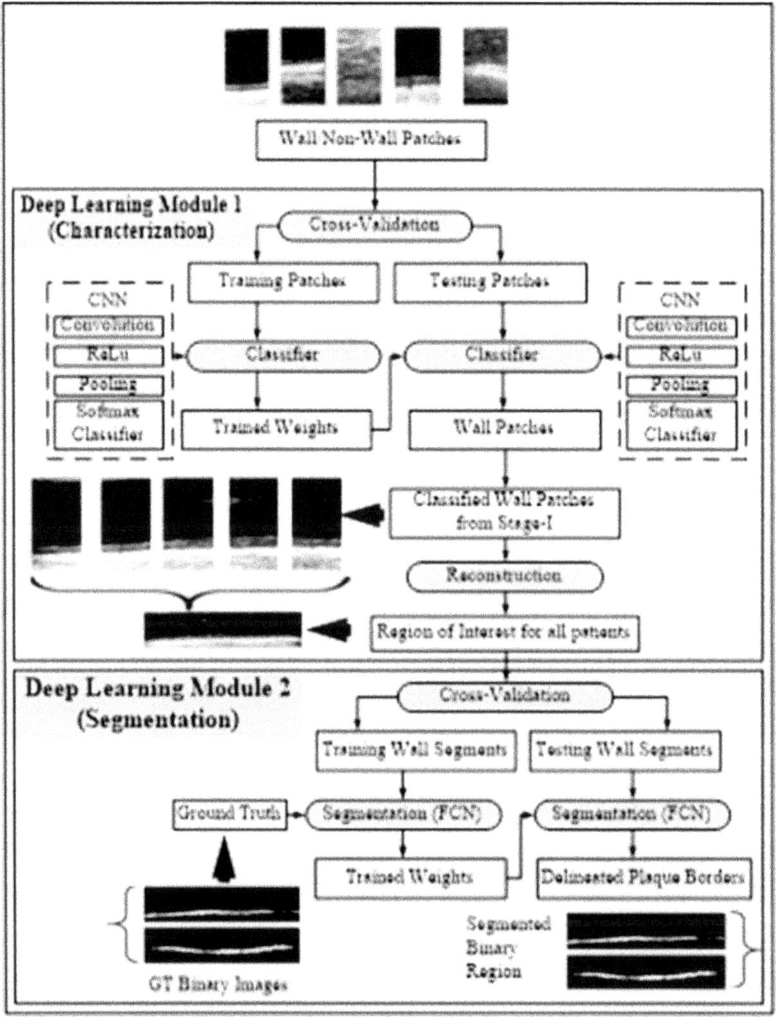


**Figure 7.4.** The two-stage DL model's training/testing approach. The first module 1 is utilized for the categorization of wall and non-wall patches, whereas the second module 2 is used for interface segmentation. (Reproduced with permission from [84]. Copyright 2020 Elsevier.)

*Stage 1: Characterization component*

Two comparative strategies were followed: one with augmentation and one without augmentation. Data augmentation is a strategy that augments the dataset size by geometrical transformations: (i) a 30° rotation, (ii) a length and breadth shift by 0.1, (iii) a shear range of 0.2, (iv) a zoomed range of 0.2, and (v) a flip horizontally. The data augmentation resulted in a sum of 24 000 patches. This constituted 4000 original and 20 000 augmented patches. The K10 cross-validation (90% training and 10% testing) was applied to the ROI and non-ROI patches for dividing the patches into training and testing. Two training/testing systems were employed separately for with and without segmentation. The output patches were fed into the DL architecture [70] comprising 22 layers (two convolutions + nine inception layers + one linear layer). The DL network is shown in figure 7.5. The categorized wall and non-wall patches are shown in figure 7.6. Once characterization was performed, the ROI patches were reconstructed consisting of the wall region, which was passed onto stage 2, or the segmentation phase.

*Stage 2: Segmentation component*

The LI and MA borders were traced and outlined by an qualified radiologist using the ImgTracer™ [20, 26, 53, 71, 72] software which allows zooming and a marking facility to trace the exact wall on the LI and MA borders. The ImgTracer™ has been used in several studies in the past [29, 46, 51, 52, 73–75]. The ROI segments built up from the patches along with the binary GT are fed into the stage 2 segmentation module. The corresponding images of the ROI segments along with their GT counterparts are shown in figures 7.7 and 7.8. As used before in stage 1, the K10 cross-validation

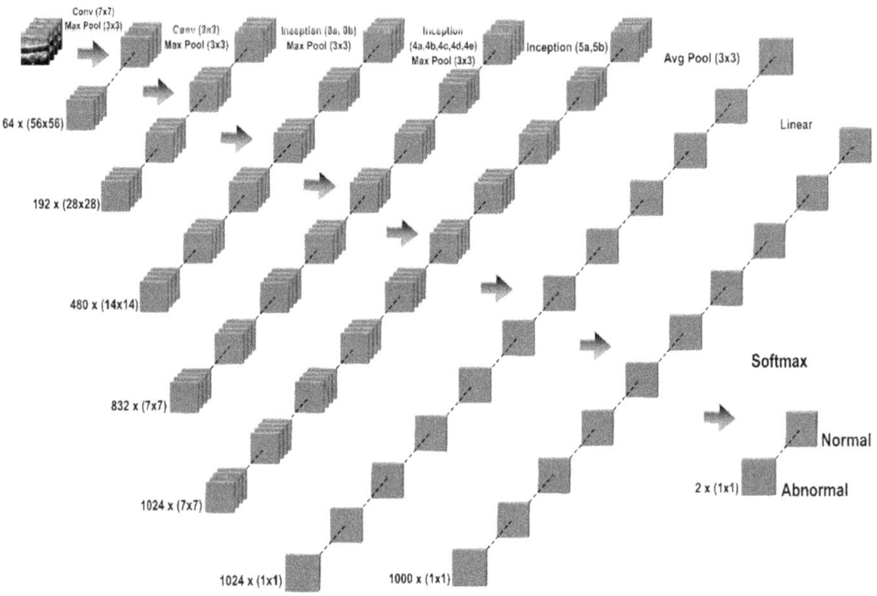


**Figure 7.5.** The stage 1 deep learning network for classification. (Image courtesy of AtheroPoint™. Reproduced with permission from [84]. Copyright 2020 Elsevier.)

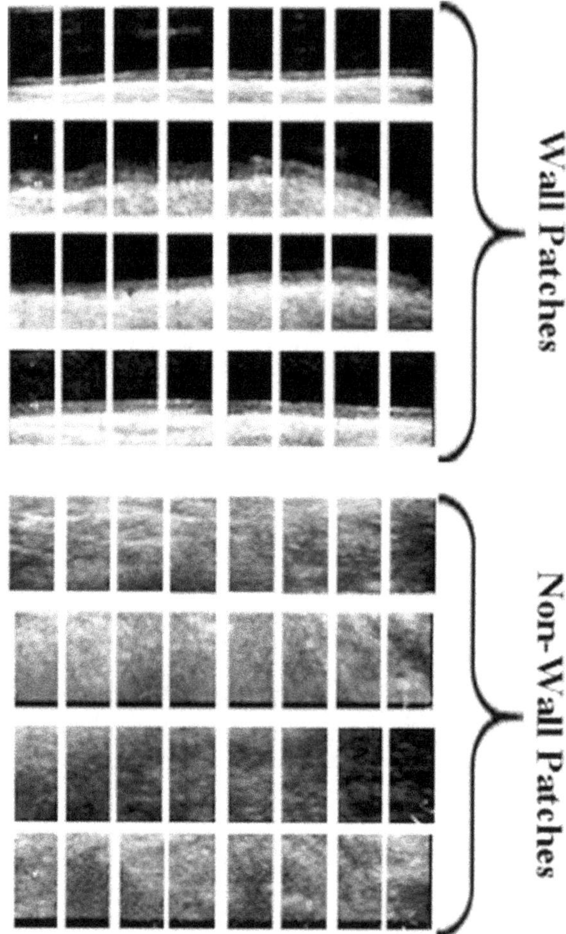


**Figure 7.6.** Categorized wall and non-wall patches from the stage 1 DL module. (Reproduced with permission from [84]. Copyright 2020 Elsevier.)

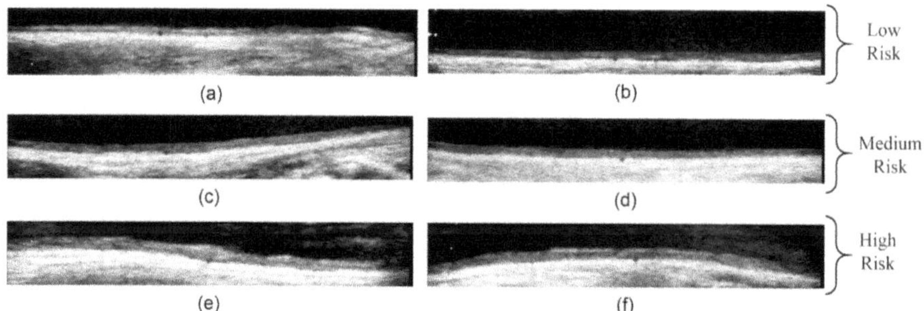


**Figure 7.7.** Patch combination for input to the segmentation module. (Reproduced with permission from [84]. Copyright 2020 Elsevier.)

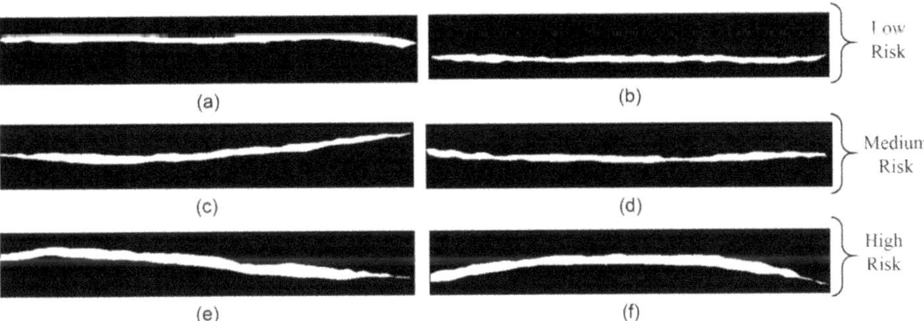



**Figure 7.8.** Binary GT images analogous to low, medium, and high-risk patients. (Reproduced with permission from [84]. Copyright 2020 Elsevier.)

protocol was applied for dividing the dataset into training and testing datasets. The images and their GT counterpart are fed into the FCN. The FCN is made up of two parts: an encoder and a decoder. There are 13 convolution layers and five pooling layers in the encoder. This is also known as the contraction path, as the high-level features are mined and down-sampled along the way. The decoder is known as the expansion path and consists of three up-sample layers. The decoder up-samples the feature maps mined from the encoder to the original size of the image, where a pixel-to-pixel characterization is performed with the corresponding binary GT. The FCN employs two skipping layers to merge features from intermediary layers of the encoder and decoder to recover spatial information lost during down-sampling. Summarizing, the discriminative feature maps mapped from the input images are up-sampled using up-sampling and skipping layers resulting in characterization of the pixels. The FCN model is displayed in figure 7.9. Finally, the LI and MA borders are delineated from the segmented images and cIMT and TPA are computed from it using a polyline distance method (PDM) [72]. The segmented images with the original GT and corresponding AI delineated borders are shown in figure 7.10.

*Workflow of the AI architecture*
Here, the two-stage AI based model was used where each stage used two independent models. The methodology used was patching where image patches were first created from the original image. The reasons for this were better ROI detection and achieving higher segmentation accuracy in the form of low cIMT and TPA bias error. The initial stage comprises a CNN model for patch categorization and the subsequent stage is the actual plaque segmentation. The patching process consists of removing the background information, and splitting the original CCA image into near lumen and far lumen regions. The far lumen region is further split horizontally into the ROI and non-ROI region. The ROI and non-ROI region are split vertically into eight equal halves, resulting in 16 image patches. Two strategies were followed while training/testing the images: one without augmentation and one with augmentation. The augmentation policy followed the geometrical transformation of the image patches to artificially inflate the dataset size for better training. After this stage the patches were divided into training/testing sets using K10

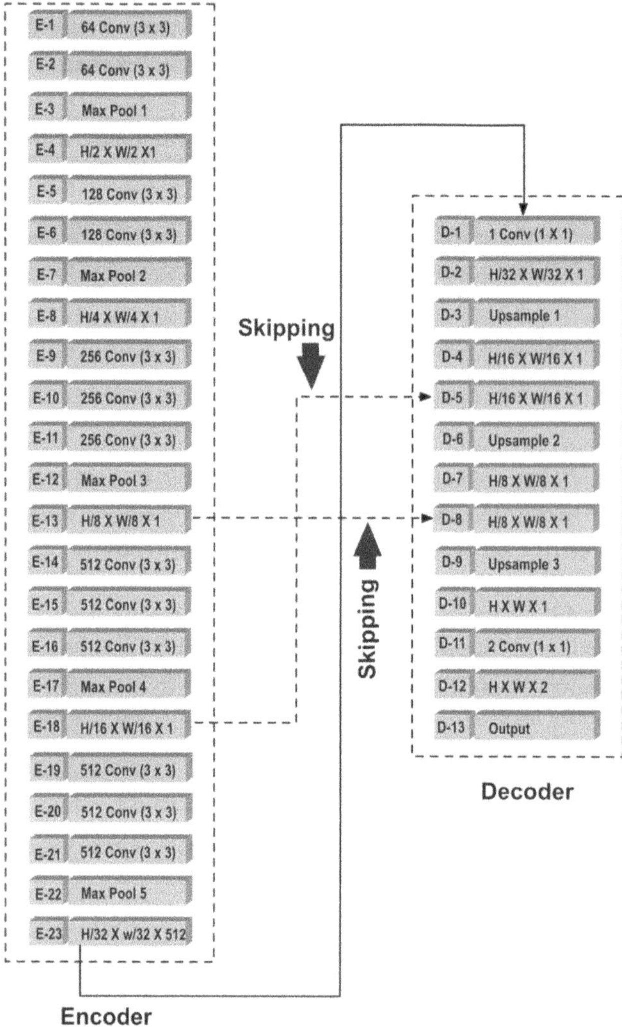


**Figure 7.9.** Stage 2 segmentation DL network. (Reproduced with permission from AtheroPoint™. Reproduced with permission from [84]. Copyright 2020 Elsevier.)

cross-validation and input to the first stage of the AI model. The first stage AI model consists of the CNN model to characterize the input patches into ROI and non-ROI patches. To mine high-level characteristics from the CCA images, the CNN model uses many layers of convolution and pooling. The goal of convolution is to multiply the original image pixels using filters (kernel matrices). The output is a convolved version of the original image, often known as a feature map, that is decreased in dimension via pooling. To generate high-level features, the convolution and pooling operations are repeated layer by layer. For characterization, the features are sent to a fully connected network. After the ROI patches have been identified, the ROI segment is formed by joining them together. For segmentation and border

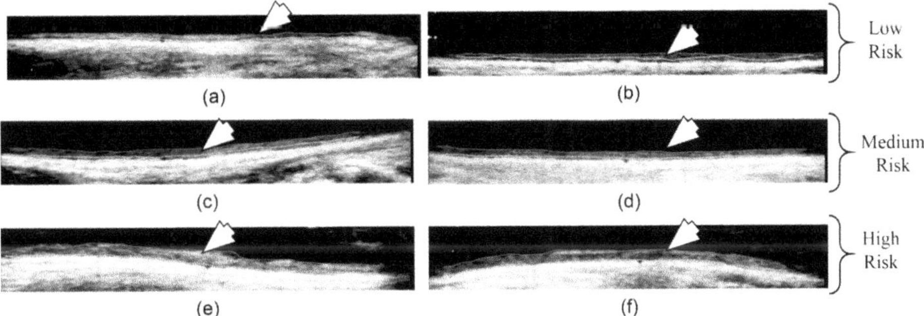


**Figure 7.10.** The low risk patients are shown in row 1 (a) and (b), medium risk patients are shown in row 2 (c) and (d), and the high-risk patients are shown in row 3 (e) and (f). The red and green lines represent the DL LI and MA boundaries, respectively. The DL output is depicted in a yellowish-green hue. (Reproduced with permission from [84]. Copyright 2020 Elsevier.)

delineation, the ROI segment is transmitted to the second step. The FCN makes up the second stage. The FCN is comparable to CNN except that it does not have a completely connected layer. Instead, the FCN uses skipping and up-sampling layers to provide dense characterization. The FCN consist of two parts: the encoder and decoder. The encoder is a 13 layer CNN which outputs the feature map. In the decoder phase, the output feature map is up-sampled and merged with feature maps with intermediate layers of the encoder. Finally, segmented/bifurcated output is obtained. The LI and MA walls are delineated and cIMT and PA values are computed.

## 7.5 Experimental protocol and results

The K10 cross-validation was used to divide the image dataset into training and testing volumes. The K10 protocol follows 90% training and 10% testing in both stages of the AI model. The K10 is a standard protocol followed globally for AI based systems. In this protocol the images are randomly divided into ten blocks. From them, nine blocks are used for training and one block is used for testing. Each block is used alternately for testing while others are used for training. The results for each training/testing phase are compiled to generate the final result.

The characterization module in the initial stage ran for 16 000 iterations for both with and without augmentation. The resultant output image is shown in figure 7.11. Without augmentation, the stage 1 characterization was 89%. However, with augmentation the characterization accuracy was overwhelmingly 99%. The accuracy plot with respect to iterations is shown in figure 7.12. Therefore, the data input to stage 2 was with augmentation. After characterization of the ROI patches and their integration, the ROI segments are fed into stage 2 for segmentation and LI and MA border delineation. The cIMT and TPA bias error is 0.0935 ± 0.0637 mm and 2.7939 ± 2.3702 mm$^2$, respectively. The regression plot showed a high degree of correlation for both cIMT (0.99, $p$-value < 0.0001) and TPA (0.89, $p$-value < 0.0001. The plots are shown in figures 7.13(a) and (b).

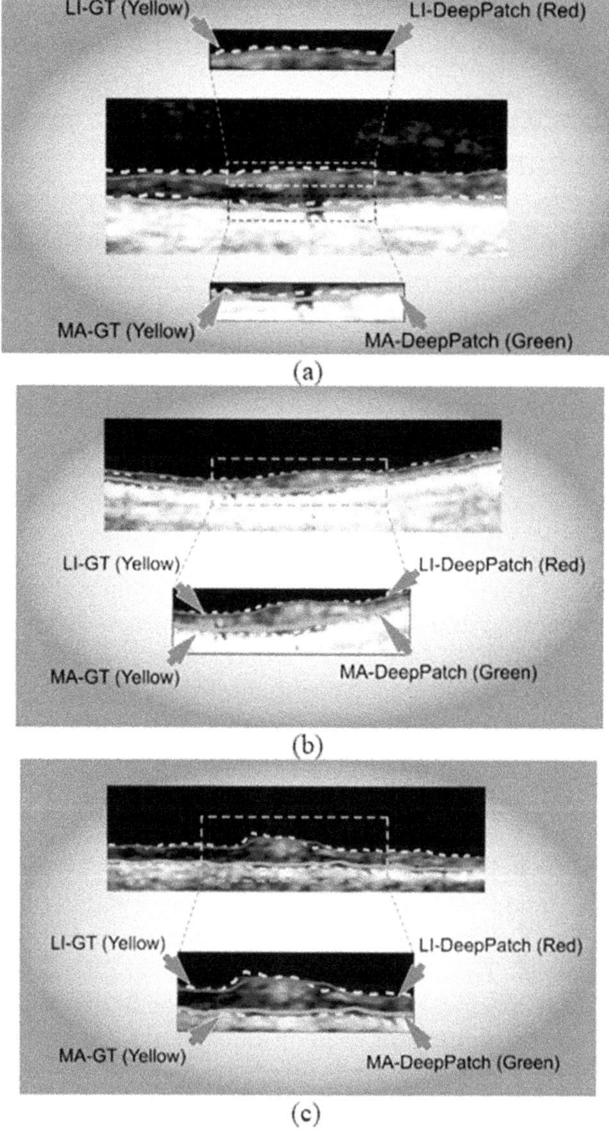



**Figure 7.11.** The two-stage DL based system outputs with magnified outlined borders. The dashed yellow lines indicate the GT LI and MA borders. The red and green lines represent the DL based LI and MA boundaries, respectively. (Reproduced with permission from AtheroPoint™. Reproduced with permission from [84]. Copyright 2020 Elsevier.)

*Bias error statistics*

According to the absolute cIMT bias error plot, 90% of the image collection has an bias error < 0.19 mm. The signed cIMT bias error plot corroborates that 90% of photos have an bias error < 0.18 mm, while 90% of the image collection has an bias error > 0.13 mm. Figures 7.14(a) and (b) illustrate the absolute and signed bias error

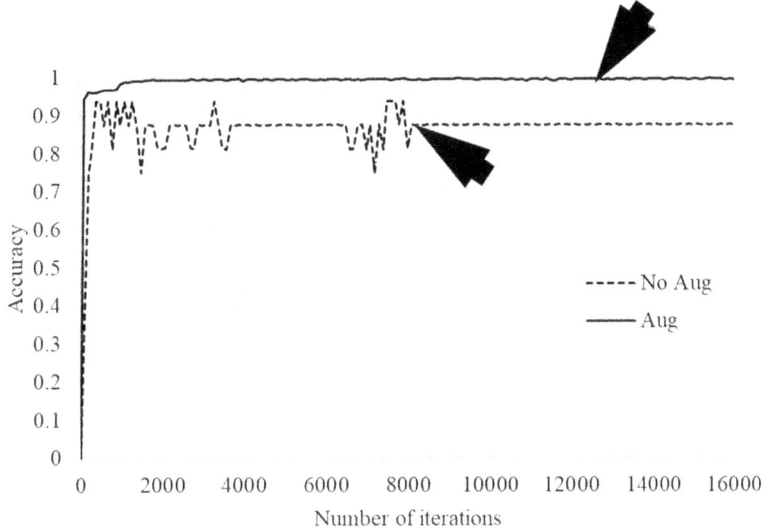



**Figure 7.12.** Stage 1 classification: accuracy versus iterations using and not using augmentation. (Reproduced with permission from [84]. Copyright 2020 Elsevier.)

plots for cIMT, respectively. Similarly for TPA, the absolute bias error for 90% images $< 5.75$ mm$^2$. The signed bias error plot for TPA shows that 90% of images have bias error $< 4.75$ mm$^2$ and 90% of images have TPA bias error $> -4.5$ mm$^2$. The absolute and signed TPA bias error plots are displayed in figures 7.14(c) and (d), respectively.

*Benchmarking*
The work presented here has been compared with an earlier application of DL for cIMT estimation [21]. A comparative analysis was performed for 250 CCA images in terms of cIMT bias error, regression, bias error analysis etc. In the prior application, two sets of GTs were used, which will be known as GT1 and GT2, while the gold standard used here will only be known as GT hereafter. The cIMT bias error for the GT for the current work is 0.0935 $\pm$ 0.0637 mm, which is better than the prior implementation using GT1 by 20% (0.1164 $\pm$ 0.1122 mm) and for GT2 by 18% (0.1146 $\pm$ 0.0940 mm). The GT1 and GT2 regression plots are shown in figure 7.15 for the 250 images. The performance of the current work 0.99 ($p < 0.0001$) for the GT, is better than for GT1 0.83 ($p < 0.0001$)) and GT2 0.87 ($p < 0.0001$). The bias error analysis of the current work when compared to the previous work showed that for 90% of images, the performance of absolute bias error for GT ($<0.19$ mm) is less than for GT1 ($<0.24$ mm) and GT2 ($<0.23$ mm) for the previous work. The details are given in table 7.2 and bias error plots are shown in figure 7.16. For signed bias error, in the current work the performance for 90% of the images for the GT ($-0.13$ mm $<$ signed error $< 0.18$ mm) was better than GT1 ($-0.15$ mm $<$ signed error $< 0.18$ mm) and GT2 ($-0.15$ mm $<$ signed error $< 0.19$ mm) for the previous technique. The resultant plots are shown in figure 7.17.

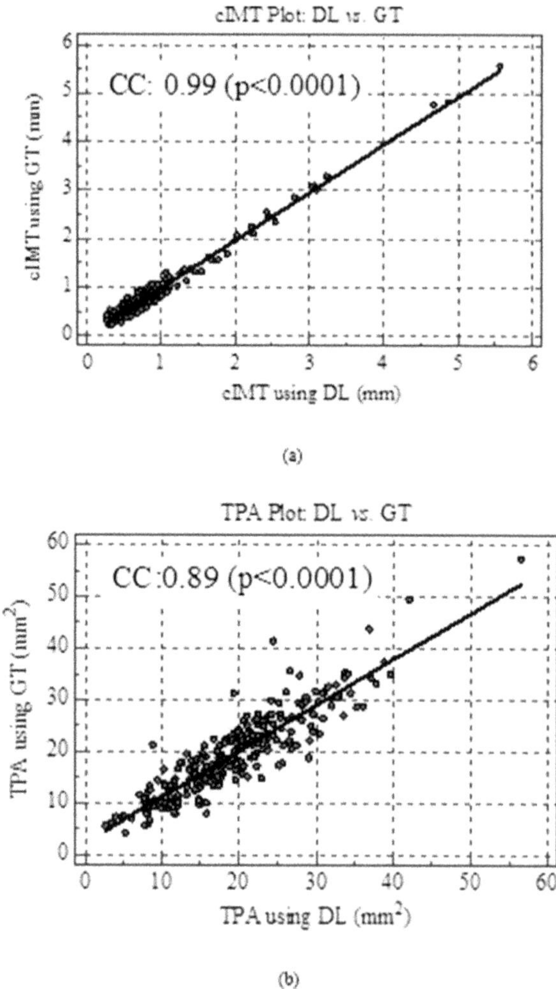


**Figure 7.13.** (a) cIMT and (b) TPA regression plots. (Reproduced with permission from [84]. Copyright 2020 Elsevier.)

*cIMT and TPA correlation*
The link between cIMT and TPA are given in the regression plots in figure 7.18. The cIMT versus TPA with respect to DL is given in figure 7.18(a), while for the GT it is displayed in figure 7.18(b). Both plots show positive correlations of 0.19 ($p = 0.03$) and 0.18 ($p = 0.006$) for DL and the GT, respectively.

## 7.6 Statistical tests

*Statistical tests*
The Bland–Altman, Mann–Whitney, and Wilcoxon plots for the tow-stage DL based system are shown in figures 7.19, 7.20, and 7.21, respectively. The *p*-values for the Mann–Whitney and Wilcoxon tests are 0.2798 and 0.0005, respectively, which

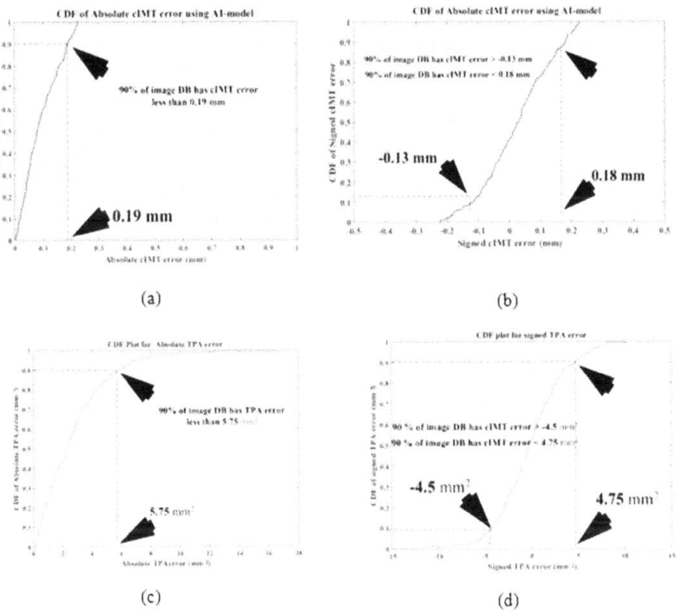

Figure 7.14. Two-stage DL based method for (a) absolute cIMT bias error, (b) signed cIMT bias error, (c) absolute TPA bias error, and (d) signed TPA bias error. (Reproduced with permission from [84]. Copyright 2020 Elsevier.)

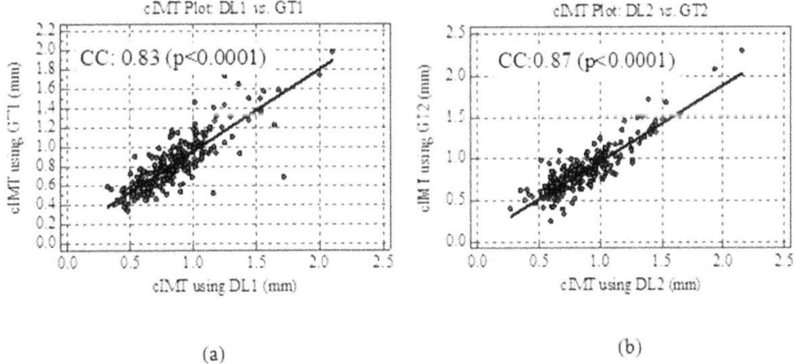

Figure 7.15. Predicted versus actual cIMT values correlation plot [21] for (a) GT1 and (b) GT2. (Reproduced with permission from [84]. Copyright 2020 Elsevier.)

are statistically significant. The corresponding statistical data for the Mann–Whitney and Wilcoxon plots are given in tables 7.3 and 7.4, respectively.

*ROC analysis*

In a recent study involving 7865 women and 6349 men in the age group 45–64 it was seen that the CVD risk for women occurs at cIMT values > 1.0 mm with a hazard ratio of 8.5 [76]. The hazard ratio for men was 3.6. In another study on 4476 subjects [77] the risk was lower when cIMT was < 0.87 mm. However, the CVD incidence increased when cIMT values were greater than 1.18 mm. In this work we have taken risk thresholds of 0.85 mm and

**Table 7.2.** Absolute correlation coefficient (CC) with the DL based comparative analysis of [21]. (Reproduced with permission from [84]. Copyright 2020 Elsevier.)

| Methodology | CC | Improvement of CC | 90% of images in the database have cIMT error less than | Improvement of CDF cIMT bias error |
|---|---|---|---|---|
| Two-stage DL model for GT | 0.99 ($p < 0.0001$) | | 0.19 mm | |
| DL model for GT1[21] | 0.83 ($p < 0.0001$) | 19% | 0.24 mm | 21% |
| DL model for GT2[21] | 0.87 ($p < 0.0001$) | 14% | 0.23 mm | 17% |

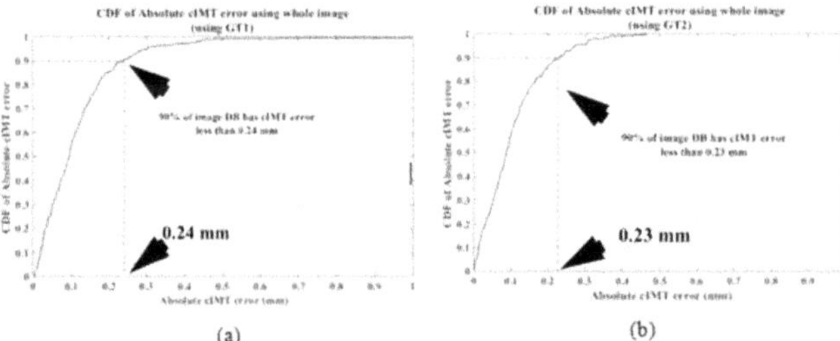

Figure 7.16. CDF absolute cIMT bias error plot for (a) GT1 and (b) GT2 using the proposed method [21]. (Reproduced with permission from [84]. Copyright 2020 Elsevier.)

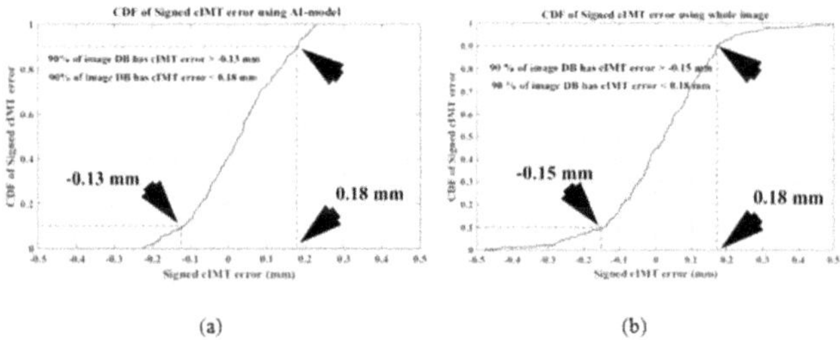

Figure 7.17. CDF signed cIMT bias error plot for (a) GT1 and (b) GT2 using the proposed method [21]. (Reproduced with permission from [84]. Copyright 2020 Elsevier.)

0.90 mm to characterize most risky patients. The ROC plots are shown in figures 7.22(a) and (b), respectively. The AUCs for risk thresholds of 0.85 mm and 0.90 mm are 0.85 and 0.87 respectively, authenticating the DL based system for clinical review.

## 7.7 Discussion

In this work we have developed a two-stage AI model for detection of the ROI and segmentation of the plaque. Initially the CCA is divided into patches and passed to the first stage. Stage 1 characterizes the ROI patches from the non-ROI ones, combines the

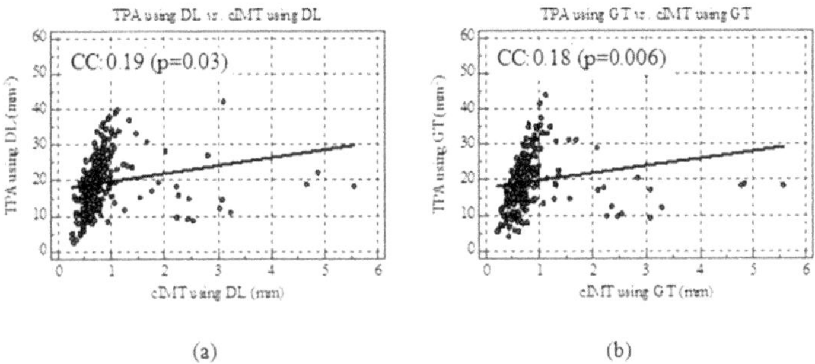

(a)        (b)

**Figure 7.18.** TPA versus cIMT regression plots using (a) DL and (b) the GT. (Reproduced with permission from [84]. Copyright 2020 Elsevier.)

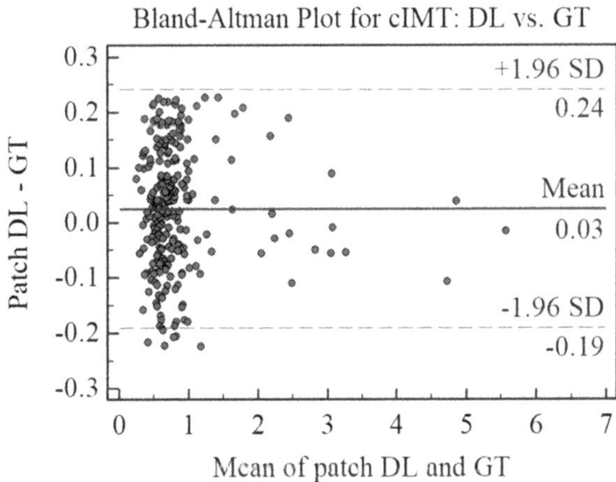

**Figure 7.19.** DL w.r.t GT Bland–Altman plot. (Reproduced with permission from [84]. Copyright 2020 Elsevier.)

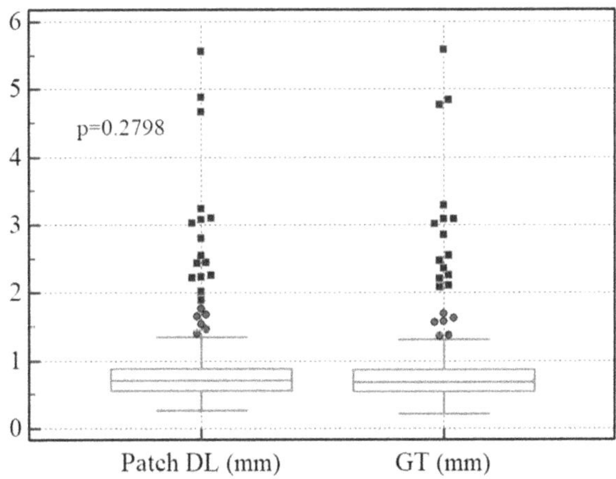

**Figure 7.20.** Mann–Whitney test. (Reproduced with permission from [84]. Copyright 2020 Elsevier.)

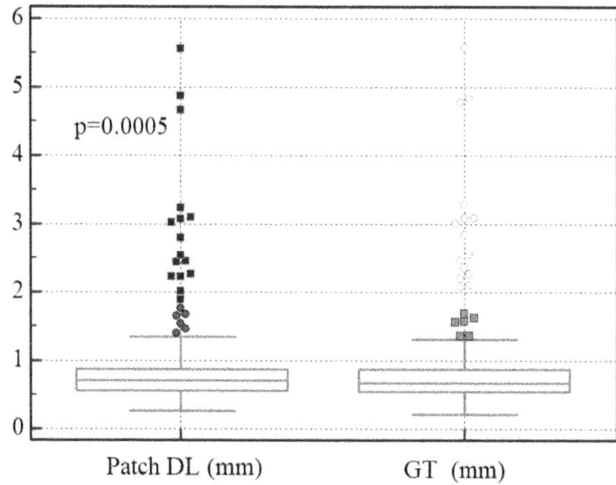


**Figure 7.21.** Wilcoxon test. (Reproduced with permission from [84]. Copyright 2020 Elsevier.)

**Table 7.3.** Mann–Whitney test parameters. (Reproduced with permission from [84]. Copyright 2020 Elsevier.)

| Parameters | Values |
| --- | --- |
| Mean rank of first group | 257.4840 |
| Mean rank of second group | 243.5160 |
| Mann–Whitney $U$ | 29504.00 |
| Large sample test statistic $Z$ | 1.081 |
| Two-tailed probability | $p = 0.2798$ |

**Table 7.4.** Wilkinson test parameters. (Reproduced with permission from [84]. Copyright 2020 Elsevier.)

| Parameters | Values |
| --- | --- |
| Large sample test statistic $Z$ | 3.464797 |
| Two-tailed probability | $p = 0.0005$ |
| Number of +ve differences | 102 |
| Number of −ve differences | 148 |

ROI patches, and passes them onto the second stage. The second stage the DL network segments the plaque area from the ROI. The LI and MA borders are delineated from the segmented plaque and the cIMT and TPA are computed. As already discussed in the previous sections, the cIMT bias error was 0.0935 ± 0.0637 mm for the given GT, which was far less, by 20%, than for GT1 (i.e. 0.1164 ± 0.1122 mm) and by 18% for GT2 for a previous implementation. The TPA bias error was 2.7939 ± 2.3702 mm$^2$. The regression plots for AI versus GT for both cIMT and TPA showed high correlation at 0.99 ($p$-value < 0.0001) and 0.89 ($p$-value < 0.0001), respectively. The cIMT regression CC value of 0.99 ($p$-value < 0.0001) for the current method was better

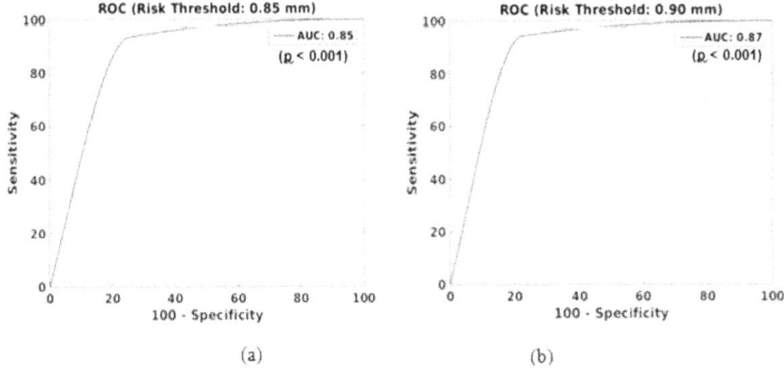

**Figure 7.22.** ROC curves: (a) a risk threshold of 0.85 mm (AUC: 0.85) and (b) a risk threshold of 0.90 mm (AUC: 0.87). (Reproduced with permission from [84]. Copyright 2020 Elsevier.)

**Table 7.5.** Benchmarking. (BEP: bulb edge point.) (Reproduced with permission from [84]. Copyright 2020 Elsevier.)

| Reference | Data size | Method | cIMT bias error (mm) |
|---|---|---|---|
| Molinari *et al* (2012) [30] | 200 | Morphological image processing, level set | 0.144 ± 0.179 mm |
| Molinari *et al* (2014) [79] | 665 | CALEX 1.0 | 0.191 ± 0.217 mm |
| Molinari *et al* (2011) [32] | 647 | CARES | 0.172 ± 0.222 mm |
| Molinari *et al* (2011) [28] | 657 | CAMES 1.0 | 0.154 ± 0.227 mm |
| Molinari *et al* (2011) [33] | 630 | CAUDLES | 0.224 ± 0.252 mm |
| Molinari *et al* (2011) [80] | 665 | FOAM | 0.150 ± 0.169 mm |
| Ikeda *et al* (2017) [81] | 649 | BEP detection | 0.0106 ± 0.00310 mm |
| Rosa *et al* (2014) [65] | 55 | Autoencoder and ANN | 0.0499 ± 0.0498 mm |
| Maria *et al* (2019) [67] | 331 | FCN | 0.02 mm |
| Biswas *et al* (2018) [21] | 396 | FCN based (with GT1) | 0.126 ± 0.134 mm |
| Biswas *et al* (2018) [21] | 396 | FCN based (with GT2) | 0.124 ± 0.100mm |
| **Proposed** | 250 | Artificial intelligence | **0. 093 ± 0.0637mm** |

than for GT1 (0.83 ($p < 0.0001$)) and GT2 (0.87 ($p < 0.0001$)). The improved performance is attributes to better ROI detection, the omission of similar echogenic structures, and reduced noise, which affected the output accuracy in the previous implementation. This is evident from the better performance in the case of absolute bias error which showed a decrease of 21% and 17% compared to ground truths GT1 and GT2, respectively. One of the key takeaways from the study is the use of augmentation, which improved the characterization accuracy in stage 1 to 99% from 89% (without augmentation). The cross-validation protocol used was K10, which has been used in several other studies [21, 54, 70, 78]. The benchmarking is provided next.

*Benchmarking*
The benchmarking is given in table 7.5.

*Strengths and weaknesses of the patch process*

When segmenting whole images, similar echogenic structures such as the jugular vein and noise affect the performance of the characterization module. Therefore, accurate ROI detection becomes a necessity. In this regard, the two-stage model provides an accurate ROI excluding noise and similar structures, resulting in better accuracy and performance. Therefore DL based studies are better in terms of performance. However, DL networks use multiple layers consisting of millions of learning parameters which require huge computation times to converge, leading to delays and cost escalation. GPUs which contain thousands of cores can reduce the computation times significantly but increase the overall cost. Therefore a proper balance is required between time and cost. The use of AI can be also applied to Big Data [82] and other applications [83]. One of the key issues is CVD risk estimation using DL, which has been done using ML to date [48].

## 7.8 Conclusion

A novel two-stage model was used for plaque volume (cIMT and TPA) computation from ultrasound CCA images. The initial stage was used for ROI detection while the next stage was used for segmentation and border delineation. ROI detection is based on patching, which eliminates similar echogenic structure such as the lumen and noise. The two-stage model used two independent DL models for cIMT measurement. The results show that the patching based two-stage DL model works better than all the previous models.

## References

[1] World Health Organization *Cardiovascular diseases (CVDs)* https://www.who.int/news-room/fact-sheets/detail/cardiovascular-diseases-(cvds)

[2] Benjamin E J, Blaha M J, Chiuve S E, Cushman M, Das S R, Deo R, Floyd J, Fornage M, Gillespie C and Isasi C 2017 Heart disease and stroke statistics—2017 update: a report from the American Heart Association *Circulation* **135** e146–603

[3] Suri J S, Kathuria C and Molinari F 2010 *Atherosclerosis Disease Management* (Berlin: Springer)

[4] Luca S, Jamthikar A, Gupta D, Khanna N N, Viskovic K, Suri H S, Gupta A, Mavrogeni S, Turk M and Laird J R 2019 Global perspective on carotid intima-media thickness and plaque: should the current measurement guidelines be revisited? *Int. Angiol.* **38** 451–65

[5] Saba L, Sanches J M, Pedro L M and Suri J S 2014 *Multi-modality Atherosclerosis Imaging and Diagnosis* (Berlin: Springer)

[6] Ross R 1999 Atherosclerosis—an inflammatory disease *New Engl. J. Med.* **340** 115–26

[7] Jonasson L, Holm J, Skalli O, Bondjers G and Hansson G K 1986 Regional accumulations of T cells, macrophages, and smooth muscle cells in the human atherosclerotic plaque *Arteriosclerosis* **6** 131–8

[8] Das P, van de Berg Bentz D and Becker A 1989 Atherosclerotic lesions in humans. *In situ* immunophenotypic analysis suggesting an immune mediated response *Lab Invest.* **61** 166–70

[9] Tedgui A and Mallat Z 2006 Cytokines in atherosclerosis: pathogenic and regulatory pathways *Physiol. Rev.* **86** 515–81

[10] Falk E, Shah P K and Fuster V 1995 Coronary plaque disruption *Circulation* **92** 657–71

[11] Suri J S and Laxminarayan S 2003 *Angiography and Plaque Imaging: Advanced Segmentation Techniques* (Boca Raton, FL: CRC Press)

[12] Seabra J and Sanches J 2012 *Ultrasound Imaging: Advances and Applications* (New York: Springer)

[13] Delsanto S, Molinari F, Giustetto P, Liboni W, Badalamenti S and Suri J S 2007 Characterization of a completely user-independent algorithm for carotid artery segmentation in 2-D ultrasound images *IEEE Trans. Instrum. Meas.* **56** 1265–74

[14] Molinari F, Zeng G and Suri J S 2010 Intima-media thickness: setting a standard for a completely automated method of ultrasound measurement *IEEE Trans. Ultrason. Ferroelectr. Freq. Control* **57** 1112–24

[15] Acharya U R, Faust O, Sree S V, Molinari F, Saba L, Nicolaides A and Suri J S 2011 An accurate and generalized approach to plaque characterization in 346 carotid ultrasound scans *IEEE Trans. Instrum. Meas.* **61** 1045–53

[16] Peters S, Den Ruijter H, Palmer M, Grobbee D, Crouse J III, O'Leary D, Evans G, Raichlen J, Lind L and Bots M 2012 Manual or semi-automated edge detection of the maximal far wall common carotid intima–media thickness: a direct comparison *J. Iint. Med.* **271** 247–56

[17] Polak J F and O'Leary D H 2015 Edge-detected common carotid artery intima–media thickness and incident coronary heart disease in the multi-ethnic study of atherosclerosis *J. Am. Heart Assoc.* **4** e001492

[18] Molinari F, Meiburger K M, Saba L, Acharya U R, Ledda G, Zeng G, Ho S Y S, Ahuja A T, Ho S C and Nicolaides A 2012 Ultrasound IMT measurement on a multi-ethnic and multi-institutional database: our review and experience using four fully automated and one semi-automated methods *Comput. Methods Programs Biomed.* **108** 946–60

[19] Molinari F, Zeng G and Suri J S 2011 Inter-greedy technique for fusion of different segmentation strategies leading to high-performance carotid IMT measurement in ultrasound images *Atherosclerosis Disease Management* (Berlin: Springer) pp 253–79

[20] Kumar P K, Araki T, Rajan J, Saba L, Lavra F, Ikeda N, Sharma A M, Shafique S, Nicolaides A and Laird J R 2017 Accurate lumen diameter measurement in curved vessels in carotid ultrasound: an iterative scale-space and spatial transformation approach *Med. Biol. Eng. Comput.* **55** 1415–34

[21] Biswas M, Kuppili V, Araki T, Edla D R, Godia E C, Saba L, Suri H S, Omerzu T, Laird J R and Khanna N N 2018 Deep learning strategy for accurate carotid intima-media thickness measurement: an ultrasound study on Japanese diabetic cohort *Comput. Biol. Med.* **98** 100–17

[22] Krizhevsky A, Sutskever I and Hinton G E 2012 ImageNet characterization with deep convolutional neural networks *Adv. Neural Inf. Process. Syst.* **60** 84–90

[23] Molinari F M, Meiburger K, Zeng G, Acharya U R, Liboni W, Nicolaides A and Suri J S 2012 Carotid artery recognition system: a comparison of three automated paradigms for ultrasound images *Med. Phys.* **39** 378–91

[24] Long J, Shelhamer E and Darrell T 2015 Fully convolutional networks for semantic segmentation *Proc. of the IEEE Conf. on Computer Vision and Pattern Recognition* 3431–40

[25] Molinari F, Meiburger K M, Saba L, Acharya U R, Ledda M, Nicolaides A and Suri J S 2012 Constrained snake vs conventional snake for carotid ultrasound automated IMT measurements on multi-center data sets *Ultrasonics* **52** 949–61

[26] Molinari F, Meiburger K M, Saba L, Zeng G, Acharya U R, Ledda M, Nicolaides A and Suri J S 2012 Fully automated dual-snake formulation for carotid intima-media thickness measurement: a new approach *J. Ultrasound Med.* **31** 1123–36

[27] Saba L, Molinari F, Meiburger K, Piga M, Zeng G, Rajendra U A, Nicolaides A and Suri J 2012 What is the correct distance measurement metric when measuring carotid ultrasound intima-media thickness automatically? *Int. Angiol.* **31** 483–9

[28] Molinari F, Pattichis C S, Zeng G, Saba L, Acharya U R, Sanfilippo R, Nicolaides A and Suri J S 2011 Completely automated multiresolution edge snapper—a new technique for an accurate carotid ultrasound IMT measurement: clinical validation and benchmarking on a multi-institutional database *IEEE Trans. Image Process.* **21** 1211–22

[29] Molinari F, Zeng G and Suri J S 2010 A state of the art review on intima–media thickness (IMT) measurement and wall segmentation techniques for carotid ultrasound *Comput. Methods Programs Biomed.* **100** 201–21

[30] Molinari F, Krishnamurthi G, Acharya U R, Sree S V, Zeng G, Saba L, Nicolaides A and Suri J S 2012 Hypothesis validation of far-wall brightness in carotid-artery ultrasound for feature-based IMT measurement using a combination of level-set segmentation and registration *IEEE Trans. Instrum. Meas.* **61** 1054–63

[31] Meiburger K M, Molinari F, Acharya U R, Saba L, Rodrigues P, Liboni W, Nicolaides A and Suri J S 2011 Automated carotid artery intima layer regional segmentation *Phys. Med. Biol.* **56** 4073

[32] Molinari F, Acharya U R, Zeng G, Meiburger K M and Suri J S 2011 Completely automated robust edge snapper for carotid ultrasound IMT measurement on a multi-institutional database of 300 images *Med. Biol. Eng. Comput.* **49** 935–45

[33] Molinari F, Meiburger K M, Zeng G, Nicolaides A and Suri J S 2011 CAUDLES-EF: carotid automated ultrasound double line extraction system using edge flow *J. Digital Imag.* **24** 1059–77

[34] Molinari F, Liboni W, Pantziaris M and Suri J 2011 CALSFOAM-completed automated local statistics based first order absolute moment' for carotid wall recognition, segmentation and IMT measurement: validation and benchmarking on a 300 patient database *Int. Angiol.* **30** 227–41

[35] Molinari F, Meiburger K M, Saba L, Acharya U R, Famiglietti L, Georgiou N, Nicolaides A, Mamidi R S, Kuper H and Suri J S 2014 Automated carotid IMT measurement and its validation in low contrast ultrasound database of 885 patient Indian population epidemiological study: results of AtheroEdge® software *Multi-modality Atherosclerosis Imaging and Diagnosis* (Berlin: Springer) pp 209–19

[36] Spence J D, Eliasziw M, DiCicco M, Hackam D G, Galil R and Lohmann T 2002 Carotid plaque area: a tool for targeting and evaluating vascular preventive therapy *Stroke* **33** 2916–22

[37] Cuadrado-Godia E, Maniruzzaman M, Araki T, Puvvula A, Rahman M J, Saba L, Suri H S, Gupta A, Banchhor S K and Teji J S 2018 Morphologic TPA (mTPA) and composite risk score for moderate carotid atherosclerotic plaque is strongly associated with HbA1c in diabetes cohort *Comput. Biol. Med.* **101** 128–45

[38] Puvvula A, Jamthikar A D, Gupta D, Khanna N N, Porcu M, Saba L, Viskovic K, Ajuluchukwu J N, Gupta A and Mavrogeni S 2020 Morphological carotid plaque area is associated with glomerular filtration rate: a study of South Asian Indian patients with diabetes and chronic kidney disease *Angiology* **71** 520–35

[39] Khanna N N, Jamthikar A D, Gupta D, Araki T, Piga M, Saba L, Carcassi C, Nicolaides A, Laird J R and Suri H S 2019 Effect of carotid image-based phenotypes on cardiovascular risk calculator: AECRS1.0 *Med. Biol. Eng. Comput.* **57** 1553–66

[40] Ikeda N, Gupta A, Dey N, Bose S, Shafique S, Arak T, Godia E C, Saba L, Laird J R and Nicolaides A 2015 Improved correlation between carotid and coronary atherosclerosis SYNTAX score using automated ultrasound carotid bulb plaque IMT measurement *Ultrasound Med. Biol.* **41** 1247–62

[41] Ikeda N, Araki T, Sugi K, Nakamura M, Deidda M, Molinari F, Meiburger K M, Acharya U R, Saba L and Bassareo P P 2014 Ankle–brachial index and its link to automated carotid ultrasound measurement of intima–media thickness variability in 500 Japanese coronary artery disease patients *Curr. Atheroscler. Rep.* **16** 393

[42] Saba L, Ikeda N, Deidda M, Araki T, Molinari F, Meiburger K M, Acharya U R, Nagashima Y, Mercuro G and Nakano M 2013 Association of automated carotid IMT measurement and HbA1c in Japanese patients with coronary artery disease *Diabetes Res. Clin. Pract.* **100** 348–53

[43] Araki T, Ikeda N, Molinari F, Dey N, Acharjee S, Saba L and Suri J 2014 Link between automated coronary calcium volumes from intravascular ultrasound to automated carotid IMT from B-mode ultrasound in coronary artery disease population *Int. Angiol.* **33** 392–403

[44] Acharya U R, Sree S V, Molinari F, Saba L, Nicolaides A and Suri J S 2015 An automated technique for carotid far wall characterization using grayscale features and wall thickness variability *J. Clin. Ultrasound* **43** 302–11

[45] Saba L, Mallarini G, Sanfilippo R, Zeng G, Montisci R and Suri J 2012 Intima media thickness variability (IMTV) and its association with cerebrovascular events: a novel marker of carotid therosclerosis *Cardiovasc. Diagn. Ther.* **2** 10

[46] Saba L, Meiburger K M, Molinari F, Ledda G, Anzidei M, Acharya U R, Zeng G, Shafique S, Nicolaides A and Suri J S 2012 Carotid IMT variability (IMTV) and its validation in symptomatic versus asymptomatic Italian population: can this be a useful index for studying symptomaticity? *Echocardiography* **29** 1111–9

[47] Lucatelli P, Raz E, Saba L, Argiolas G M, Montisci R, Wintermark M, King K S, Molinari F, Ikeda N and Siotto P 2016 Relationship between leukoaraiosis, carotid intima-media thickness and intima-media thickness variability: preliminary results *Eur. Radiol.* **26** 4423–31

[48] Khanna N N, Jamthikar A D, Araki T, Gupta D, Piga M, Saba L, Carcassi C, Nicolaides A, Laird J R and Suri H S 2019 Nonlinear model for the carotid artery disease 10-year risk prediction by fusing conventional cardiovascular factors to carotid ultrasound image phenotypes: a Japanese diabetes cohort study *Echocardiography* **36** 345–61

[49] Khanna N N, Jamthikar A D, Gupta D, Nicolaides A, Araki T, Saba L, Cuadrado-Godia E, Sharma A, Omerzu T and Suri H S 2019 Performance evaluation of 10-year ultrasound image-based stroke/cardiovascular (CV) risk calculator by comparing against ten conventional CV risk calculators: a diabetic study *Comput. Biol. Med.* **105** 125–43

[50] Cuadrado-Godia E, Jamthikar A D, Gupta D, Khanna N N, Araki T, Maniruzzaman M, Saba L, Nicolaides A, Sharma A and Omerzu T 2019 Ranking of stroke and cardiovascular risk factors for an optimal risk calculator design: logistic regression approach *Comput. Biol. Med.* **108** 182–95

[51] Saba L, Montisci R, Molinari F, Tallapally N, Zeng G, Mallarini G and Suri J S 2012 Comparison between manual and automated analysis for the quantification of carotid wall by using sonography. A validation study with CT *Eur. J. Radiol.* **81** 911–8

[52] Saba L, Banchhor S K, Araki T, Viskovic K, Londhe N D, Laird J R, Suri H S and Suri J S 2018 Intra-and inter-operator reproducibility of automated cloud-based carotid lumen diameter ultrasound measurement *Indian Heart J.* **70** 649–64

[53] Saba L, Banchhor S K, Suri H S, Londhe N D, Araki T, Ikeda N, Viskovic K, Shafique S, Laird J R and Gupta A 2016 Accurate cloud-based smart IMT measurement, its validation and stroke risk stratification in carotid ultrasound: a web-based point-of-care tool for multicenter clinical trial *Comput. Biol. Med.* **75** 217–34

[54] Biswas M, Kuppili V, Saba L, Edla D R, Suri H S, Sharma A, Cuadrado-Godia E, Laird J R, Nicolaides A and Suri J S 2019 Deep learning fully convolution network for lumen characterization in diabetic patients using carotid ultrasound: a tool for stroke risk *Med. Biol. Eng. Comput.* **57** 543–64

[55] Saba L, Biswas M, Suri H S, Viskovic K, Laird J R, Cuadrado-Godia E, Nicolaides A, Khanna N, Viswanathan V and Suri J S 2019 Ultrasound-based carotid stenosis measurement and risk stratification in diabetic cohort: a deep learning paradigm *Cardiovasc. Diagn. Ther.* **9** 439

[56] Saba L, Dey N, Ashour A S, Samanta S, Nath S S, Chakraborty S, Sanches J, Kumar D, Marinho R and Suri J S 2016 Automated stratification of liver disease in ultrasound: an online accurate feature characterization paradigm *Comput. Methods Programs Biomed.* **130** 118–34

[57] Cover T and Hart P 1967 Nearest neighbor pattern characterization *IEEE Trans. Inf. Theory* **13** 21–7

[58] Cortes C and Vapnik V 1995 Support-vector networks *Mach. Learn.* **20** 273–97

[59] Quinlan J R 1986 Induction of decision trees *Mach. Learn.* **1** 81–106

[60] Majumder M 2015 Artificial neural network *Impact of Urbanization on Water Shortage in Face of Climatic Aberrations* (Berlin: Springer) pp 49–54

[61] Ding S, Zhao H, Zhang Y, Xu X and Nie R 2015 Extreme learning machine: algorithm, theory and applications *Artif. Intell. Rev.* **44** 103–15

[62] Hinton G E 2009 Deep belief networks *Scholarpedia* **4** 5947

[63] Baldi P 2012 Autoencoders, unsupervised learning, and deep architectures *Proc. of ICML Workshop on Unsupervised and Transfer Learning* pp 37–49

[64] He K, Zhang X, Ren S and Sun J 2016 Deep residual learning for image recognition *Proc. of the IEEE Conf. on Computer Vision and Pattern Recognition* pp 770–8

[65] Menchón-Lara R-M and Sancho-Gómez J-L 2015 Fully automatic segmentation of ultrasound common carotid artery images based on machine learning *Neurocomputing* **151** 161–7

[66] Shin J, Tajbakhsh N, Todd Hurst R, Kendall C B and Liang J 2016 Automating carotid intima-media thickness video interpretation with convolutional neural networks *Proc. of the IEEE Conf. on Computer Vision and Pattern Recognition* pp 2526–35

[67] del Mar Vila M, Remeseiro B, Grau M, Elosua R, Betriu À, Fernandez-Giraldez E and Igual L 2020 Semantic segmentation with DenseNets for carotid artery ultrasound plaque segmentation and CIMT estimation *Artif. Intell. Med.* **103** 101784

[68] Suri J S, Liu K, Reden L and Laxminarayan S N 2002 White and black blood volumetric angiographic filtering: ellipsoidal scale-space approach *IEEE Trans. Inf. Technol. Biomed.* **6** 142–58

[69] Suri J S 2001 White matter/gray matter boundary segmentation using geometric snakes: a fuzzy deformable model *Int. Conf. on Advances in Pattern Recognition* (Berlin: Springer) pp 333–40

[70] Biswas M, Kuppili V, Edla D R, Suri H S, Saba L, Marinhoe R T, Sanches J M and Suri J S 2018 Symtosis: a liver ultrasound tissue characterization and risk stratification in optimized deep learning paradigm *Comput. Methods Programs Biomed.* **155** 165–77

[71] Banchhor S K, Araki T, Londhe N D, Ikeda N, Radeva P, Elbaz A, Saba L, Nicolaides A, Shafique S and Laird J R 2016 Five multiresolution-based calcium volume measurement techniques from coronary IVUS videos: a comparative approach *Comput. Methods Programs Biomed.* **134** 237–58

[72] Saba L, Banchhor S K, Londhe N D, Araki T, Laird J R, Gupta A, Nicolaides A and Suri J S 2017 Web-based accurate measurements of carotid lumen diameter and stenosis severity: an ultrasound-based clinical tool for stroke risk assessment during multicenter clinical trials *Comput. Biol. Med.* **91** 306–17

[73] Araki T, Ikeda N, Shukla D, Londhe N D, Shrivastava V K, Banchhor S K, Saba L, Nicolaides A, Shafique S and Laird J R 2016 A new method for IVUS-based coronary artery disease risk stratification: a link between coronary and carotid ultrasound plaque burdens *Comput. Methods Programs Biomed.* **124** 161–79

[74] Araki T, Ikeda N, Shukla D, Jain P K, Londhe N D, Shrivastava V K, Banchhor S K, Saba L, Nicolaides A and Shafique S 2016 PCA-based polling strategy in machine learning framework for coronary artery disease risk assessment in intravascular ultrasound: a link between carotid and coronary grayscale plaque morphology *Comput. Methods Programs Biomed.* **128** 137–58

[75] Suri J S, Haralick R M and Sheehan F H 2000 Greedy algorithm for bias error correction in automatically produced boundaries from low contrast ventriculograms *Pattern Anal. Appl.* **3** 39–60

[76] Chambless L E, Folsom A R, Clegg L X, Sharrett A R, Shahar E, Nieto F J, Rosamond W D and Evans G 2000 Carotid wall thickness is predictive of incident clinical stroke: the Atherosclerosis Risk in Communities (ARIC) study *Am. J. Epidemiol.* **151** 478–87

[77] O'Leary D H, Polak J F, Kronmal R A, Manolio T A, Burke G L and Wolfson S K Jr 1999 Carotid-artery intima and media thickness as a risk factor for myocardial infarction and stroke in older adults *New Engl. J. Med.* **340** 14–22

[78] Kuppili V, Biswas M, Sreekumar A, Suri H S, Saba L, Edla D R, Marinhoe R T, Sanches J M and Suri J S 2017 Extreme learning machine framework for risk stratification of fatty liver disease using ultrasound tissue characterization *J. Med. Syst.* **41** 152

[79] Molinari F, Acharya U R, Saba L, Nicolaides A and Suri J S 2014 Hypothesis validation of far wall brightness in carotid artery ultrasound for feature-based IMT measurement using a combination of level set segmentation and registration *Multi-Modality Atherosclerosis Imaging and Diagnosis* (Berlin: Springer) pp 255–67

[80] Molinari F and Suri J S 2011 Automated measurement of carotid artery intima-media thickness *Ultrasound and Carotid Bifurcation Atherosclerosis* (Berlin: Springer) pp 177–92

[81] Ikeda N, Dey N, Sharma A, Gupta A, Bose S, Acharjee S, Shafique S, Cuadrado-Godia E, Araki T and Saba L 2017 Automated segmental-IMT measurement in thin/thick plaque with bulb presence in carotid ultrasound from multiple scanners: stroke risk assessment *Comput. Methods Programs Biomed.* **141** 73–81

[82] El-Baz A and Suri J S 2019 *Big Data in Multimodal Medical Imaging* (Boca Raton, FL: CRC Press)

[83] Sinha G and Suri J S 2020 Introduction to cognitive science, informatics, and modeling *Cognitive Informatics, Computer Modelling, and Cognitive Science* (Amsterdam: Elsevier) pp 1–12

[84] Biswas M *et al* 2020 Two-stage artificial intelligence model for jointly measurement of atherosclerotic wall thickness and plaque burden in carotid ultrasound: a screening tool for cardiovascular/stroke risk assessment *Comput. Biol. Med.* **123** 103847

# Chapter 8

# Stenosis measurement from ultrasound carotid artery images in the deep learning paradigm

**Mainak Biswas and Jasjit S Suri**

Stroke happens to 15 million people around the globe annually, killing 5 million instantly and leaving another 5 million disabled. Precautionary steps, including carotid artery assessment, leading to an effective treatment plan could save many lives globally. However such clinical assessments are invasive, costly, and risky. Therefore, non-invasive methods, e.g. ultrasound (US), are much preferred. In recent years, machine intelligence paradigms such as machine learning (ML) and deep learning (DL) have been used in image characterization. These tools have been used for characterizing diseases and segmenting infected areas from different imaging modalities. In this chapter, a new DL model is proposed for stenosis measurement from carotid US scans. Three ground truths (GTs) developed by three radiologists were used to train three instances of the DL network. The ground truths were traced manually for the lumen–intima (LI) and media–adventitia (MA) walls from the US scans. Each trained DL network delineates the borders from raw carotid scans, and stenosis is measured. For the experiment, 407 carotid scans from both sides of the neck were collected from 204 patients (one right-side scan was not available). The lumen diameter mean error biases for the three datasets were $0.19 \pm 0.27$ mm ($p < 0.001$), $0.23 \pm 0.23$ mm ($p < 0.001$), and $0.21 \pm 0.19$ mm ($p < 0.001$), respectively. The precisions-of-merit for the three trained DL systems were 96.7%, 96.1%, and 96.4% for GT1, GT2, and GT3, respectively. The areas-under-the-curve (AUCs) corresponding to the three trained DL systems for GT1, GT2, and GT3 were 0.90, 0.94, and 0.86, respectively. The DL network showed better accuracy in providing stable, robust SSI computation.

## 8.1 Introduction

Annually, stroke affects around 15 million people globally, killing 5 million people and disabling another 5 million [1]. In the USA, 795 000 people suffer from stroke annually, with one American experiencing it every 40 s. This leads to a direct economic cost of USD 33 billion dollars and indirect costs of USD 20.6 billion dollars [2]. One of the reasons for stroke is stenosis, i.e. narrowing of the blood vessels to the brain. This narrowing happens due to the accumulation of plaque within the walls of the carotid artery, as shown in figure 8.1. One of the main reasons for the formation of plaque is the inflammatory disease called atherosclerosis [3, 4]. Atherosclerosis leads to epithelial dysfunction leading to the accumulation to monocytes, macrophages, and fats and other lipids within the arterial walls. Over the years the plaque accumulates, forming a necrotic core with a thin fibrous cap. Finally, the cap ruptures which leads to thrombosis, blockage of blood flow, and then stroke [5, 6]. There are multiple causes of atherosclerosis, chief among them are genetics [7–9], age [10, 11], obesity [12], diet [13], smoking [14], diabetes mellitus [15], alcohol, physical inactivity [16], hypertension [17], pollution [18] etc.

There are different types of imaging modalities, such as ultrasound (US) [19], magnetic resonance imaging (MRI) [20], and computed tomography (CT) [21, 22], for effective carotid scans. Although MRI is able to capture soft tissue information, it has several disadvantages, such as the gantry size and limitations in scanning patients who have metal objects in their bodies, e.g. a pacemaker. CT produces non-ionizing radiation which can be harmful to tissues, even leading to cancer, and hence should be avoided [23]. In such a scenario, ultrasound provides a radiation free, low-cost, non-invasive, and portable tool for carotid imaging [24–26]. There are different ways in which ultrasound imaging is carried out, mainly color Doppler imaging and B-mode ultrasound imaging. Doppler imaging detects blood pressure and flow to diagnose several complications such as clots and blood blockages [26]. There are

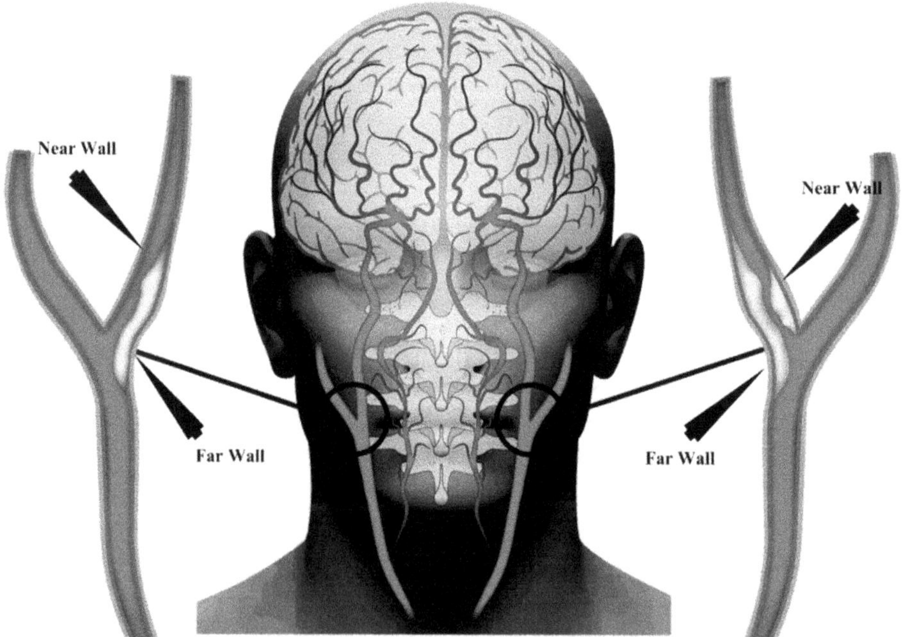


**Figure 8.1.** The anatomy of the common carotid artery and the stenosis caused by plaque development. (Image courtesy of AtheroPoint™, Roseville, CA. Reproduced with permission from [66].)

several disadvantages to Doppler imaging, such as a dependence on blood velocity, arterial wall stiffness, and low resolution due to distortion caused by impedance mismatch between the fluid and vessel walls [27–32]. B-mode ultrasound generates two-dimensional images based on acoustic impedance from the tissue area. The resolution of B-mode ultrasound images is also very high due to the employment of harmonic as well as compound imaging by US original equipment manufacturer. Thus B-mode ultrasound imaging is preferred to Doppler imaging since different tissues have unique acoustic impedances, resulting in better clarity [33, 34].

Several works have been performed in the area of automated diagnosis in carotid imaging. Most of these were on the recognition of the carotid artery and the quantification of the lumen diameter (LD) [35, 36] and intima–media complex [37, 38], and morphology based characterization of tissue [39, 40] etc. Some works have also linked the plaque burden in the carotid artery to cardio-vascular risk in the form of ankle brachial index [41], HbA1c [42], and coronary risk score [43]. In this study we propose employing a deep learning based mechanism for (a) delineation of the far and near lumen–intima borders, (b) stenosis measurement, and (c) risk characterization. Stenosis is measured following the NASCAT criteria [44–46].

There have been several works done previously in the area of lumen characterization and management [47, 48]. Most of the works follow similar trajectory. Several image features based on texture, brightness, and contrast of the image are collected in relation to the lumen. Based on these features the LD is computed and stenosis measurement is performed. However, the feature mining algorithms developed are manually constructed and cannot be generalized for all scenarios. Moreover, carotid scans may contain noise and therefore may lead to erroneous results. These methods of LD and stenosis measurement lack intelligence based solutions. Any intelligent solution requires memory and learning which again involve training and testing. In the past decade deep learning (DL) has gained favor in the field of medical imaging. DL has provided considerable improvements in the characterization and segmentation of medical images. DL models, unlike conventional models, perform both internal feature mining and characterization. DL consists of several layers of brain-mimicking neural networks capable of learning the complex relationships existing within patterns. In this regard a DL based solution is proposed for lumen recognition and stenosis measurement. The DL based model for stenosis measurement consists of four stages. Stage 1 prepares the carotid scans where they are cropped to remove non-tissue regions and dimensionality reduction is carried out for faster processing. Stage 2 consists of the DL network for lumen characterization using a fully convolutional network (FCN) [49]. Stage 3 performs LI far and near border delineation from the characterization and up-sampling to original size. Stage 4 is the risk analysis and performance estimation. The process model is shown in figure 8.2.

**Figure 8.2.** Pre-processing, lumen identification, LI near/far border mining system, stenosis measurement system, and performance evaluation comprise the global DL system. Reproduced with permission from [66].

The rest of the paper is organized as follows. Sections 8.2 and 8.3 provide the patient demographics and the methodology, respectively. Sections 8.4, 8.5, and 8.6 provide the results, statistical analysis, and discussion. Finally the paper concludes in section 8.7.

## 8.2 Patient demographics and image acquisition

A total of 204 patients took part in the experiment, including 157 male and 47 female. The average age of all the patients was 69 ± 11 years. A total of 407 B-mode ultrasound images were collected for the right and left carotid arteries (one right carotid scan was not available). The IRB of Toho University approved the data collection and the consent of all patients was taken prior. The other details of the patients are specified in table 8.1. Amongst the patients 92 were smokers, 93 patients were on statins, and 84 patients had renin angiotensin. The equipment used was an Aplio XV, Aplio KG, Xario scanner (Toshiba Inc). The scanner had a transducer of 7.5 MHz to examine the left and right carotid arteries. Scans were performed by a sonographer with 15 years' experience. The mean pixel resolution was 0.05 ± 0.01 mm/pixel.

## 8.3 Methodology

For tracing the walls of the carotid artery, the three radiologists used ImgTracer™, a user-friendly software [50]. The tracing was performed on 15–25 edge points close to the border, and performed across the entire carotid artery. For better visualization, the ImgTracer™ has a zoom feature. The output resulted in an array of $(x, y)$ points of traced boundaries. Now we discuss the AI methodology displayed in figure 8.2. In this work, the methodology was divided into three phases—phase 1: multiresolution; phase 2: lumen detection; phase 3: LI and MA border delineation; and phase 4: stenosis estimation and prognosis. The detailed global system is shown in figure 8.3.

**Table 8.1.** Patient demographic details. (Reproduced with permission from [66].)

| Parameter | Value |
|---|---|
| Gender (male) | 77.0% |
| Mean age | 69 ± 11 years |
| Low-density lipoprotein cholesterol | 99.80 ± 31.30 mg dl$^{-1}$ |
| High-density lipoprotein cholesterol | 50.40 ± 15.40 mg dl$^{-1}$ |
| Total cholesterol | 174.6 ± 37.7 mg dl$^{-1}$ |
| HbA1c | 5.8 ± 1.0 mg dl$^{-1}$ |
| Cr | 1.6 ± 2.1 mg dl$^{-1}$ |
| CKD stage | 3.1 ± 0.9 |
| eGFR | 45.4% |
| History of CVD (%) | 12.1% |
| Smokers | 45.1% |
| Total patients undergoing dyslipidemia | 57.3% |
| Total patients taking statins | 45.6% |

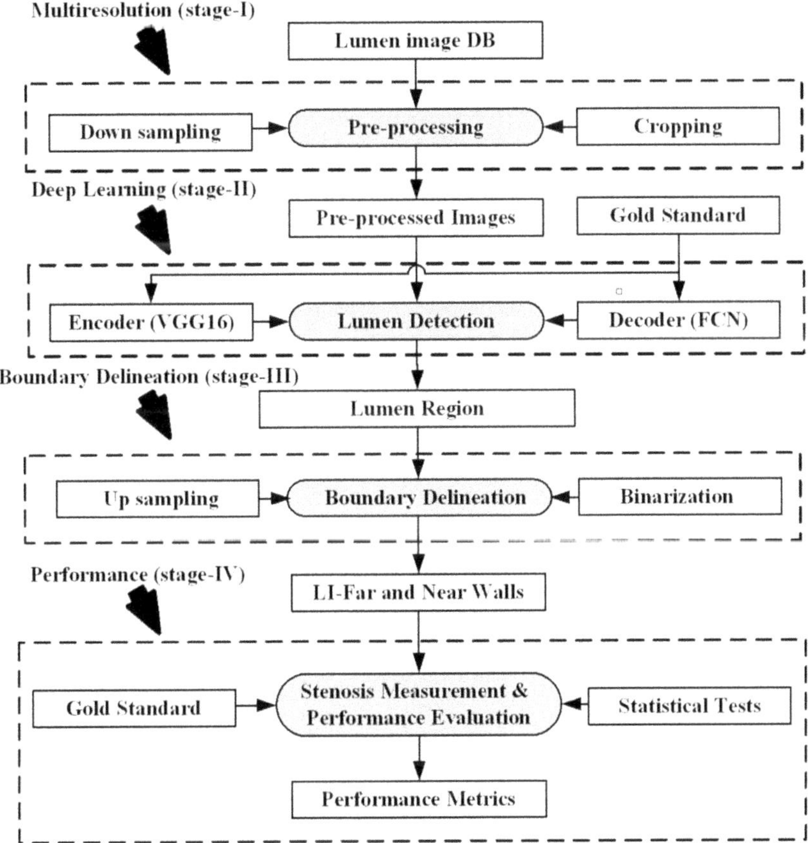


**Figure 8.3.** Stenosis estimation global model.

The local system is shown in figure 8.4. In the multiresolution stage [51, 52] the input images are pre-processed for the removal of non-tissue regions. The pre-processing steps are divided into four steps, including image cropping, binarization, reduction, and down-sampling. The image cropping task is performed to remove all background information except for the tissue region. In the binarization, the traced boundary images are converted into binary ground truth. The tissue information generated has noisy left and right edges due to the shape of the probe, which provides the best resolution at the center. Therefore, in the third step reduction is performed of 10% from each side to remove the left and right noise. The last step performs 50% down-sampling in each image for faster computation.

The DL system constitutes two phases: the encoder and the decoder [53]. The encoder is used for higher level feature mining. The decoder takes these features as input and performs lumen segmentation through feature enhancement and merging (discussed later). The deeper layers help in better understanding and learning of complex relationships within the data. The encoder, once trained, generates features which represent a miniature predicted segmentation output. The decoder up-samples

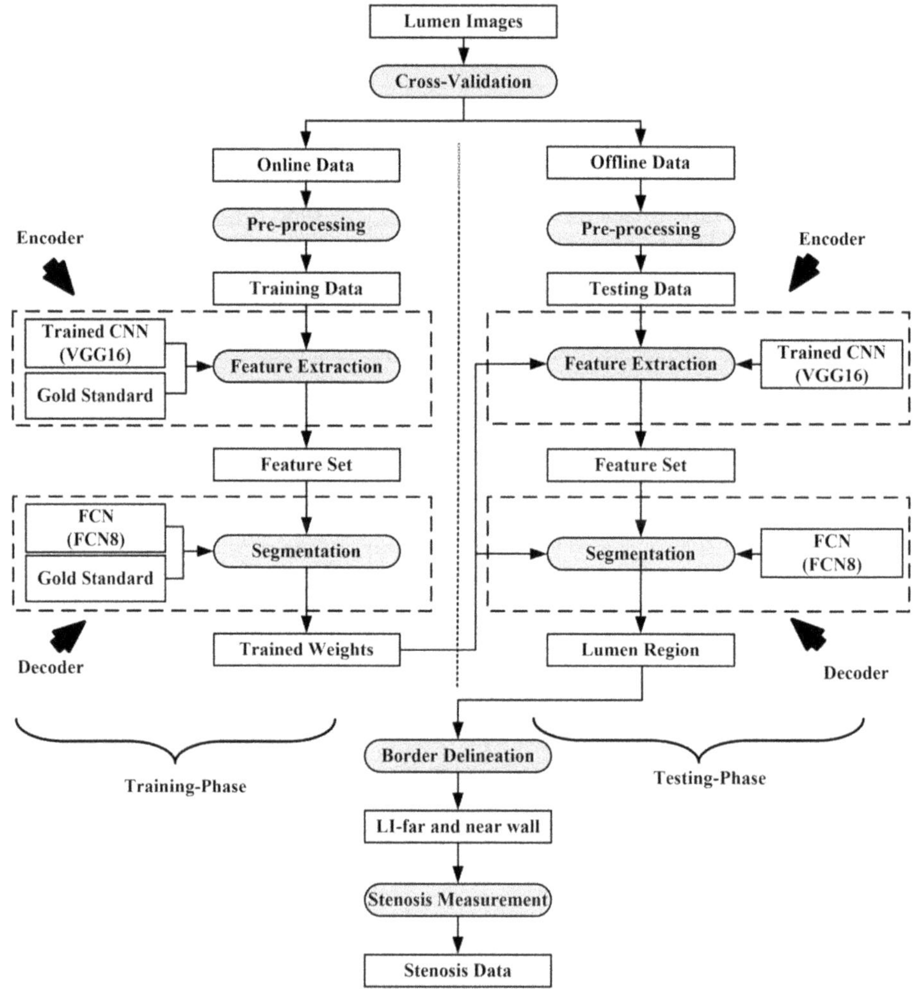



**Figure 8.4.** The training and testing paradigm of the DL based stenosis measurement model. (Reproduced with permission from [66].)

this output to the same dimension as the ground truth using up-sampling and merging operations. The predicted and the actual GTs are compared pixel-to-pixel and, based on the loss function parameters of the DL networks, they are modified until they reach convergence. Our model is inspired by FCNs [54]. FCNs are convolutional neural networks having upsampling and skipping layers in place of fully connected network for semantic segmentation [55]. The DL based encoder–decoder model is shown in figure 8.5. A detailed analysis of the model is given next. The mining of features is performed by a model similar to the first 13 layers of the VGG16 network [49] which is known as the encoder.

The layers have been already discussed in previous chapters. We used the first 13 layers of the VGG16 layers and removed the fully connected network. One of

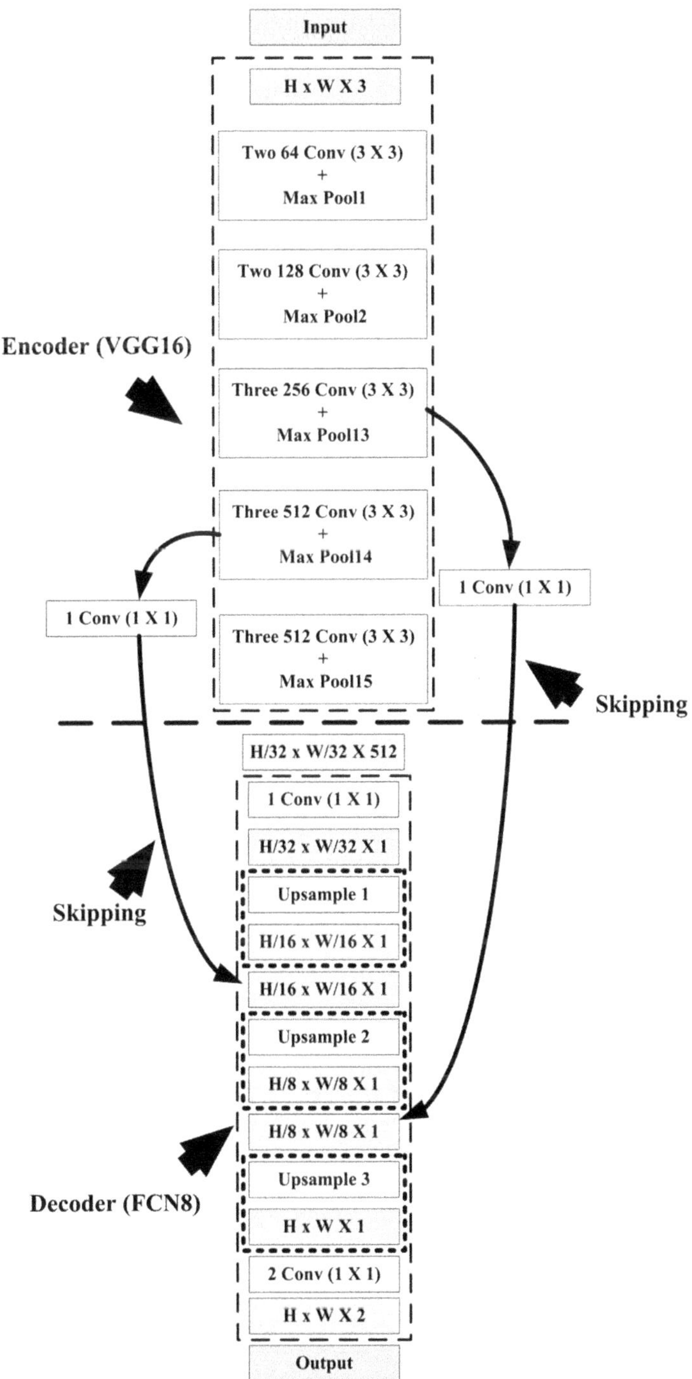



**Figure 8.5.** The DL architecture employed for stenosis measurement. (Reproduced with permission from [66].)

advantages of the VGG16 network is that it uses small filter sizes and hence retains better quality features than other CNN models. The weights of the VGG16 network are initialized using pre-trained weights. The decoder consists of three up-sampling layers, and uses two intermediate skipping layers. The up-sample layers restore the size of features extracted during the encoder phase to the same dimension as the GT. The skipping layers on the other hand merge features from the intermediate encoder layers.

This helps in the recovery of the structural data lost during the encoder phase. Thereafter, a pixel-to-pixel comparison occurs with the GT, using the loss function which is specified by

$$\partial_L(\beta_1, \beta_2) = \frac{1}{|N|} \sum_{n \in N} \sum_{l \in C} \beta_{2n}(l) \log \beta_{1n}(l), \tag{8.1}$$

where $\beta_1$ is the prediction, $\beta_2$ is the GT, $L$ is the label, C is the total number of classes, and $N$ is the quantity of images. Once the loss is decided it is used to update the weights of the network. A dropout strategy of 50% is used to avoid overfitting [56]. After the training/testing phase is completed, the LI far and near walls are delineated automatically from the segmented image. Finally, the NASCET criterion is applied to compute the stenosis severity index (SSI) which is the ratio of the LD minimum and LD normal. This is specified by

$$SSI(\%) = \left(1 - \frac{LD_{minimum}}{LD_{normal}}\right) \times 100. \tag{8.2}$$

## 8.4 Experimental protocol, performance parameters, and results

The cross-validation protocol used here is the K10 protocol (90% images for training + 10% images for testing). The mechanism follows partitioning the dataset into ten sets. Training and testing the model therefore happens ten times for full overall learning. In this work three tracers were employed for preparing the GT. Hence, three models of the same architecture were trained independently for the three GTs. The predicted boundary lines were derived separately for the three GTs and performance parameters were collected. The different performance and measurement parameters in addition to SSI are specified in the following. The first performance parameter is the precision of merit (PoM) for the SSI or POM$_{SSI}$ which is specified by

$$PoM_{SSI}(\%) = 100 - \left[\left(\frac{\sum_{i=1}^{N} \frac{\left|SSI_{dl(i)} - SSI_{gt(i)}\right|}{SSI_{gt(i)}}}{N}\right) \times 100\right], \tag{8.3}$$

where $N$ is the number of images in the cohort, and $i$ is the image number. We have computed $\text{POM}_{\text{LD}}$ for $\text{LD}_{\text{dl}(i)}$ and $\text{LD}_{\text{gt}(i)}$ which is specified by

$$\text{PoM}_{\text{LD}}(\%) = 100 - \left[\left(\frac{\sum\limits_{i=1}^{N}\dfrac{\left|\text{LD}_{\text{dl}(i)} - \text{LD}_{\text{gt}(i)}\right|}{\text{LD}_{\text{gt}(i)}}}{N}\right) \times 100\right]. \tag{8.4}$$

The polyline distance metric (PDM) is used for computing the stenosis and LD error bias. Let the LI far and LI near borders be denoted as $C_1$ and $C_2$, respectively. Let the reference point on $C_1$ be vertex $R_1$ and the segment in $C_2$ be defined by vertices $R_2$ and $R_3$. Let the distance between $R_1$ and $R_2$ be $d_1$ and the distance between $R_1$ and $R_3$ be denoted as $d_2$. Let $D(R_1, L)$ be the PD between vertex $R_1$: $(x_1, y_1)$ on $C_1$ and line segment $L$ formed by two points $R_2$: $(x_2, y_2)$ and $R_3$: $(x_3, y_3)$. Let delta $(\delta)$ be the distance of the reference point $R_1$ towards the line segment $L$. The perpendicular distance between the line segment $L$ and the reference point, $R_1$, is specified by $d_p$. Then, the polyline distance $D(R_1, L)$ can be defined as

$$D(R_1, L) = \begin{cases} |d_p| & 0 < \delta < 1 \\ \min(d_1, d_2) & \delta < 0,\ \delta > 1 \end{cases}, \tag{8.5}$$

where

$$d_1 = \sqrt{(x_1 - x_2)^2 + (y_1 - y_2)^2} \tag{8.6}$$

$$d_2 = \sqrt{(x_1 - x_3)^2 + (y_1 - y_3)^2} \tag{8.7}$$

$$\delta = \frac{(y_3 - y_2)(y_1 - y_2) + (x_3 - x_2)(x_1 - x_2)}{(x_3 - x_2)^2 + (y_3 - y_2)^2} \tag{8.8}$$

and

$$d_p = \frac{(y_3 - y_2)(x_2 - x_1) + (x_3 - x_2)(y_1 - y_2)}{\sqrt{(x_3 - x_2)^2 + (y_3 - y_2)^2}}. \tag{8.9}$$

The process to obtain $D(R_1, L)$ is repeated for the remaining points of the outline $C_j$ and is specified by

$$D(C_1, C_2) = \sum_{i=1}^{N} D(R_i, S_{C_2}), \tag{8.10}$$

where $N$ represents all points on $C_1$ and $S_{C_2}$ is the section on outline of $C_2$. This algorithm is recurring in reverse, where $C_2$ becomes the reference outline and $C_1$ becomes the segment outline. The reverse is symbolized as $D(C_2, C_1)$. Finally, by combining both $D(C_1, C_2)$ and $D(C_2, C_1)$, we obtain the PDM which is specified by

$$D_{\text{PDM}}(C_1: C_2) = \frac{D(C_1, C_2) + D(C_2, C_1)}{(\#\text{points} \in C_1 + \#\text{points} \in C_2)}. \qquad (8.11)$$

*Results for LD*

For the first DL1, the corresponding PoMs for stenosis and the LD error bias are specified in tables 8.2 and 8.3, respectively. Similarly, the stenosis and LD error bias tables for DL2 and DL3 are shown in tables 8.4 and 8.5 and tables 8.6 and 8.7, respectively. The corresponding plots for DL1, DL2, and DL3 are shown in figures 8.6, 8.7, and 8.8, respectively. Each of the DL systems was trained for 4000 iterations.

*Estimation of LD/SSI when trained using GT1*

Approximately 40% of images showed stenosis was present irrespective of the ground truth, confirming the accuracy of the DL. For 407 images, the mean stenosis DL1 and GT1 were 20.5% and 18.6%, respectively, as shown in table 8.2. The mean LD error bias between DL1 and GT1 was $0.19 \pm 0.27$ mm ($p < 0.001$) as displayed in table 8.3. Figure 8.6 shows the stenosis plot, where high stenosis was noted

**Table 8.2.** Mean PoM stenosis (%) of DL1 w.r.t. GT1. (Reproduced with permission from [66].)

| # Images | Mean DL1 stenosis (%) | Mean GT1 stenosis (%) | Mean PoM1 (stenosis) (%) |
|---|---|---|---|
| 10 | 57.2 | 57.2 | 93.7 |
| 50 | 42.1 | 40.8 | 94.6 |
| 100 | 35.2 | 33.1 | 95.6 |
| 150 | 31.2 | 28.9 | 96.1 |
| 200 | 28.4 | 26.0 | 96.3 |
| 250 | 26.1 | 23.8 | 96.2 |
| 300 | 24.2 | 21.9 | 96.4 |
| 350 | 22.5 | 20.3 | 96.5 |
| 407 | 20.5 | 18.6 | 96.7 |

**Table 8.3.** LD absolute mean error bias (mm) and PoM (%) computed between DL1 and GT1. (Reproduced with permission from [66].)

| Range (mm) | # Images w.r.t. GT1 | # Images w.r.t. DL1 | Mean LD using GT1 (mm) | Mean LD using DL1 (mm) | LD absolute mean error bias (mm) | Chi-squared test w.r.t. LD error bias | PoM1 (LD) (%) |
|---|---|---|---|---|---|---|---|
| <8.5 | 406 | 406 | $6.07 \pm 0.91$ | $6.09 \pm 0.94$ | $0.19 \pm 0.27$ | $p < 0.001$ | 96.71 |
| <8.0 | 399 | 396 | $6.00 \pm 0.85$ | $6.04 \pm 0.87$ | $0.20 \pm 0.27$ | $p < 0.001$ | 96.73 |
| <7.5 | 382 | 377 | $5.90 \pm 0.78$ | $5.96 \pm 0.80$ | $0.19 \pm 0.27$ | $p < 0.001$ | 96.76 |
| <7.0 | 346 | 341 | $5.82 \pm 0.70$ | $5.82 \pm 0.71$ | $0.19 \pm 0.27$ | $p < 0.001$ | 96.73 |
| <6.5 | 282 | 278 | $5.61 \pm 0.61$ | $5.63 \pm 0.63$ | $0.18 \pm 0.27$ | $p < 0.001$ | 96.72 |
| <6.0 | 190 | 200 | $5.31 \pm 0.49$ | $5.38 \pm 0.55$ | $0.20 \pm 0.31$ | $p < 0.001$ | 96.32 |
| <5.5 | 113 | 110 | $5.01 \pm 0.42$ | $5.07 \pm 0.51$ | $0.21 \pm 0.37$ | $p < 0.001$ | 95.71 |
| <5.0 | 49 | 45 | $4.64 \pm 0.37$ | $4.71 \pm 0.54$ | $0.27 \pm 0.53$ | $p < 0.001$ | 94.15 |

**Table 8.4.** Mean PoM stenosis (%) of DL2 w.r.t. GT2. (Reproduced with permission from [66].)

| # Images | Mean DL2 stenosis (%) | Mean GT2 stenosis (%) | Mean PoM2 (stenosis) (%) |
| --- | --- | --- | --- |
| 10 | 48.9 | 54.6 | 92.8 |
| 50 | 37.4 | 39.7 | 94.1 |
| 100 | 31.0 | 32.2 | 95.0 |
| 150 | 26.9 | 27.9 | 95.6 |
| 200 | 24.1 | 25.1 | 95.9 |
| 250 | 21.9 | 22.8 | 96.0 |
| 300 | 20.2 | 21.0 | 96.0 |
| 350 | 18.7 | 19.4 | 96.1 |
| 407 | 17.0 | 17.7 | 96.1 |

**Table 8.5.** lD absolute mean error bias (mm) and PoM (%) computed between DL2 and GT2. (Reproduced with permission from [66].)

| Range (mm) | # Images w.r.t. GT2 | # Images w.r.t. DL2 | Mean LD using GT2 (mm) | Mean LD using DL2 (mm) | LD absolute mean error bias (mm) | Chi-squared test w.r.t. LD error bias | PoM2 (LD) (%) |
| --- | --- | --- | --- | --- | --- | --- | --- |
| <8.5 | 406 | 406 | $5.91 \pm 0.88$ | $6.05 \pm 0.91$ | $0.23 \pm 0.23$ | $p < 0.001$ | 96.09 |
| <8.0 | 403 | 400 | $5.89 \pm 0.84$ | $6.01 \pm 0.85$ | $0.23 \pm 0.23$ | $p < 0.001$ | 96.08 |
| <7.5 | 387 | 384 | $5.81 \pm 0.77$ | $5.93 \pm 0.78$ | $0.23 \pm 0.23$ | $p < 0.001$ | 96.04 |
| <7.0 | 364 | 354 | $5.72 \pm 0.70$ | $5.82 \pm 0.70$ | $0.23 \pm 0.23$ | $p < 0.001$ | 95.90 |
| <6.5 | 311 | 293 | $5.56 \pm 0.60$ | $5.64 \pm 0.62$ | $0.24 \pm 0.24$ | $p < 0.001$ | 95.76 |
| <6.0 | 230 | 203 | $5.31 \pm 0.51$ | $5.38 \pm 0.54$ | $0.25 \pm 0.25$ | $p < 0.001$ | 95.31 |
| <5.5 | 130 | 113 | $4.98 \pm 0.42$ | $5.08 \pm 0.51$ | $0.26 \pm 0.28$ | $p < 0.001$ | 94.72 |
| <5.0 | 54 | 45 | $4.59 \pm 0.38$ | $4.71 \pm 0.50$ | $0.31 \pm 0.33$ | $p < 0.001$ | 93.23 |

**Table 8.6.** Mean PoM stenosis (%) of DL3 w.r.t. GT3. (Reproduced with permission from [66].)

| # Images | Mean DL3 stenosis (%) | Mean GT3 stenosis (%) | Mean PoM3 (stenosis) (%) |
| --- | --- | --- | --- |
| 10 | 58.4 | 52.6 | 91.2 |
| 50 | 43.6 | 38.3 | 93.7 |
| 100 | 36.3 | 32.2 | 95.4 |
| 150 | 32.2 | 28.5 | 95.7 |
| 200 | 29.3 | 26.0 | 95.7 |
| 250 | 26.9 | 23.9 | 95.9 |
| 300 | 25.0 | 22.2 | 96.1 |
| 350 | 23.3 | 20.7 | 96.2 |
| 407 | 21.3 | 19.0 | 96.4 |

**Table 8.7.** LD absolute mean error bias (mm) and PoM (%) computed between DL3 and GT3. (Reproduced with permission from [66].)

| Range (mm) | # Images w.r.t. GT3 | # Images w.r.t. DL3 | Mean LD using GT3 (mm) | Mean LD using DL3 (mm) | LD absolute mean error bias (mm) | Chi-squared test w.r.t. LD error bias | PoM3 (LD) (%) |
|---|---|---|---|---|---|---|---|
| <8.5 | 406 | 406 | 6.00 ± 0.90 | 6.01 ± 0.94 | 0.21 ± o.19 | $p < 0.001$ | 96.40 |
| <8.0 | 401 | 400 | 5.97 ± 0.85 | 5.97 ± 0.89 | 0.21 ± 0.21 | $p < 0.001$ | 96.38 |
| <7.5 | 387 | 380 | 5.90 ± 0.79 | 5.88 ± 0.81 | 0.22 ± 0.21 | $p < 0.001$ | 96.33 |
| <7.0 | 350 | 349 | 5.76 ± 0.69 | 5.77 ± 0.74 | 0.21 ± 0.20 | $p < 0.001$ | 96.34 |
| <6.5 | 293 | 292 | 5.58 ± 0.60 | 5.59 ± 0.66 | 0.21 ± 0.20 | $p < 0.001$ | 96.31 |
| <6.0 | 209 | 207 | 5.31 ± 0.50 | 5.33 ± 0.58 | 0.22 ± 0.21 | $p < 0.001$ | 95.92 |
| <5.5 | 127 | 119 | 5.01 ± 0.41 | 5.02 ± 0.51 | 0.20 ± 0.18 | $p < 0.001$ | 95.98 |
| <5.0 | 50 | 56 | 4.60 ± 0.35 | 4.67 ± 0.42 | 0.19 ± 0.15 | $p < 0.001$ | 95.97 |

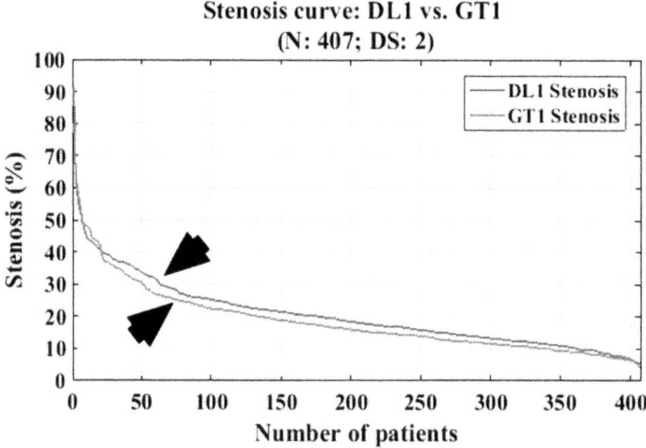


**Figure 8.6.** Stenosis plot for DL1 and GT1. (Reproduced with permission from [66].)

initially for 20–25 patients. The DL1 curve rises above the GT1 curve for 380 patients meeting the latter at the end. The PoM is 96.7%.

*Estimation of LD/SSI when trained using GT2*
Table 8.4 shows stenosis for both DL2 and GT2 at 17.0% and 17.7%, respectively. The PoM value is 96.1%. The mean LD error bias between DL2 and GT2 is 0.23 ± 0.23 mm ($p < 0.001$) shown in table 8.5. The stenosis curve in figure 8.7 shows strong resemblance between DL2 and GT2.

*Estimation of LD/SSI when trained using GT3*
The mean stenosis for DL3 and GT3 were 21.3% and 19.0%, respectively, as shown in table 8.6. The PoM value was 96.4%. Table 8.7 provides the mean LD error bias at 0.21 ± 0.19 mm ($p < 0.001$). The stenosis curve for DL3 and GT3 in figure 8.8 shows a strong correlation.

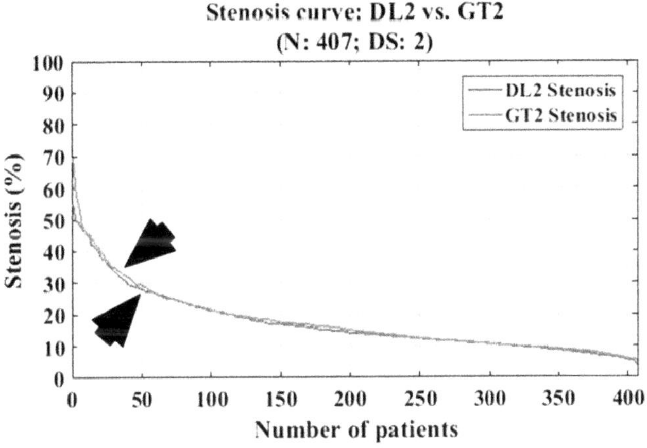


**Figure 8.7.** Stenosis plot for DL2 and GT2. (Reproduced with permission from [66].)

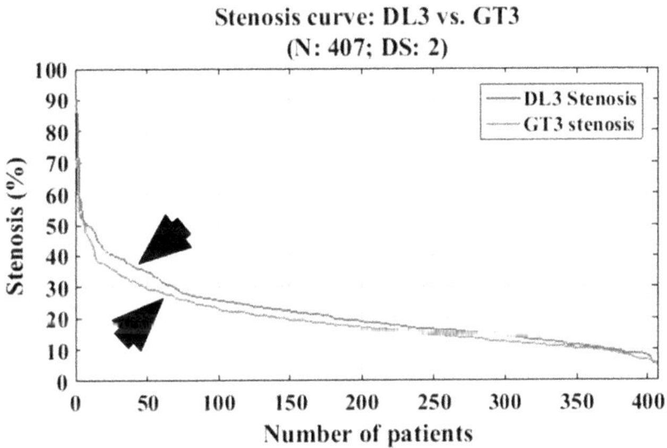


**Figure 8.8.** Stenosis plot for DL3 and GT3. (Reproduced with permission from [66].)

The results indicate that the cohort of 204 subjects had mild stenosis and subclinical atherosclerosis. The high stenosis where endarterectomy or stenting occurs is typically in the range of 60%–80% [57, 58].

*Results: correlation coefficient of DL based stenosis against GT stenosis*
The regression plots for DL and GT for the three instances can be seen in figures 8.9(a), (b), and (c), respectively. All plots have high CC values at 0.93, 0.94, and 0.93, respectively. On the whole, the results were good enough to indicate that the DL system was accurate in predicting stenosis.

*Results: Bland–Altman plots for DL stenosis against GT stenosis*
The Bland–Altman plots for DL1 with respect to GT1, DL2 with respect to GT2, and DL3 with respect to GT3 are shown in figures 8.10(a), (b), and (c), respectively. The bias values with respect to these plots were −1.8, 0.2, and −1.9, respectively, with GT showing the lowest bias.

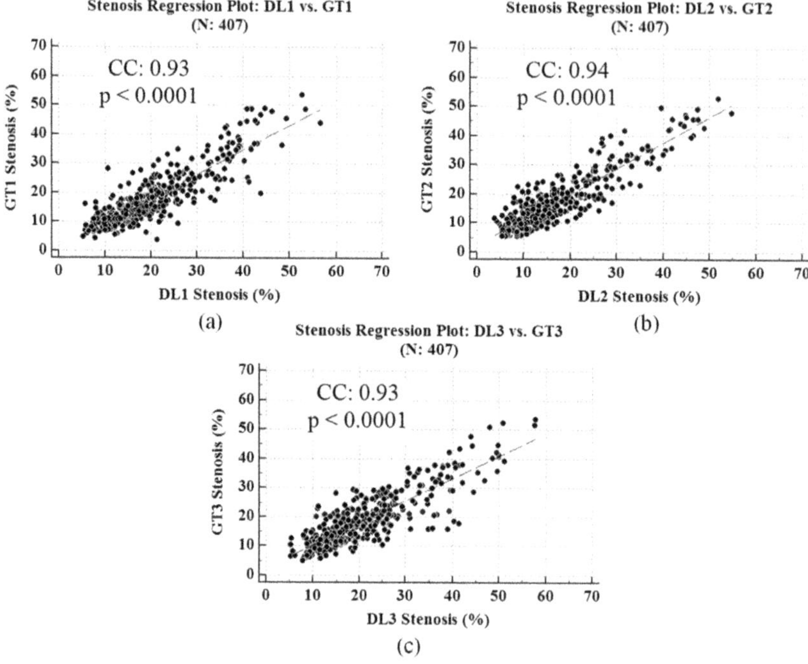


**Figure 8.9.** (a) Regression plot for DL1 and GT1, (b) regression plot for DL2 and GT2, and (c) regression plot for DL3 and GT3. (Reproduced with permission from [66].)

## 8.5 Statistical tests, variability and error bias analysis, and risk characterization

Statistical tools are an important element of data analysis. The tests performed as a part of the analysis were Wilcoxon, Mann–Whitney, paired $t$-test, ANOVA, Kruskal–Wallis, and Friedman. We also performed inter-operator variability analysis. The overall performance analysis is discussed at the end. The $p$-values for the Wilcoxon and paired-test irrespective of the DL and GT values are below 0.0001. For Mann–Whitney, the $p$-values for the three sets of DL were $p = 0.0005$, $p = 0.0012$, and $p = 0.0017$, respectively. Hence, for Mann–Whitney the null hypothesis showed that the data distribution could not be retained for DL1, DL2, and DL3. The box-plots for the three DL systems for the Wilcoxon test, paired $t$-test, and Mann–Whitney test are shown in figures 8.11(a)–(c), 8.12(a)–(c), and 8.13(a)–(c), respectively. The corresponding values are also specified in tables 8.8, 8.9, and 8.10, respectively. Further, ANOVA and Kruskal–Wallis tests were performed. The ANOVA test $p$-values were $p < 0.001$ for DL1, $p < 0.001$ for DL2, and $p < 0.001$ for DL3, respectively. The Kruskal–Wallis $p$-values were $p < 0.466956$ for DL1, $p < 0.448880$ for DL2, $p < 0.449992$ for DL3, respectively, all accepting the null hypothesis. The Friedman test was performed with $p < 0.00001$ for DL1, $p < 0.00001$ for DL2, $p < 0.00001$ for DL3, respectively, rejecting the null hypothesis. The results are listed in tables 8.11, 8.12, and 8.13.

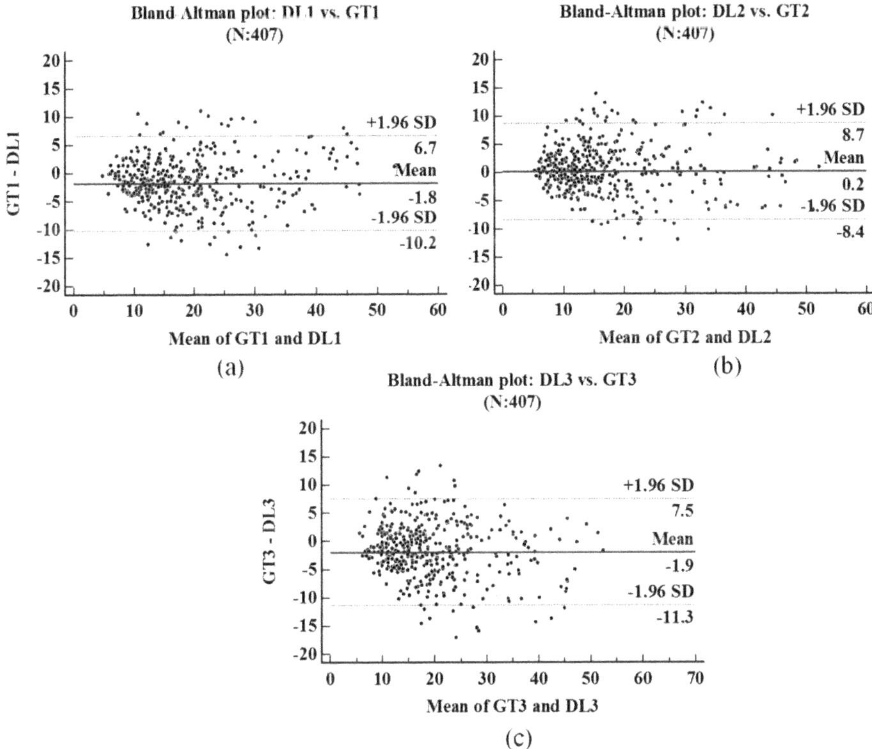


**Figure 8.10.** (a) Bland–Altman plot for DL1 and GT1, (b) DL2 and GT2, and (c) DL3 and GT3. (Reproduced with permission from [66].)

*Inter-operator variability*

The inter-operator variability analysis shows the DL and GT predictions against each other. The corresponding regression plots are shown in figures 8.14(a), (b), and (c), respectively. The CC values for DL1 and DL2, DL1 and DL3, and DL2 and DL3 are 0.93, 0.93, and 0.92, respectively. Also, the CC values for GT1 and GT2, GT1 and GT3, and GT2 and GT3 are 0.95, 0.96, and 0.95, respectively.

*Performance evaluation*

The CDF plot shown in figure 8.15 clearly shows that 90% of patients had a stenosis error bias of less than 10%. Also for 70% patients the stenosis error bias was < 5%. The DL1, DL2, and DL3 signed CDF plots are shown in figures 8.16(a), (b), and (c), respectively. For DL1, 90% of patients had stenosis error bias < 8% and > −5%. Similarly for DL2 90% of patients has a stenosis error bias < 5% and >−5%. Finally, for DL3, 90% of patients had stenosis error bias < 9% and > −5%. All values show that the DL system is robust and ready for clinical trials.

*Risk categorization and the ROC study*

Based on the stenosis, the patients were categorized into low, medium, and high risk. The cohort had a maximum stenosis of 70%. Therefore, the low risk patients had stenosis less than 25%, medium risk patients had stenosis between 25% and 50%, and

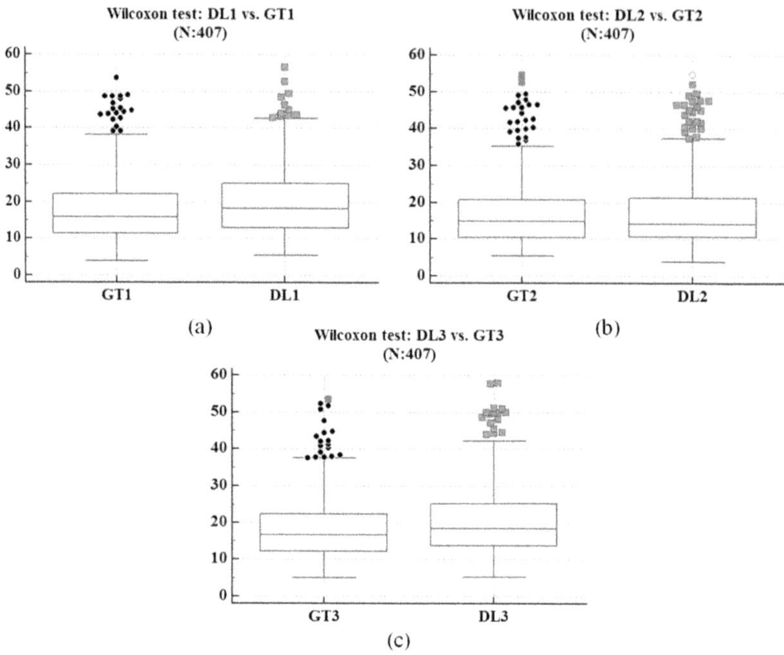

**Figure 8.11.** (a) Wilcoxon box plot for DL1 and GT1, (b) DL2 and GT2, and (c) DL3 and GT3. (Reproduced with permission from [66].)

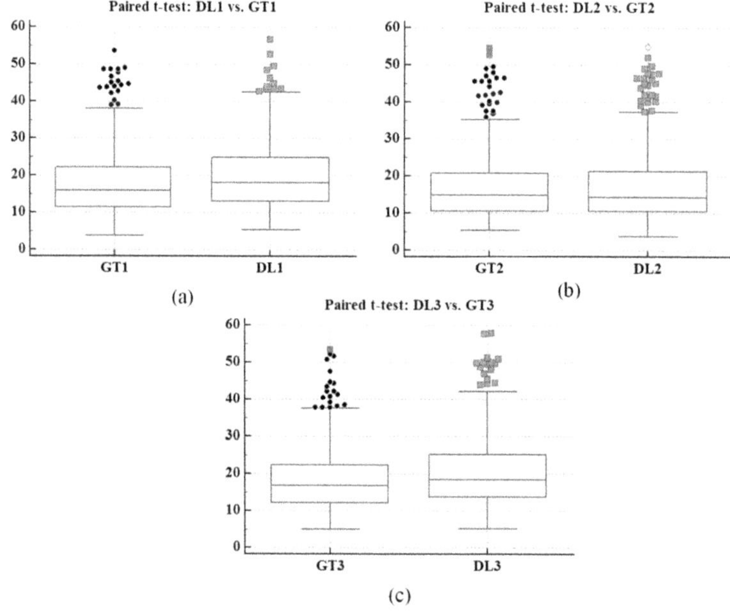

**Figure 8.12.** (a) Paired *t*-test box plot for DL1 and GT1, (b) DL2 and GT2, and (c) DL3 and GT3. (Reproduced with permission from [66].)

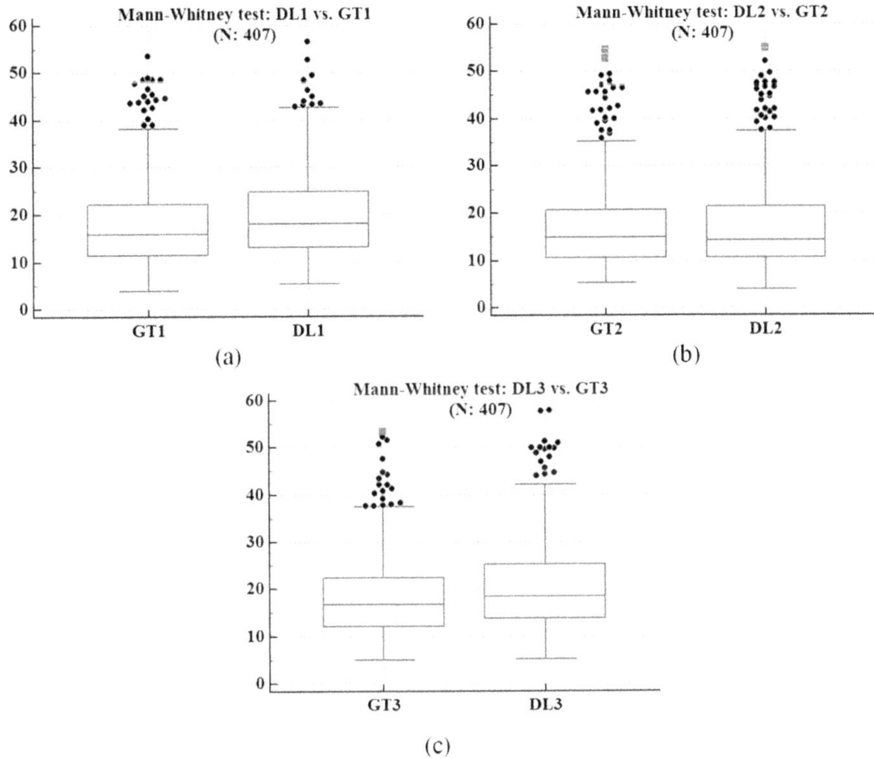

**Figure 8.13.** (a) Mann–Whitney test box plot for DL1 and GT1, (b) DL2 and GT2, and (c) Mann–Whitney DL3 and GT3. (Reproduced with permission from [66].)

**Table 8.8.** Wilcoxon test parameters. (Reproduced with permission from [66].)

| Parameters | DL1 | DL2 | DL3 |
|---|---|---|---|
| Large sample test statistic $Z$ | −8.165608 | −6.319204 | −7.882409 |
| Two-tailed probability | $p < 0.0001$ | $p < 0.0001$ | $p < 0.0001$ |
| Number of +ve differences | 278 | 271 | 267 |
| Number of -ve differences | 129 | 136 | 140 |

**Table 8.9.** Paired $t$-test parameters. (Reproduced with permission from [66].)

| Parameters | DL1 | DL2 | DL3 |
|---|---|---|---|
| Mean difference | 1.9223 | 1.9224 | 2.0314 |
| Standard deviation of differences | 4.8163 | 4.8832 | 4.8928 |
| Standard error bias of mean difference | 0.2387 | 0.2472 | 0.2425 |
| 95% CI | 1.4530 to 2.3916 | 1.5388 to 2.3544 | 1.5547 to 2.5082 |
| Test statistic $t$ | 8.052 | −0.406 | 8.376 |
| Degrees of freedom (DF) | 407 | 407 | 407 |
| Two-tailed probability | $p < 0.0001$ | $p < 0.0001$ | $p < 0.0001$ |

**Table 8.10.** Mann–Whitney test parameters. (Reproduced with permission from [66].)

| Parameters | DL1 | DL2 | DL3 |
| --- | --- | --- | --- |
| Average rank of first group | 378.7445 | 389.2162 | 381.5774 |
| Average rank of second group | 436.2555 | 433.7838 | 433.4226 |
| Mann–Whitney $U$ | 71121.00 | 71312.00 | 72274.00 |
| Test statistic $Z$ (corrected for ties) | 3.489 | 3.451 | 3.146 |
| Two-tailed probability | $p = 0.0005$ | $p = 0.0012$ | $p = 0.0017$ |

**Table 8.11.** ANOVA test parameters. (Reproduced with permission from [66].)

| Parameters | DL1 | DL2 | DL3 |
| --- | --- | --- | --- |
| Sum of squares | 37096.4027 | 38596.4637 | 38749.5063 |
| DF | 406 | 406 | 406 |
| $F$-ratio | 3.405 | 310.764 | 2.581 |
| Significance | $p < 0.001$ | $p < 0.001$ | $p < 0.001$ |

**Table 8.12.** Kruskal–Wallis test parameters. (Reproduced with permission from [66].)

| Parameters | DL1 | DL2 | DL3 |
| --- | --- | --- | --- |
| Test statistic | 405.6928 | 405.9889 | 405.9087 |
| Corrected for ties $H_t$ | 405.6930 | 405.9889 | 405.9088 |
| DF | 404 | 403 | 403 |
| Significance level | $p = 0.466956$ | $p = 0.448880$ | $p = 0.449992$ |

**Table 8.13.** Friedman test parameters. (Reproduced with permission from [66].)

| Parameters | DL1 | GT1 | DL2 | GT2 | DL3 | GT3 |
| --- | --- | --- | --- | --- | --- | --- |
| N | 407 | 407 | 407 | 407 | 407 | 407 |
| Minimum | 5.3298 | 4.0761 | 3.8359 | 5.323 | 5.2161 | 5.0032 |
| 25th percentile | 13.063 | 11.416 | 10.705 | 10.597 | 13.698 | 12.126 |
| Median | 18.114 | 15.893 | 14.548 | 15.086 | 18.391 | 16.8 |
| 75th percentile | 24.88 | 22.16 | 21.434 | 21.398 | 25.231 | 22.562 |
| Maximum | 52.644 | 53.519 | 54.808 | 52.663 | 53.611 | 52.189 |
| $F$ | 60.9085 | | 49.0025 | | 40.8077 | |
| Significance | $p < 0.00001$ | | $p < 0.00001$ | | $p < 0.00001$ | |

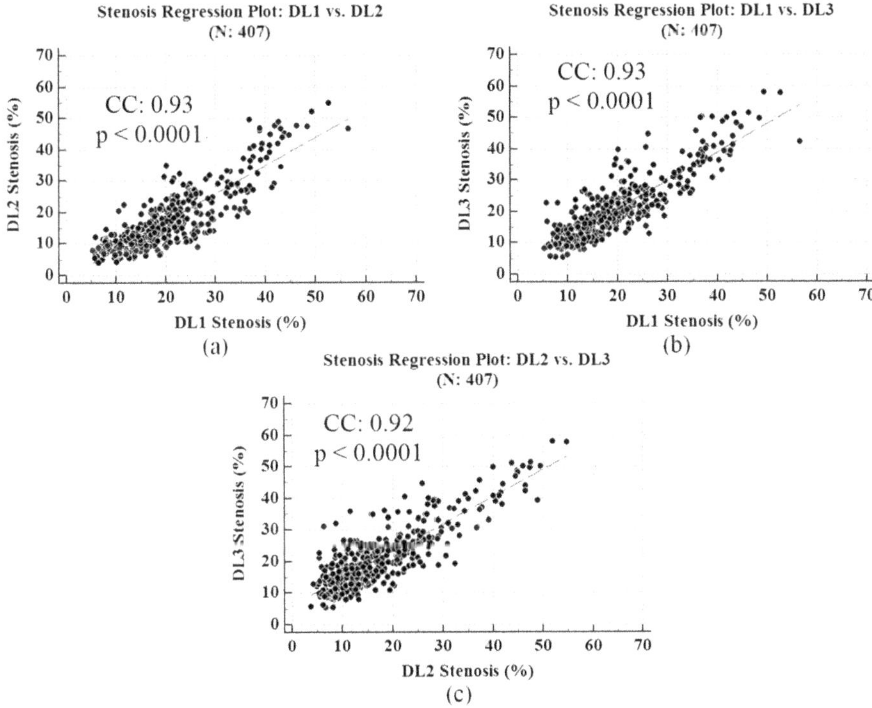



**Figure 8.14.** The regression plot for (a) DL1 and DL2, (b) DL2 and DL3, and (c) DL2 and DL3. (Reproduced with permission from [66].)

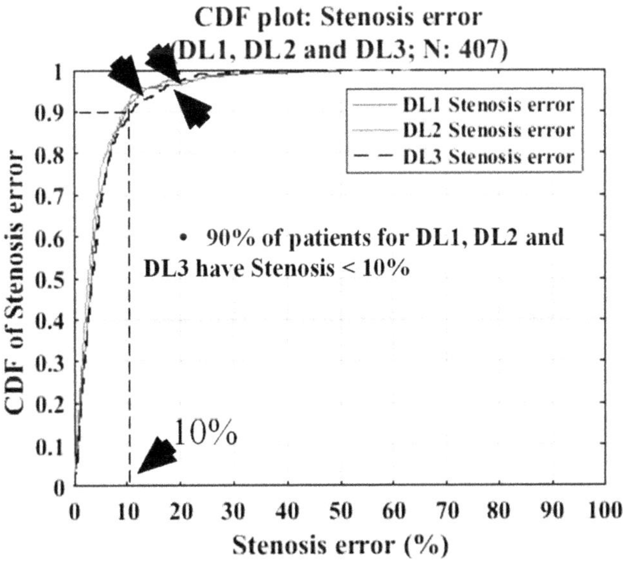



**Figure 8.15.** CDF of stenosis error bias for DL1, DL2, and DL3. (Reproduced with permission from [66].)

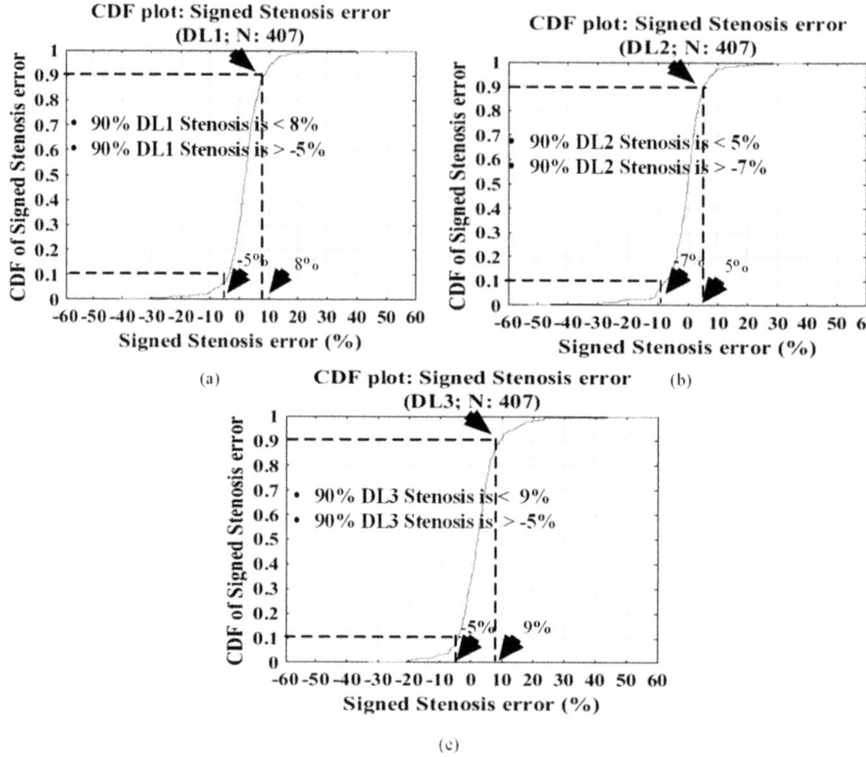


**Figure 8.16.** (a) DL, (b) DL2, and (c) DL3 CDF of signed stenosis error bias. (Reproduced with permission from [66].)

high risk patients had stenosis >50%. The DL1, DL2, and DL3 ROC plot is shown in figure 8.17. A detailed analysis is provided in the discussion section.

## 8.6 Discussion

In this study DL based stenosis measurement was performed from ultrasound carotid scans. The DL based model showed great accuracy and robustness in detecting stenosis. The DL model consisted of a 13-layer encoder for high level feature mining. It contained a further three-layer decoder for up-scaling the features to the dimension of the GT for training and testing. The DL model used merging of feature sets from the intermediate encoder and decoder layers to retain spatial information lost during down-sampling. Three radiologists were used for tracing the GT. The system consisted four stages: pre-processing, DL segmentation, boundary delineation, and stenosis measurement with the DL being the chief part. The stenosis was measured by employing the NASCET criteria, and risk analysis was performed. The AUC values for risk categorization for the three instances of DL were 0.90, 0.94, and 0.86, respectively. The examples of low, medium, and high risk classes are shown in figure 8.18. Benchmarking is presented in the next section.

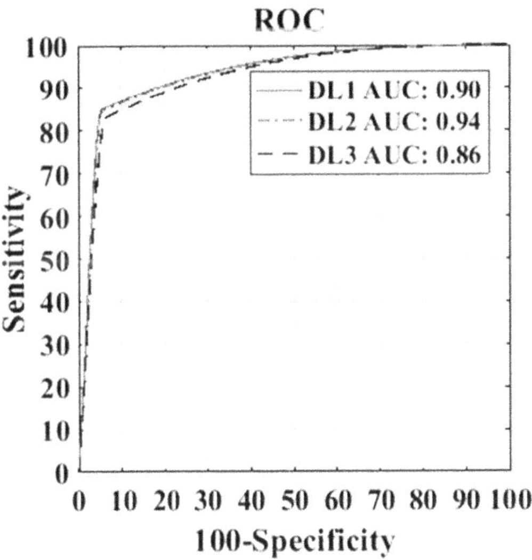

**Figure 8.17.** DL1, DL2, and DL3 ROC plot. (Reproduced with permission from [66].)

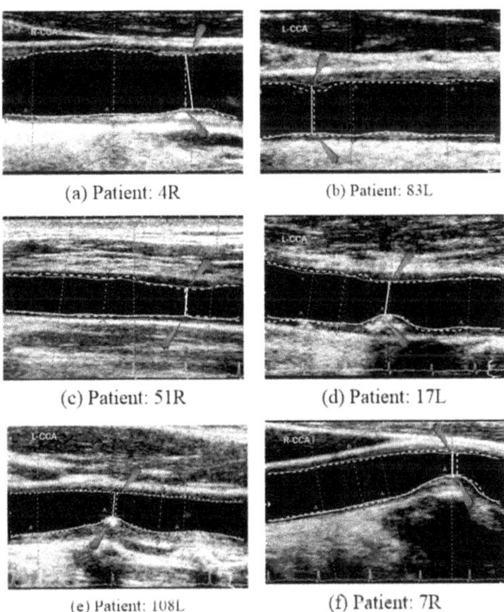

**Figure 8.18.** The GT is represented by yellow dashed lines, while the deep learning is represented by a red line and green line for the LI near and far walls, respectively. The white dashed line shows GT stenosis, and the white line shows DL stenosis. Row #1 depicts patients with low risk stenosis (a), (b); row #2 depicts patients with intermediate risk stenosis (c), (d); and row #3 depicts patients with high risk stenosis (e), (f). (Reproduced with permission from [66].)

**Table 8.14.** LD error bias benchmarking table. (Reproduced with permission from [66].)

| Authors | Method | # Images | LD error bias (mm) | PoM (%) |
|---|---|---|---|---|
| Saba *et al* (2015) [60] | Gaussian filter + spectral analysis (GSA) | 404 | GSA1: 0.25 ± 0.24<br>GSA2: 0.27 ± 0.26 | |
| Araki *et al* (2016) [47] | Region-based (RB) | 300 | | RB: 97.9 |
| Araki *et al* (2016) [47] | Boundary-based (BB) | 300 | | BB: 85.2 |
| Araki *et al* (2016) [61] | Spectral analysis | 404 | SA1: 0.26 ± 0.28<br>SA2: 0.27 ± 0.26 | |
| Krishna *et al* (2017) [35] | Scale space | 404 | SS1: 0.25 ± 0.24<br>SS2: 0.27 ± 0.25 | SS1: 95.9<br>SS2: 95.1 |
| Saba *et al* (2017) [48] | Level set + region + boundary (LSRB) | 100 | | LSRB: 96.2 |
| Proposed | Deep learning | 407 | DL1: 0.19 ± 0.27<br>DL2: 0.23 ± 0.23<br>DL3: 0.21 ± 0.19 | DL1: 96.7<br>DL2: 96.1<br>DL3: 96.4 |

*LD, stenosis, and risk benchmarking*

Earlier models used various image features such as texture, brightness, contrast, and color for delineating boundaries. In this respect, El Barghouty *et al* [59] used a grayscale median (GSM) to characterize low and high risk. In another work, Acharya *et al* [40] used an ML based model for approximation of risk from plaque images. DL inspired stenosis based risk investigation was presented for the first time in that paper. Regarding lumen detection and stenosis based stroke risk analysis, some works were performed, which are listed in tables 8.14 and 8.15, respectively.

## 8.7 Conclusion

In this work, several new aspects of DL based LD and stenosis measurement were presented. CVD risk analysis was also performed. The study showed that DL based lumen detection is a powerful instrument for both stenosis estimation and risk analysis. In the future DL based tools can be employed in diagnostic centers for clinical diagnosis and prognosis. These tools can help clinicians to make faster and more accurate decisions and prescribe a treatment plan for quick recovery and CVD avoidance.

**Table 8.15.** Stenosis benchmarking table. (Reproduced with permission from [66].)

| Authors | Method | # Patients | Stenosis cut-offs |
|---|---|---|---|
| Bruno *et al* (2001) [62] | Manual | 22 | • Risk: SSI < 50% |
| Nicholaidas *et al* (2005) [63] | Manual | 1115 | • Low risk: SSI <30%<br>• Mild risk: 30%< SSI < 49%<br>• High risk: SSI >50% |
| Kakkos *et al* (2008) [64] | Manual | 821 | • Risk: 60%<SSI<99% |
| Schieneider *et al* (2010) [65] | Manual | | • Risk: 60%<SSI<99% |
| Proposed | Deep learning | 204 | • Low risk: SSI <25%<br>• Mild risk: 25%< SSI <50%<br>• High risk: SSI >50% |

# References


[1] Virani S S, Alonso A, Benjamin E J, Bittencourt M S, Callaway C W, Carson A P, Chamberlain A M, Chang A R, Cheng S and Delling F N 2020 Heart disease and stroke statistics—2020 update: a report from the American Heart Association *Circulation* E139–596

[2] https://www.cdc.gov/nchs/hus/data-finder.htm?&subject=Heart%20disease

[3] Suri J S, Kathuria C and Molinari F 2010 *Atherosclerosis Disease Management* (Berlin: Springer)

[4] Saba L, Sanches J M, Pedro L M and Suri J S 2014 *Multi-modality Atherosclerosis Imaging and Diagnosis* (Berlin: Springer)

[5] Spence J D 2015 Management of asymptomatic carotid stenosis *Neurol. Clin.* **33** 443–57

[6] Carr S, Farb A, Pearce W H, Virmani R and Yao J S 1996 Atherosclerotic plaque rupture in symptomatic carotid artery stenosis *J. Vasc. Surg.* **23** 755–66

[7] Orho-Melander M 2015 Genetics of coronary heart disease: towards causal mechanisms, novel drug targets and more personalized prevention *J. Int. Med.* **278** 433–46

[8] Prabhakaran D, Jeemon P and Roy A 2016 Cardiovascular diseases in India: current epidemiology and future directions *Circulation* **133** 1605–20

[9] Pandey A K, Blaha M J, Sharma K, Rivera J, Budoff M J, Blankstein R, Al-Mallah M, Wong N D, Shaw L and Carr J 2014 Family history of coronary heart disease and the incidence and progression of coronary artery calcification: Multi-Ethnic Study of Atherosclerosis (MESA) *Atherosclerosis* **232** 369–76

[10] Saba L, Mallarini G, Sanfilippo R, Zeng G, Montisci R and Suri J 2012 Intima media thickness variability (IMTV) and its association with cerebrovascular events: a novel marker of carotid therosclerosis? *Cardiovasc. Diag. Ther.* **2** 10

[11] Molinari F, Zeng G and Suri J S 2010 A state of the art review on intima–media thickness (IMT) measurement and wall segmentation techniques for carotid ultrasound *Comput. Methods Programs Biomed.* **100** 201–21

[12] Ayer J, Charakida M, Deanfield J E and Celermajer D S 2015 Lifetime risk: childhood obesity and cardiovascular risk *Eur. Heart J.* **36** 1371–6

[13] D'Alessandro A and De Pergola G 2015 Mediterranean diet and cardiovascular disease: a critical evaluation of *a priori* dietary indexes *Nutrients* **7** 7863–88

[14] Critchley J A and Capewell S 2003 Mortality risk reduction associated with smoking cessation in patients with coronary heart disease: a systematic review *JAMA* **290** 86–97

[15] Keating S T, Plutzky J and El-Osta A 2016 Epigenetic changes in diabetes and cardiovascular risk *Circ. Res.* **118** 1706–22

[16] Haskell W L, Lee I-M, Pate R R, Powell K E, Blair S N, Franklin B A, Macera C A, Heath G W, Thompson P D and Bauman A 2007 Physical activity and public health: updated recommendation for adults from the American College of Sports Medicine and the American Heart Association *Circulation* **116** 1081

[17] Hyman L, Schachat A P, He Q and Leske M C 2000 Hypertension, cardiovascular disease, and age-related macular degeneration *Arch. Ophthalmol.* **118** 351–8

[18] Newby D E, Mannucci P M, Tell G S, Baccarelli A A, Brook R D, Donaldson K, Forastiere F, Franchini M, Franco O H and Graham I 2015 Expert position paper on air pollution and cardiovascular disease *Eur. Heart J.* **36** 83–93

[19] Sanches J M, Laine A F and Suri J S 2012 *Ultrasound Imaging* (Berlin: Springer)

[20] Das K, Singh P, Chawla Y, Duseja A, Dhiman R K and Suri S 2008 Magnetic resonance imaging of brain in patients with cirrhotic and non-cirrhotic portal hypertension *Dig. Dis. Sci.* **53** 2793–8

[21] Saba L and Suri J S 2019 *Multi-Detector CT Imaging Handbook* (Boca Raton, FL: CRC Press)

[22] El-Baz A and Suri J S 2011 *Lung Imaging and Computer Aided Diagnosis* (Boca Raton, FL: CRC Press)

[23] Saba L and Suri J S 2013 *Multi-Detector CT Imaging: Principles, Head, Neck, and Vascular Systems* vol 1 (Boca Raton, FL: CRC Press)

[24] Seabra J and Sanches J 2012 *Ultrasound Imaging: Advances and Applications* (New York: Springer)

[25] Saba L, Acharya U R, Guerriero S and Suri J S 2014 *Ovarian Neoplasm Imaging* (Berlin: Springer)

[26] Suri J S 2008 *Advances in Diagnostic and Therapeutic Ultrasound Imaging* (Boston, MA: Artech House)

[27] Jensen J A 1996 *Estimation of Blood Velocities using Ultrasound: A Signal Processing Approach* (Cambridge: Cambridge University Press)

[28] Mehra S 2010 Role of duplex Doppler sonography in arterial stenoses *J. Indian Acad. Clin. Med.* **11** 294–9

[29] Jones S, Leclerc H, Chatzimavroudis G P, Kim Y, Scott N and Yoganathan A P 1996 The influence of acoustic impedance mismatch on poststenotic pulsed-Doppler ultrasound measurements in a coronary artery model *Ultrasound Med. Biol.* **22** 623–34

[30] Mitchell D G 1990 Color Doppler imaging: principles, limitations, and artifacts *Radiology* **177** 1–10

[31] Bjällmark A, Lind B, Peolsson M, Shahgaldi K, Brodin L-Å and Nowak J 2010 Ultrasonographic strain imaging is superior to conventional non-invasive measures of vascular stiffness in the detection of age-dependent differences in the mechanical properties of the common carotid artery *Eur. J. Echocardiogr.* **11** 630–6

[32] Benthin M, Dahl P, Ruzicka R and Lindström K 1991 Calculation of pulse-wave velocity using cross correlation—effects of reflexes in the arterial tree *Ultrasound Med. Biol.* **17** 461–9

[33] Tole N M and Ostensen H 2005 *Basic Physics of Ultrasonographic Imaging* (Geneva: World Health Organization)

[34] Londhe N D and Suri J S 2016 Superharmonic imaging for medical ultrasound: a review *J. Med. Syst.* **40** 279

[35] Kumar P K, Araki T, Rajan J, Saba L, Lavra F, Ikeda N, Sharma A M, Shafique S, Nicolaides A and Laird J R 2017 Accurate LD measurement in curved vessels in carotid ultrasound: an iterative scale-space and spatial transformation approach *Med. Biol. Eng. Comput.* **55** 1415–34

[36] Biswas M, Kuppili V, Saba L, Edla D R, Suri H S, Sharma A, Cuadrado-Godia E, Laird J R, Nicolaides A and Suri J S 2019 Deep learning fully convolution network for lumen characterization in diabetic patients using carotid ultrasound: a tool for stroke risk *Med. Biol. Eng. Comput.* **57** 543–64

[37] Molinari F, Meiburger K M, Saba L, Zeng G, Acharya U R, Ledda M, Nicolaides A and Suri J S 2012 Fully automated dual-snake formulation for carotid intima-media thickness measurement: a new approach *J. Ultrasound Med.* **31** 1123–36

[38] Biswas M, Kuppili V, Araki T, Edla D R, Godia E C, Saba L, Suri H S, Omerzu T, Laird J R and Khanna N N 2018 Deep learning strategy for accurate carotid intima-media thickness measurement: an ultrasound study on Japanese diabetic cohort *Comput. Biol. Med.* **98** 100–17

[39] Acharya U R, Faust O, Sree S V, Molinari F, Saba L, Nicolaides A and Suri J S 2011 An accurate and generalized approach to plaque characterization in 346 carotid ultrasound scans *IEEE Trans. Instrum. Meas.* **61** 1045–53

[40] Acharya U R, Krishnan M M R, Sree S V, Sanches J, Shafique S, Nicolaides A, Pedro L M and Suri J S 2012 Plaque tissue characterization and classification in ultrasound carotid scans: a paradigm for vascular feature amalgamation *IEEE Trans. Instrum. Meas.* **62** 392–400

[41] Ikeda N, Araki T, Sugi K, Nakamura M, Deidda M, Molinari F, Meiburger K M, Acharya U R, Saba L and Bassareo P P 2014 Ankle–brachial index and its link to automated carotid ultrasound measurement of intima–media thickness variability in 500 Japanese coronary artery disease patients *Curr. Atheroscler. Rep.* **16** 393

[42] Saba L, Ikeda N, Deidda M, Araki T, Molinari F, Meiburger K M, Acharya U R, Nagashima Y, Mercuro G and Nakano M 2013 Association of automated carotid IMT measurement and HbA1c in Japanese patients with coronary artery disease *Diabetes Res. Clin. Pract.* **100** 348–53

[43] Ikeda N, Gupta A, Dey N, Bose S, Shafique S, Arak T, Godia E C, Saba L, Laird J R and Nicolaides A 2015 Improved correlation between carotid and coronary atherosclerosis SYNTAX score using automated ultrasound carotid bulb plaque IMT measurement *Ultrasound Med. Biol.* **41** 1247–62

[44] Fox A J 1993 How to measure carotid stenosis *Radiology* **186** 316–8

[45] Beebe II G, Salles-Cunha S X, Scissons R P, Dosick S M, Whalen R C, Gale S S, Pigott J P and Seiwert A J 1999 Carotid arterial ultrasound scan imaging: a direct approach to stenosis measurement *J. Vasc. Surg.* **29** 838–44

[46] Nicolaides A N, Kakkos S K, Kyriacou E, Griffin M, Sabetai M, Thomas D J, Tegos T, Geroulakos G, Labropoulos N and Doré C J 2010 Asymptomatic internal carotid artery stenosis and cerebrovascular risk stratification *J. Vasc. Surg.* **52** 1486–96

[47] Araki T, Kumar P K, Suri H S, Ikeda N, Gupta A, Saba L, Rajan J, Lavra F, Sharma A M and Shafique S 2016 Two automated techniques for carotid LD measurement: regional versus boundary approaches *J. Med. Syst.* **40** 182

[48] Saba L, Banchhor S K, Londhe N D, Araki T, Laird J R, Gupta A, Nicolaides A and Suri J S 2017 Web-based accurate measurements of carotid LD and stenosis severity: an ultrasound-based clinical tool for stroke risk assessment during multicenter clinical trials *Comput. Biol. Med.* **91** 306–17

[49] Simonyan K and Zisserman A 2014 Very deep convolutional networks for large-scale image recognition, arXiv:1409.1556

[50] Saba L, Banchhor S K, Suri H S, Londhe N D, Araki T, Ikeda N, Viskovic K, Shafique S, Laird J R and Gupta A 2016 Accurate cloud-based smart IMT measurement, its validation and stroke risk stratification in carotid ultrasound: a web-based point-of-care tool for multicenter clinical trial *Comput. Biol. Med.* **75** 217–34

[51] Molinari F, Pattichis C S, Zeng G, Saba L, Acharya U R, Sanfilippo R, Nicolaides A and Suri J S 2011 Completely automated multiresolution edge snapper—a new technique for an accurate carotid ultrasound IMT measurement: clinical validation and benchmarking on a multi-institutional database *IEEE Trans. Image Process.* **21** 1211–22

[52] Molinari F, Meiburger K M, Saba L, Acharya U R, Ledda G, Zeng G, Ho S Y S, Ahuja A T, Ho S C and Nicolaides A 2012 Ultrasound IMT measurement on a multi-ethnic and multi-institutional database: our review and experience using four fully automated and one semi-automated methods *Comput. Methods Programs Biomed.* **108** 946–60

[53] Teichmann M, Weber M, Zoellner M, Cipolla R and Urtasun R 2018 Multinet: real-time joint semantic reasoning for autonomous driving *2018 IEEE Intelligent Vehicles Symp.* (Piscataway, NJ: IEEE) pp 1013–20

[54] Long J, Shelhamer E and Darrell T 2015 Fully convolutional networks for semantic segmentation *Proc. of the IEEE Conf. on Computer Vision and Pattern Recognition* pp 3431–40

[55] LeCun Y, Bengio Y and Hinton G 2015 Deep learning *Nature* **521** 436–44

[56] Srivastava N, Hinton G, Krizhevsky A, Sutskever I and Salakhutdinov R 2014 Dropout: a simple way to prevent neural networks from overfitting *J. Mach. Learn. Res.* **15** 1929–58

[57] Abbott A L and Nicolaides A N 2015 Improving outcomes in patients with carotid stenosis: call for better research opportunities and standards *Am. Heart Assoc.* **46** 7–8

[58] Craig D, Meguro K, Watridge C, Robertson J, Barnett H and Fox A 1982 Intracranial internal carotid artery stenosis *Stroke* **13** 825–8

[59] El-Barghouty N, Nicolaides A, Bahal V, Geroulakos G and Androulakis A 1996 The identification of the high risk carotid plaque *Eur. J. Vasc. Endovasc. Surg.* **11** 470–8

[60] Saba L, Than J C M, Noor N M, Rijal O M, Kassim R M, Yunus A, Ng C R and Suri J S 2016 Inter-observer variability analysis of automatic lung delineation in normal and disease patients *J. Med. Syst.* **40** 142

[61] Araki T, Kumar A M, Krishna Kumar P, Gupta A, Saba L, Rajan J, Lavra F, Sharma A M, Shafique S and Nicolaides A 2016 Ultrasound-based automated carotid LD/stenosis measurement and its validation system *J. Vasc. Ultrasound* **40** 120–34

[62] Randoux B, Marro B, Koskas F, Duyme M, Sahel M, Zouaoui A and Marsault C 2001 Carotid artery stenosis: prospective comparison of CT, three-dimensional gadolinium-enhanced MR, and conventional angiography *Radiology* **220** 179–85

[63] Nicolaides A, Kakkos S, Griffin M, Sabetai M, Dhanjil S, Tegos T, Thomas D, Giannoukas A, Geroulakos G and Georgiou N 2005 Severity of asymptomatic carotid stenosis and risk of ipsilateral hemispheric ischaemic events: results from the ACSRS study *Eur. J. Vasc. Endovasc. Surg.* **30** 275–84

[64] Kakkos S K, Sabetai M, Tegos T, Stevens J, Thomas D, Griffin M, Geroulakos G and Nicolaides A N 2009 Silent embolic infarcts on computed tomography brain scans and risk of ipsilateral hemispheric events in patients with asymptomatic internal carotid artery stenosis *J. Vasc. Surg.* **49** 902–9

[65] Schneider P A and Naylor A R 2010 Asymptomatic carotid artery stenosis—medical therapy alone versus medical therapy plus carotid endarterectomy or stenting *J. Vasc. Surg.* **52** 499–507

[66] Saba L 2019 Ultrasound-based carotid stenosis measurement and risk stratification in diabetic cohort: a deep learning paradigm *Cardiovasc. Diagn. Ther.* **9** 439

# Chapter 9

# A systematic review of conventional and deep learning models for the measurement of plaque burden

**Mainak Biswas and Jasjit S Suri**

Cardiovascular diseases (CVDs) are the primary cause of death in the world. CVDs are mainly caused by an artery inflammation disease called atherosclerosis. The degree of arterial damage can be assessed by two major image phenotypes, namely, carotid intima–media thickness (cIMT) and plaque area (PA). Previously image segmentation techniques were based on conventional image processing and semi-automated digital imaging solutions. These methods are unreliable, tedious, slow, and not robust. In this study we look into artificial intelligence (AI) based automated technologies for cIMT and PA measurement.

Machine learning (ML) and deep learning (DL) technologies are AI techniques generally used in the area of cIMT and PA measurement from ultrasound images. ML and DL are supervised learning techniques where learning takes place from learned sample data. In this review we look into several factors such as (i) the impact of AI based learning of cIMT and PA measurement, (ii) the formal definition of ML/DL models, and (iii) plaque segmentation techniques for cIMT/PA measurement using (a) ROI detection and (b) lumen–intima and media–adventitia border estimation.

## 9.1 Introduction

Cardiovascular diseases accounted for nearly 17.6 million deaths globally in 2016, which was almost 14.6% higher than in 2006 [1]. The increase in CVD deaths can be attributed to increased use of tobacco, physical inactivity, obesity, high blood pressure, diabetes, arthritis, coronary artery disease etc. This results in increased healthcare costs. The total estimated direct and indirect cost to the USA was 351.2 billion dollars in 2014–15, adjusted for inflation [1, 2]. Atherosclerosis, an arterial inflammation disease, is the main cause of CVDs [3, 4]. Atherosclerosis is initiated by endothelium dysfunction [4, 5], where the thin lining of cells within the interior surface of blood vessels becomes damaged. This leads to an influx of blood effluents entering the thin arterial wall leading to inflammation [1, 6–8]. Several studies over the years have shown atherosclerosis progress in the carotid arteries. In a recent study on 68 asymptomatic patients with over 50% stenosis, it was shown that the wall area increases at 2.2% per year over a period of 18 months [9]. A study on another group of 250 patients with 40%–99% stenosis, showed a high percentage of lipid and a high risk of cerebral infarction (hazard ratio = 4.4) [10]. All these studies confirm the presence of a ticking time bomb where the artery after a certain period cannot withstand the plaque burden, leading first to rupture in the fibrous cap, next to thrombosis, and finally stroke [11–13] (shown in figure 9.1). Atherosclerosis is related to neuronal diseases such as dementia [14] and Alzheimer's [17], and arthritis [20], coronary artery diseases [21], leukoaraiosis [15, 16], and renal diseases [18, 19].

The standard process of quantifying atherosclerosis is measuring the carotid intima–media thickness (cIMT) and plaque area (PA). This procedure is also known as atherosclerotic disease monitoring or vascular screening [22]. cIMT is the measured distance between the media–adventitia (MA) and lumen–intima (LI) borders [23, 24]. The area between these two borders is called the PA [25–32]. Recently, cIMT and PA measurement has been used to compute composite risk scores. The traditional risk scores did not cover these image biomarkers, i.e. Framingham [33], ASCVD [34], Reynold risk score (RRS) [35], United Kingdom Prospective Diabetes Study (UKPDS56) [36], UKPDS60 [37], QRISK2 [38], and Joint British Societies (JBS3) [39], etc. The new age risk predictors such as AtheroEdge Composite Risk Score (AECRS 1.0) [20] was developed by Suri and his team. AECRS 1.0 uses an morphology based CVD prediction tool which includes both convention and image based phenotypes such as cIMT and PA. These image based phenotypes are computed based on ultrasound scanning of carotid segments such as the common carotid artery (CCA), bulb, or internal carotid artery acquired using 2D B-mode ultrasound [5, 40, 41]. Suri and his team developed

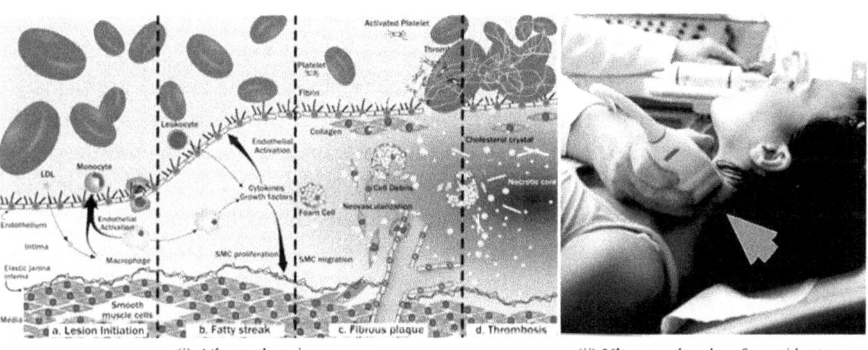

(i) Atherosclerosis progress.      (ii) Ultrasound probe of carotid artery.

**Figure 9.1.** (i) Atherosclerosis progression consisting of (a) endothelium dysfunction and lesion initiation, (b) formation of a fatty streak, (c) formation of a fibrous plaque underlying a fibrous cap, and (d) rupture and thrombosis. (ii) Ultrasound scanning of the carotid artery. (Both images courtesy of AtheroPoint™, Roseville, CA, USA. Reproduced with permission of Springer from [162]. Copyright 2021 Society for Imaging Informatics in Medicine.)

another ten-year risk calculator AECRS 2.0 based on conventional, blood-based biomarkers and image phenotypes. Since the image based phenotypes are becoming an integral part of these new age calculators, a better understanding is required to understand the evolution of automated techniques for cIMT/PA measurement in the context of artificial intelligence (AI). Several research groups have contributed their findings in the area of automated cIMT/PA measurement. These research groups have used various technique for cIMT/PA measurement. These research groups work on various facades of cIMT imaging, i.e. static (image) to dynamic (video). These models vary from basic morphology threshold based approaches to current deep learning (DL) paradigms. The research group findings can be categorized into three generations based on ROI segmentation and LI/MA border delineation technologies [42]. The first generation were primary image processing techniques to detect LI/MA borders and then use caliper solutions to compute the mean distance. The second generation used contour based methods consisting of para-metric/geometric curves. Many of these techniques are semi-automated [43–46]. Signal-based methods such as scale-space and deformable models also belong to this generation under the AtheroEdge™ models category. These models are fully automated and commercially available [47, 48]. The third-generation technologies are based on AI paradigms of machine learning (ML) [49] and DL [50] for cIMT/PA measurement. All the generations use the same distance measurement techniques such as Hausdorff distance [51] and 'Suri's polyline distance method' (PDM) [52]. In this review we study various aspect of third-generation AI based learning of cIMT/ PA measurement.

Section 9.2 covers all generations of cIMT/PA techniques in detail. Section 9.3 is dedicated to AI based automated techniques covering ML and DL based applications. Section 9.4 provides quantification techniques. The chapter is concluded in section 9.5.

## 9.2 Chronological generation of cIMT regional segmentation and cIMT measurement

cIMT is a surrogate biomarker of carotid/artery disease, as stated previously [53]. Several technologies were developed in the first generation for low-level segmentation tasks for measuring the cIMT/PA. These techniques were dynamic programming (DP) [54], the Hough transform (HT) [55], Nakagami mixture modeling [56], active contour [57], edge detection [58] and gradient-based techniques [59]. All these techniques are categorized under the class of computer-aided diagnosis [40, 60, 61– 66]. The DP approach aims to minimize the cost function. This cost function is the weighted summation of local estimations of the echo intensity, intensity gradient, and boundary continuity. Each pixel point of the polyline border is considered. The Hough transform of a line is a point in the $(s, \theta)$ plane where all points map to a single point. This basic principle is used to find line segments through different edge points. Polyline distance methods are used for computing cIMT from the boundary vertices [54]. The Hough transform has been used in various works such as by Golemati *et al* [67], Stoitis *et al* [68], Petroudi *et al* [69]. In another work by

Destrempes *et al* [56], the Nakagami distribution and motion estimation were used to estimate the cIMT. The first-generation techniques used signal processing techniques such as scale-space to detect the LI–MA boundaries [46, 70]. The second-generation technologies sometimes built on the first-generation techniques to augment their performance [24, 71–73].

Active contour models evolved in the second-generation techniques which involved fitting the contour as per local image data. Snakes is considered as an example of the active parametric contours used for LI/MA estimation and consecutively cIMT measurement [14, 74]. Curve fitting models were another legacy cIMT estimation method [75]. Level sets also belong to this generation where they propagate zero-level curves to settle at the interfaces at the LI and MA borders [76]. The first- and second-generation techniques have also been fused together for achieving better performance [46, 70]. Several second-generation techniques used augmented versions of first-generation techniques [24, 71–73]. The edge detection techniques of the second generation use a variation of gray levels and image gradient to delineate the LI and MA borders [77, 78]. In another work by Elisa *et al* [27] an automated software AtheroEdge™ system was used to delineate the LI and MA borders. The software [19, 66, 79] is based on scale-space for computing the cIMT and PA after LI and MA delineation. The third generation is based on intelligence based techniques such as ML [49] and DL [50] methods which are mainly based on neural network models. The ML models are a two-stage process where manually selected features are extracted in the first phase while the actual statistical model is applied in the second stage. The DL models on the other hand are single-stage systems where feature extraction and characterization occurs together. Both ML and DL technologies have been instrumental in developing numerous CVD risk systems with high diagnostic accuracy [80, 81]. All the generations of cIMT and PA estimation techniques are shown in figure 9.2. A detailed discussion of ML and DL technologies is provided next.

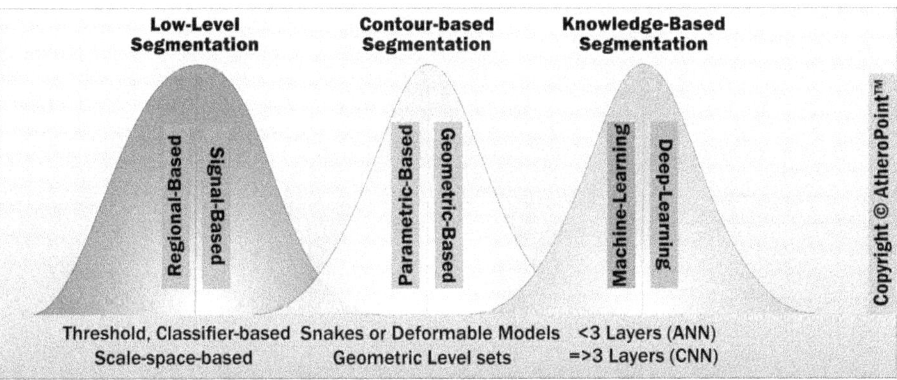



**Figure 9.2.** Three generations of cIMT/PA measurement evolution (color image). (Image courtesy of AtheroPoint™, Roseville, CA, USA. Reproduced with permission of Springer from [162]. Copyright 2021 Society for Imaging Informatics in Medicine.)

The mechanism for deriving features from instances differs between ML and DL. Feature extraction is independent of the actual model of classification in ML (details in appendix A) and done by hand (shown in figure 9.3), whereas feature extraction and characterization are indistinguishable in DL. The majority of ML classification models emerged as a single statistical learning and inference technique [82, 83]. The DL models, on the other hand, extract characteristics and produce predictions using multiple layers of statistical learning and inference approaches. As a result, when applied to a very large dataset, DL approaches are more expensive in terms of computing time and space, but they are more robust and, in certain situations, deliver superior accuracy [84]. When compared to DL, ML approaches are more time and cost efficient. For better results, these two procedures are frequently combined. The way each of these generations dealt with ultrasonic picture noise (speckle noise, scattering noise) is an interesting feature. Various denoising techniques such as Gaussian filters, anisotropic diffusion, smoothing, etc, were utilized in the first two generations [64, 65, 75, 85]. Table 9.1 provides a brief overview of technology by generation. Although technologies were split by generation, the

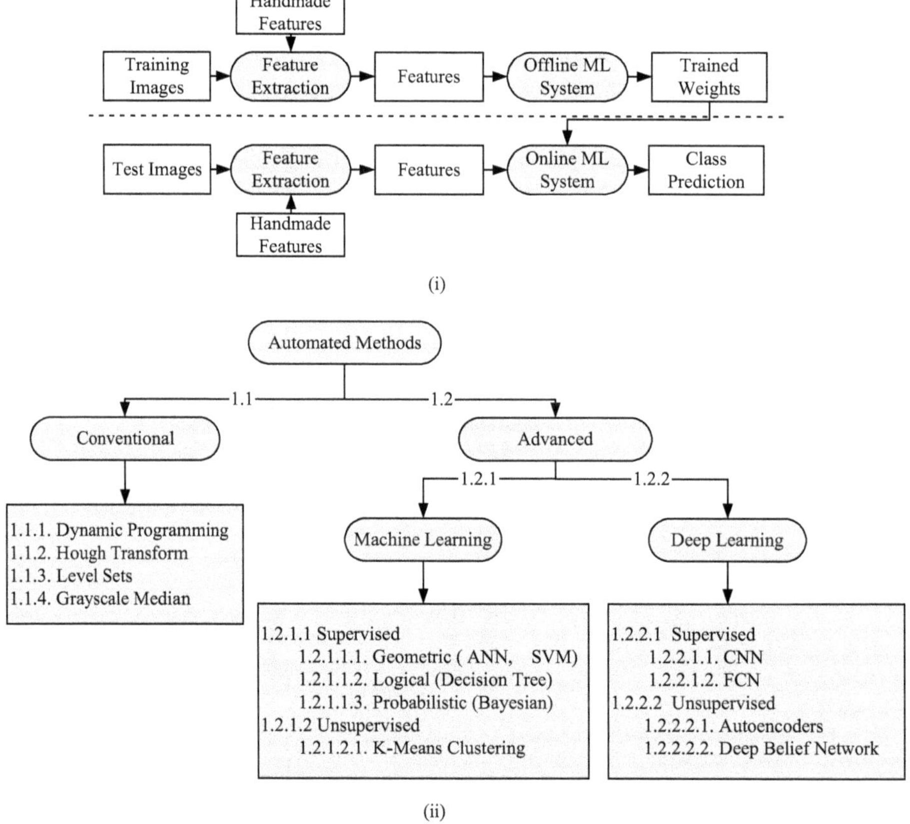

Figure 9.3. (i) Generalized ML model and (ii) classification tree of different automated models. (Reproduced with permission of Springer from [162]. Copyright 2021 Society for Imaging Informatics in Medicine.)

**Table 9.1.** The three generations of cIMT regional segmentation. (Reproduced with permission of Springer from [162]. Copyright 2021 Society for Imaging Informatics in Medicine.)

| SN Attributes | First generation | Second generation | Third generation |
|---|---|---|---|
| 1 Fields of engineering | Conventional image processing, signal processing combined with image processing | Signal processing combined with image processing | Knowledge based engineering |
| 2 Computer vision methods or fusion of methods | Dynamic programming, Hough transform | Parametric (snakes) curves, geometric curves (level sets), scale-space, conventional classifiers | Machine learning and deep learning |
| 3 Distance measurement methods or cIMT measurement methods | Vertical, shortest distance along with the columns of the image | Polyline distance method, centerline distance method | Polyline distance method, centerline distance method |

compartments were not watertight. Many models were created by combining technologies from several generations in order to improve performance. Molinari *et al* [23] employed a mixture of first- and second-generation strategies to reduce the cIMT error in this case. After studies by Spence *et al* [86, 87], Mathiesen *et al* [88], and Saba *et al* [79] firmly proved that the PA biomarker is just as essential as the cIMT, the PA biomarker has received a lot of attention. If the LI/MA borders are known, PA can be estimated along with cIMT mathematically. It is calculated by counting all pixels between the LI and MA borders and calibrating the result to millimeters squared ($mm^2$). It has also been adapted in the AECRS 1.0 [89] and AECRS 2.0 [18] systems.

## 9.3 Ml application for cIMT and PA measurement

A basic overview of machine learning techniques is provided in appendix A. Figure 9.3 depicts the various AI approaches, with descriptions provided in table 9.2. This is divided into three categories: traditional, machine learning, and deep learning. In this section we we will go through the various machine learning approaches used for cIMT/PA segmentation and measurements from CCA photos. As previously indicated, features must be retrieved before the ML model can be used. Some of the features are used to train the model, such as training data ($X_{tr}$), while the rest are used to test the model, such as test data($X_{ts}$).

The model parameters are adjusted based on the error during the training phase, which is based on the known and real outputs. This training procedure is repeated until the model parameters converge or do not change their values any longer, at

**Table 9.2.** Comparison between different cIMT regional segmentation methods for given attributes. (Reproduced with permission of Springer from [162]. Copyright 2021 Society for Imaging Informatics in Medicine.)

| Attributes | Conventional methods (1.1) | Machine learning (1.2.1) | | Deep learning (1.2.2) | |
|---|---|---|---|---|---|
| Types | Active contour, level sets, edge detection, dynamic programming (1.1.*) | Supervised (1.2.1.1) | Geometric: ANN, SVM (1.2.1.1.1) Logical: decision trees (1.2.1.1.2) | Supervised (1.2.2.1) | CNN, FCN (1.2.2.1.*) |
| | | | Probabilistic: Bayesian (1.2.1.1.3) | Unsupervised (1.2.1.2) | Autoencoders, deep belief networks (1.2.2.2.*) |
| | | Unsupervised (1.2.1.2) | $K$-means clustering (1.2.1.2.1) | | |
| Description | Techniques employing imaging features and processing for the cIMT regional segmentation. | The two-stage intelligence-based paradigm in which image features are extracted and then used by the ML model for training and testing for cIMT region identification. | | Mimics the human visual cortex with numerous hierarchical networks of neurons extracting high-level features directly from images for recognition of the cIMT region. | |
| Advantages | 1. Fastest among all the methods for region estimation as no training and testing required. 2. Simple. | 1. Uses intelligence in the form of learning from experience/training and applies to unknown instances. 2. Somewhat generalized and can be applied to instances within a similar domain. 3. Faster than deep learning. | | 1. Independent of feature extraction algorithms. 2. Can be scaled up to recognize/characterize millions of images. 3. Can be generalized and used in multiple domains. | |
| Disadvantages | 1. Task-specific and cannot be generalized. 2. Low accuracy. | 1. Dependent on the quality of features for better accuracy. 2. The accuracy curve diminishes when scaled up to a large number of instances. | | 1. Costly in terms of computation time and memory. 2. Easily overfits and therefore requires different techniques for the prevention of overfitting. | |

*Other technologies.

which point training is considered complete. As a result, the model with learned parameters is put to the test with unknown outcomes. Once the model outputs are available, they are compared to actual outputs to determine the model's performance. Arrhythmia [90] and diabetes [91] are two examples of ML applications that leverage the concept of an offline and online system. We call them supervised learning since they all use the ground truth during cross-validation, and it this used a lot in cIMT and PA computation from CCA images. Rosa *et al* [92, 93] and Molinari *et al* [23] have two different schools of thought (SOTs) for segmenting the cIMT region from ultrasonography CCA images. Before applying the ML paradigm for cIMT segmentation, both SOTs extract the ROI. The following is a description of them.

### 9.3.1 ANN model for cIMT region detection

Rosa *et al* [92] obtained cIMT values from 60 B-mode ultrasonography CCA images in three steps. The CCA images were pre-processed to extract the ROI in stage 1, and stage 2 was the AI based classification step, which separated cIMT regional pixels from non-cIMT regional pixels, resulting in a binary picture. The determination of the LI and MA boundaries from the binary image was stage 3. The lower limit of the lumen was given as the posterior wall during the preprocessing stage, which involved using the CCA's watershed transform [94–96] to detect the lumen. The final ROI in the region was where the far wall's highest point was set to 0.6 mm above the binary lumen and the bottom boundary was set to 1.5 mm below the binary mask's lowest point. As seen in figure 9.4 (i), once the ROI has been recognized its dimensions are noted and retrieved from the original image. The retrieved ROIs are then used to demarcate the cIMT border in the next stage. The following stage uses an ensemble of four artificial neural networks (ANNs) [97] to distinguish between pixels in the cIMT and non-cIMT regions. As illustrated in figure 9.4(ii), the ANN is in a form of multi-layer perceptron with three layers: input, hidden, and output. The computation is performed by the nodes in the ANN model, while the 'learning experience' is encoded in the weights between the hidden-input and hidden-output layers. These weights are typically randomly initialized and converge to a stable value as the learning advances, based on feedback on error propagation. In general, the algorithm is multiplying weights by the input values and then applying a sigmoid function $\left( \alpha = \frac{1}{1+e^{-x}} \right)$. Each computed term is also given a bias term $\beta$. The following is the output function:

$$\hat{f}(x) = \alpha(\beta^{jy} + W^{jy}(\alpha(\beta^{ij} + W^{ij}))), \tag{9.1}$$

where $i$ and $j$ signify the weights between hidden-input and hidden-output layers, respectively. The weight values ($w_{ij}$) are optimized for each error propagation using the gradient $\left( \frac{\Delta \varepsilon}{\Delta w} \right)$, which is given by

$$w_{ij} = w_{ij} - \eta \frac{\Delta \varepsilon}{\Delta w}, \tag{9.2}$$

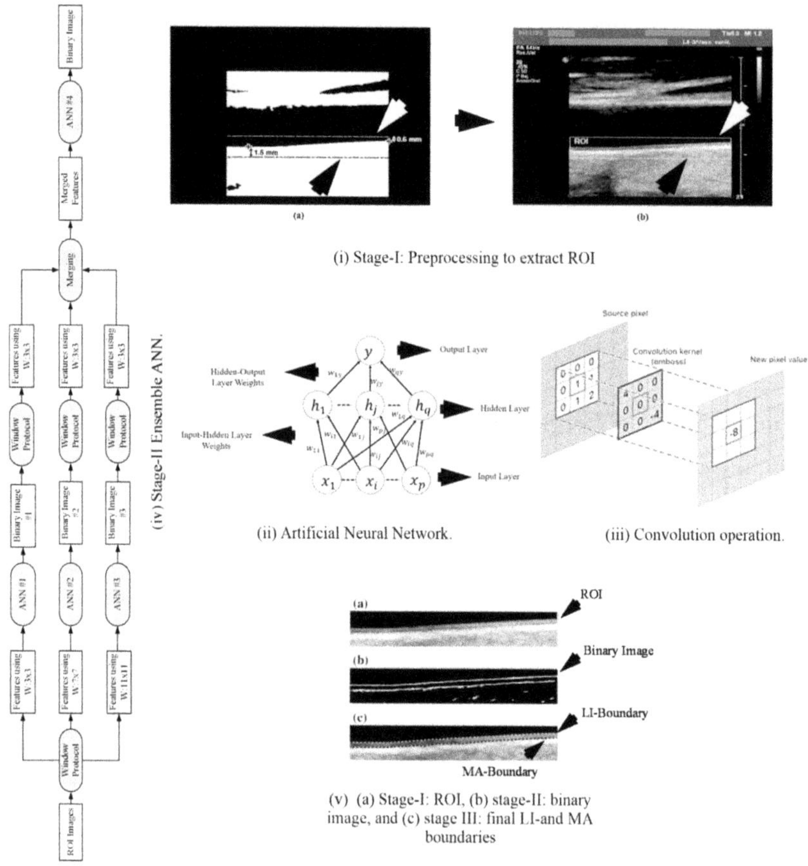



**Figure 9.4.** (i) ROI extraction, (ii) ANN, (iii) convolution operation, (iv) stage 2 ANN ensemble network, (v) outputs of [92]. ((i)–(iv) Reproduced with permission of Springer from [162]. Copyright 2021 Society for Imaging Informatics in Medicine. (v) Reproduced with permission of Springer from [92]. Copyright 2013 International Federation for Medical and Biological Engineering.

where $\eta$ is the learning parameter. The pixel intensities generated by a kernel process are the input pattern for each ANN. Three kernels of sizes $3 \times 3$, $7 \times 7$, and $11 \times 11$ were applied pixel by pixel, by shifting to collect contextual information on nearby pixel intensities for each image. Convolution is another name for the process, and figure 9.4 (iii) shows how it works with the $3 \times 3$ kernel. The ground truth was pixel class information gathered by manually segmenting CCA images and annotating whether or not each pixel represented a cIMT border. As a result, the inputs were transmitted through three ANNs for training and testing, respectively. Each of the three ANNs produces another reconstructed binary image as the output. Following that, the three output binary images are merged, and a third kernel of size $3 \times 3$ is applied to the merged picture before being fed into the fourth ANN to produce the final binary mask. Figure 9.4 (iv) depicts a representative model of the complete process. The final stage involves identifying the LI and MA borders (as shown in figure 9.4 (v)) and computing the cIMT. The polyline distance, mean absolute distance, and centerline distance were 0.03670 $\pm$ 0.02429 mm, 0.03763 $\pm$ 0.02518 mm, and 0.03683 $\pm$ 0.02450 mm, respectively.

### 9.3.2 Extreme learning machine radial basis neural network model for cIMT region detection

Rosa *et al* [93] estimated the cIMT region from 25 ultrasonography CCA images using a radial basis neural network (RBNN) [98]. RBNNs are feed-forward neural networks with a single layer of input, hidden, and output layers. The number of hidden layers in an RBNN is limited to one, which is one of the fundamental differences between it and ANNs. The basic working principle of RBNN lies in interpolating $r$ training points $x^r$ to their corresponding target variable $y^r$. The output function for an input $x^{ts}$ is given by

$$\hat{f}(x) = \sum_{r=1}^{N} w_r \varphi_r(\| x^{ts} - x^r \|), \tag{9.3}$$

where $\varphi_r$ is the Gaussian radial basis function implemented by the hidden layer, $w_r$ is the weight between the hidden and output layer, and $\varphi_r$ is the radial basis function which is given by

$$\varphi_r(\| x^{ts} - x^r \|) = \exp\left(\frac{\| x^{ts} - x^r \|^2}{2\vartheta^2}\right) \tag{9.4}$$

where $\vartheta$ represents the width. Equation (9.3) is reduced into a matrix notation,

$$W\varphi = T, \tag{9.5}$$

where $T$ is the target vector. The unknown weights can be found by using standard matrix inversion,

$$W = \varphi^{-1}T, \tag{9.6}$$

The optimally pruned extreme learning machine is used to initialize the Gaussian parameters of hidden nodes (the number of hidden nodes, centers, and the deviation of each radial unit) (OP-ELM) [99, 100]. Using the techniques described in subsection 9.3.1, the ground truth information and ROI of the 25 CCA images are recovered. The kernel generates the input pattern, just as in the prior technique [92]. Comparative research was conducted using kernel sizes ranging from 3 to 23. The kernel window size that was optimized was 19 $\times$ 19. Experienced radiologists manually traced the ground truth. For each CCA image, pixels from the LI boundary, pixels from the region between the LI and MA boundaries, and pixels from the MA wall were evaluated. Finally, as seen in figure 9.4(vi), each class of pixels was recovered and placed on the original image. The cIMT bias for this experiment was 0.065 $\pm$ 0.046 mm.

### 9.3.3 Fuzzy *K*-means classifier for cIMT region extraction

Molinari *et al* [23] proposed the idea of unsupervised fuzzy *K*-means clustering (FKMC) [101] to divide a CCA image into three parts: the plaque region, and the LI and MA borders. CULEXia is another name for this approach. The FKMC is unsupervised in the sense that the pixels have no ground truth. The FKMC algorithm is similar to the *K*-means algorithm [102] in that it initializes *K* random points that represent *K* cluster centers, and then assigns all data points in the dataset to each of the *K* clusters based on the nearest mean. The mean of all points for each cluster is represented by $c_j$. Each point $a_i$ has a membership function $b_{ij}$ that determines the degree of its membership in each cluster $c_j$:

$$b_{ij} = \frac{1}{\left( \sum_{k=1}^{K} \frac{\| a_i - c_j \|}{\| a_i - c_k \|} \right)^{\frac{2}{m-1}}} \quad 0 \leqslant b_{ij} \leqslant 1, \tag{9.7}$$

where *m* is a hyper-parameter controlling fuzziness.

The ROI is derived from each CCA image in this study by tracing the MA wall. The MA wall is traced by starting at the bottom of the image and finding the brightest local maximum for each column. Figure 9.5(i) shows the upper and lower

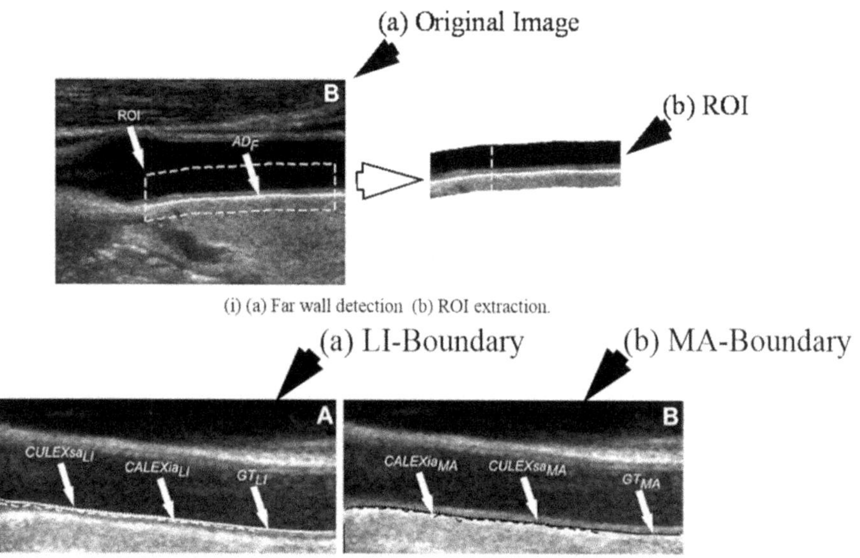


**Figure 9.5.** (i) Far wall detection and (ii) segmentation of the LI (white) and MA (black) boundaries [23]. (Reproduced with permission of Springer from [162]. Copyright 2021 Society for Imaging Informatics in Medicine. Reproduced with permission from [23]. Copyright 2010 IEEE.)

limits of the ROI, which are 1.25 mm and 0.625 mm above and below the MA wall, respectively. Following that, for each column in the ROI column-wise, three clusters of pixels, namely lumen, LI wall, and MA wall, are investigated in the next stage. Using the FKMC method, the pixels in each column are automatically assigned to each cluster. The delineation of individual walls in the ROI is marked by the sequence of LI and MA centroids for each column, as illustrated in figure 9.5 (ii). For 200 ultrasound CCA images, this technique was repeated. The method had a cIMT error of 0.054 $\pm$ 0.035 mm. The ML models presented briefly covered cIMT measurement, but they did not consider plaque area until Elisa *et al*'s [103] DL work in 2018 and Biswas *et al*'s [104] simultaneous cIMT and PA measurement in 2020. The next section gives a quick overview of DL in cIMT and PA measurement.

## 9.4 Deep learning application for cIMT and PA extraction

CNNs [50], DBNs [106], autoencoders [107], residual neural networks [108], and other DL approaches are employed for image classification and segmentation [82, 105]. The fully convolutional network (FCN) [109] is a CNN variant that does not include the connected network and is used for segmentation only. Three distinct SOTs are working on ultrasound CCA images for cIMT measurement. Before using ANNs for LI and MA boundary delineation, Rosa *et al* [110] used autoencoders for feature extraction. Suri and his colleagues employed an FCN for cIMT and PA [103] measurement on entire [111] and patch [104] CCA ultrasound images, whereas Del Mar *et al* [112] employed an FCN for full CCA images for cIMT estimation. The next few subsections present these works in detail for plaque burden measurement on CCA images.

### 9.4.1 ANN autoencoder based cIMT region segmentation

Rosa *et al* [110] employed artificial neural networks to characterize LI and MA pixels in 55 ultrasonography CCA images. To extract features from the LI and MA interface, the authors used trained autoencoders [107]. Autoencoders are neural networks that are programmed to replicate the input. They are commonly employed for unsupervised learning to decipher complex relationships in input photos. Given an input $\in [0,1]^{dx}$, using one or more hidden layers, the autoencoder converts it to a compressed representation, $Y \in [0,1]^{dy}$, with the mapping function similar to equation 9.3. Finally, the compressed image is resized to its original dimensions. This is done in order to comprehend the input vector's basic structures and relationships, as well as the training required to regenerate it. The characteristics are represented by a compressed representation of neighboring pixels. Figure 9.6(i) shows a representation of the autoencoder. The ROI extraction method is similar to that used by [92]. The authors used five ground truth images to train two autoencoders. The two autoencoders were trained using the LI and MA interface's neighborhood pixels. Two ANNs for pixel characterization were

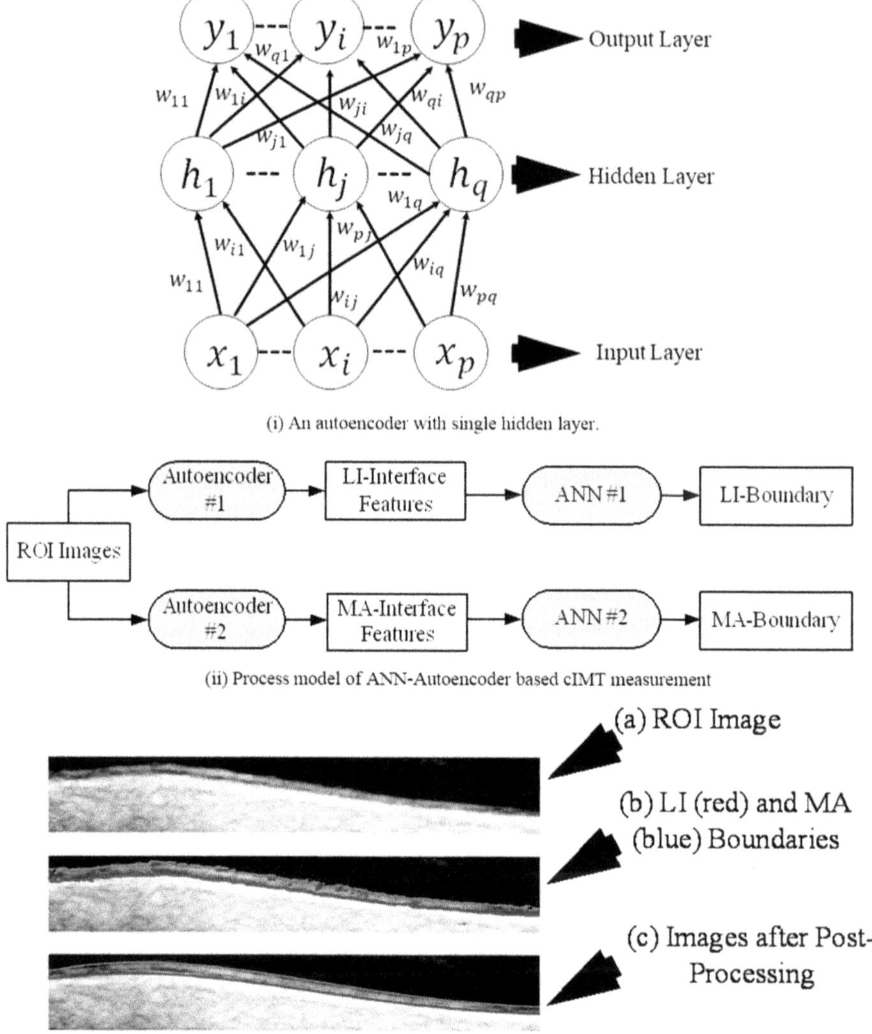



(i) An autoencoder with single hidden layer.



(ii) Process model of ANN-Autoencoder based cIMT measurement



(iii) Image outputs (a) ROI, (b) LI-and MA boundaries and (c)LI-and MA boundaries after postprocessing

**Figure 9.6.** (i) An autoencoder, (ii) process model, and (iii) image outputs. ((i) and (ii) Reproduced with permission of Springer from [162]. Copyright 2021 Society for Imaging Informatics in Medicine. (iii) Reproduced with permission from [110]. Copyright 2015 Elsevier.)

trained using the LI and MA characteristics. Finally, the LI and MA boundaries were extracted from the offline CCA images. The superfluous LI and MA pixels were removed using post-processing. The classification's cIMT bias was 0.0499 $\pm$ 0.0498 mm. The model is shown in figure 9.6 (ii) while the outcomes are shown in figure 9.6(iii).

## 9.4.2 Fully convolutional network for cIMT region estimation

The LI and MA boundaries were recovered in three steps from 396 ultrasonography CCA images by Biswas *et al* [111]. Binary images obtained from tracings of the LI and MA boundaries by two experienced radiologists using the general-purpose tracing software AtheroEdge™ [66, 113, 114] served as the ground truth. The images were cropped 10% from each side in the first multiresolution stage to guarantee that the low-contrast and non-relevant portions of the images did not impair learning in the second stage. This is usually caused by a shortage of gel or insufficient probe-to-neck contact during picture capture. An FCN was used in the second stage to separate the cIMT region from the rest of the image. The FCN based system was composed of two subsystems: the encoder and decoder, as shown in figure 9.7 (i). The encoder is 13 layers of convolution (two layers of 64 (window size = 3× 3) + two layers of 128 (window size = 3 × 3) + three layers of 256 (window size = 3 × 3) + three layers of 512 (window size = 3 × 3) + three layers of 512 (window size = 3 × 3) kernels) and five max-pooling layers to draw a down-

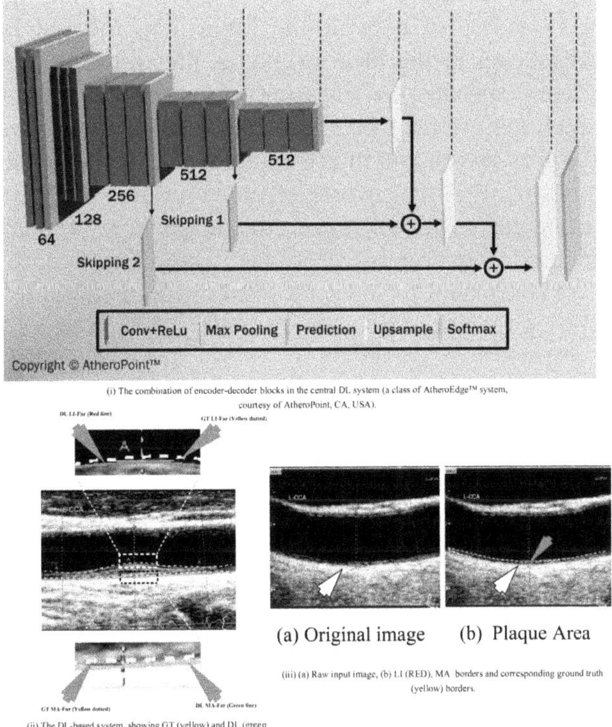



(i) The combination of encoder-decoder blocks in the central DL system (a class of AtheroEdge™ system, courtesy of AtheroPoint, CA, USA).

(a) Original image     (b) Plaque Area

(iii) (a) Raw input image, (b) LI (RED), MA borders and corresponding ground truth (yellow) borders.

(ii) The DL-based system showing GT (yellow) and DL (green and red) outputs (image courtesy AtheroPoint™, CA, USA).

**Figure 9.7.** (i) FCN model, (ii) IMT output using the FCN, and (iii) TPA output using the FCN. ((i) Reproduced with permission of Springer from [162]. Copyright 2021 Society for Imaging Informatics in Medicine. (ii) Reproduced with permission from [111]. Copyright 2018 Elsevier. (iii) Reproduced with permission from [103]. Copyright 2020 Elsevier.)

sampled representation or feature map of the image. Rosa *et al* [92] address the convolution procedure in subsection 9.3.1, which applies kernel filters on the image pixel by pixel to extract contextual information and position invariant properties of the cIMT wall region. A 1 × 1 convolution operation is used before each max-pooling step to compress the feature maps to a single map. Three up-sampling layers are used to up-sample the down-sampled feature map, two intermediate skipping operations are used to merge intermediate feature maps, and one softmax layer is used to compare pixels to pixels with the ground truth. The convolution operation is defined mathematically as

$$h(a, b) = I(a, b) \otimes k(a, b) = \sum_{s=-\frac{m}{2}}^{\frac{m}{2}} \sum_{t=-\frac{m}{2}}^{\frac{m}{2}} I(a + s, b + t) \times k(a, b), \qquad (9.8)$$

where $h$ is the output, $I$ is the image, $k$ is the kernel of size $m \times m$, $(a, b)$ represents the location of the pixel, and $(s, t)$ are dummy variables. The feature map is down-sampled using the max-pooling procedure, which keeps the most significant information from each block of the image. Figure 9.7 (i) depicts the FCN model that was used. To create a feature map of (1/32) the size of the original image, the encoder performs 13 layers of convolution (shown in the red boxes) and five max-pooling layers (shown in the blue boxes), as illustrated in figure 9.7 (i). The decoder performs dense softmax classification (shown in the orange box) with the ground truth for each feature map pixel utilizing a sequence of three up-sampling layers (shown in the gray boxes) and intermediate skipping operations (shown in the green boxes). Up-sampling is the inverse of convolution, which is applied to the feature map at the encoder's end. To restore the image to its original size, three up-sampling layers are used. The skipping procedures integrated feature maps from the encoder's contracted levels with the decoder's intermediate layers to recover spatial information lost during the encoder's down-sampling. For pixel-to-pixel character-ization, the cross-entropy loss function is

$$\theta_{\text{class}}(\alpha_1, \alpha_2) = \frac{1}{|N|} \sum_{n \in N} \sum_{l \in L} \alpha_{2n}(l) \log \alpha_{1n}(l), \qquad (9.9)$$

where $\alpha_1$ represents the prediction, $\alpha_2$ represents the gold standard (GT), $L$ represents the total number of classes, and $N$ represents the total number of images. The softmax layer is used for final characterization, which assigns a pixel to the class with the highest probability, as follows:

$$P(z_i) = \frac{e^{z_i}}{\sum_{j=1}^{N} e^{z_j}}, \qquad (9.10)$$

where $z_i$ represents the normalized score of the instance for the $i$th class. Fixing the DL iterations to 20 000 yielded the segmented images. Finally, the MA boundary was refined in the third stage utilizing calibration and a matrix inverse operation [115]. Figure 9.7 (ii) depicts the LI and MA borders, as well as their ground truth counterparts. The cIMT bias for the two ground truth values were 0.126 ± 0.134 and 0.124 ± 0.10 mm, respectively.

### 9.4.3 FCN for PA measurement

Elisa *et al* [103] used a similar technique [111] to obtain the PA values for the same cohort using the same ground truth and found some interesting results. For the two ground truth values, the PA error values were 20.52 mm$^2$ and 19.44 mm$^2$, respectively. For the two ground truths, the coefficients of correlation between the PA and cIMT using the outputs of DL were 0.92 ($p < 0.001$) and 0.94 ($p < 0.001$), respectively. Figure 9.7 (iii) depicts the final image.

### 9.4.4 Two-stage patching based AI model for cIMT and PA measurement

Biswas *et al* [104] recently used a two-stage DL based network to measure cIMT and PA. A CNN was employed in the first stage to extract the ROI, and an FCN identical to [111] was employed in the second stage to delineate the LI and MA boundaries and assess cIMT. A CNN differs from an FCN in that it uses a fully connected neural network for classification, whereas a trained FCN merely uses convolution and then up-sampling to build a feature map that represents semantic segmentation of the real image. Figure 9.8 (i) depicts a typical diagram of the two-layer CNN model. The image was finally characterized using the softmax classification function (equation 9.10). The scientists employed 22 deep layers [116] to extract high-level characteristics from the images in their study. The diabetic cohort initially had 250 CCA images collected. The AtheroEdge™ (AtheroPoint™, Roseville, CA) system was used to crop the image to remove background information and guarantee that the lumen region was in the center of the cropped CCA image. The cropped CCA image was split horizontally into two parts as a result. The bottom half of the image, which included the far wall, was removed and split horizontally. The wall information is in the upper half of the bottom half, while the tissue information is in the lower half. Both the upper and lower splits were separated into sixteen equal-sized patches on a column basis. The patches were used as input to the DL based two-stage system. To categorize the input images into wall and non-wall patches, an independent 22-layer CNN network [116] was used in the first stage. The accuracy of the CNN's characterization performance was 99.5%. After the patches had been defined, the ROI segment was created by combining the wall patches patient by patient. Figure 9.8 (ii) depicts the preprocessing and DL stage 1 inputs and outputs. The second stage DL is trained using these ROI segments and their binary ground facts. The second stage of the DL architecture is comparable to the first [111]. The second stage DL system divides the ROI segment into plaque and non-plaque regions after it has been trained.

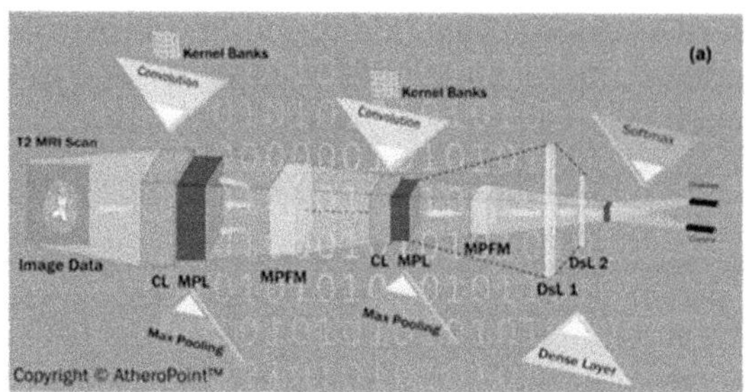

(i) A convolution neural network (courtesy of AtheroPoint™, CA, USA).

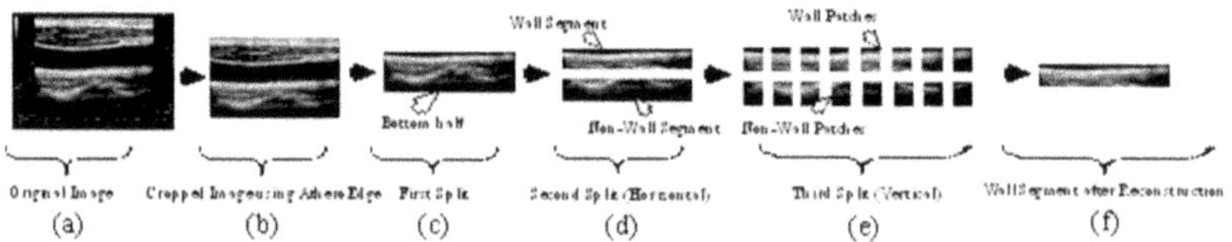

(ii) Preprocessing and characterization input and output, preprocessing: (a) original CCA image. (b) cropping using AtheroEdge™. (c) first horizontal partition. (d) second horizontal partition. (e) patching, stage-I characterization and reconstruction: (f) extracted ROI segment (image courtesy: AtheroPoint™)

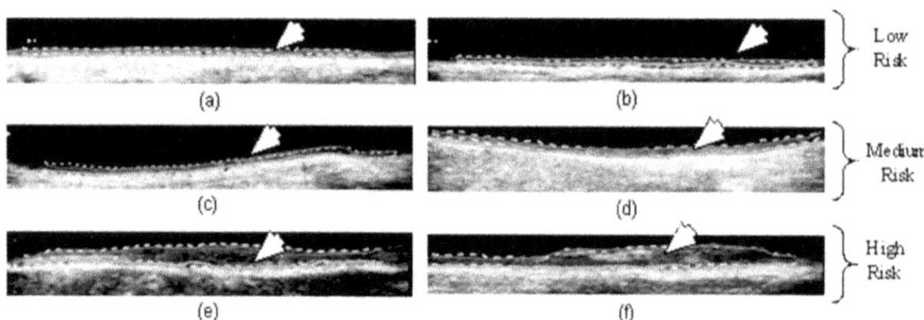

(iii) Stage-II DL output. yellow dotted line represents the ground truth LI-and MA borders. red and green line marks the deep learning LI-and MA-walls respectively (image courtesy: AtheroPoint™)

**Figure 9.8.** (i) A CNN model, (ii) patching and reconstruction process, and (iii) outputs of the two-stage DL model. ((i) and (ii) Reproduced with permission of Springer from [162]. Copyright 2021 Society for Imaging Informatics in Medicine. (iii) Reproduced with permission from [104]. Copyright 2020 Elsevier.)

Following that, the plaque region's LI and MA borders are identified, and cIMT and TPA are computed, as shown in figure 9.8 (iii). $0.0935 \pm 0.0637$ mm, $21.5794 \pm 7.9975$ mm$^2$, and $2.7939 \pm 2.3702$ mm$^2$ were found to be the cIMT error, PA, and PA error, respectively. With a cIMT mean error of 0.02 mm, a similar technique [112] was applied to the entire CCA images using DenseNet [108] for all 331 photos. Although DL techniques are more accurate, resilient, and scalable than ML techniques, they have substantially greater storage and compute costs. The multiple layers of neurons necessitate a massive number of concurrent computations that may not be achievable with desktop CPU architectures but require graphic processing units (GPUs). Furthermore, storing intermediate values necessitates a large amount of memory. As a result, machine learning is better suited to tiny systems where speed is more important than accuracy. DL, on the other hand, is best suited for industrial medical imaging applications with enormous patient volumes and shows excellent accuracy.

## 9.5 Discussion

In this study, we looked at different state-of-the-art AI algorithms that have been adopted in recent years for plaque burden estimation in the form of cIMT and PA using carotid ultrasound images. Although AI is a relatively new idea, its robustness and accuracy have had a considerable impact on the medical imaging sector. Learning from training images was introduced by ML approaches, which have lately been supplanted by DL techniques. Even when future deep learning systems outperform earlier systems, there will be a cost in terms of hardware or processing time. More thorough comparisons in terms of speed, accuracy, cohort size, and variability in carotid scan resolution, applied to both CCA and bulb segments, are needed. The presence of noise or redundant information has been shown to considerably impair the effectiveness of deep learning in various prior research, such as the use of DL in the liver [117]. As a consequence, cropping was used in order to reach considerable performance levels. Regardless of the cross-validation methodology, the amount of training data has a considerable impact on training performance. A big training pool is always better for training DL models since it captures more of the data pool's nuances and complexity. When applied to unknown data, this aids in greater performance. In addition, this paper examines three generations of cIMT/PA developments, as well as a detailed examination of several major ML and DL paradigms that computed cIMT and PA values directly from carotid ultrasound images without the need for human interaction. There was no attempt to study cIMT/PA in motion images because the scope of this study was limited to static (or frozen) ultrasound images. Traditional (non-AI) approaches for cIMT estimation in selected frames of the cardiac cycle have been used in research to explain plaque migration [118–120]. Figure 9.9 illustrates a visual comparison of AtheroEdge™, a patented software [121], the scale-space method [46], and the AI based model (Biswas *et al* [111]). The AtheroEdge™ model, which is based on splines and elastic contours, produced reasonably acceptable results with obvious border tracing. However, it falls short in noisy areas. The AI model, on the other hand, achieves a sharp and unambiguous demarcation when contrasted to the scale-space model. For the same patient, the proposed AI model findings are more aligned with the ground truth than the scale-space model. Over numerous cycles, the AI model parameters are trained to align along the LI and MA far walls, resulting in improved learning. Because of the improved training, the parameters now include believable noise

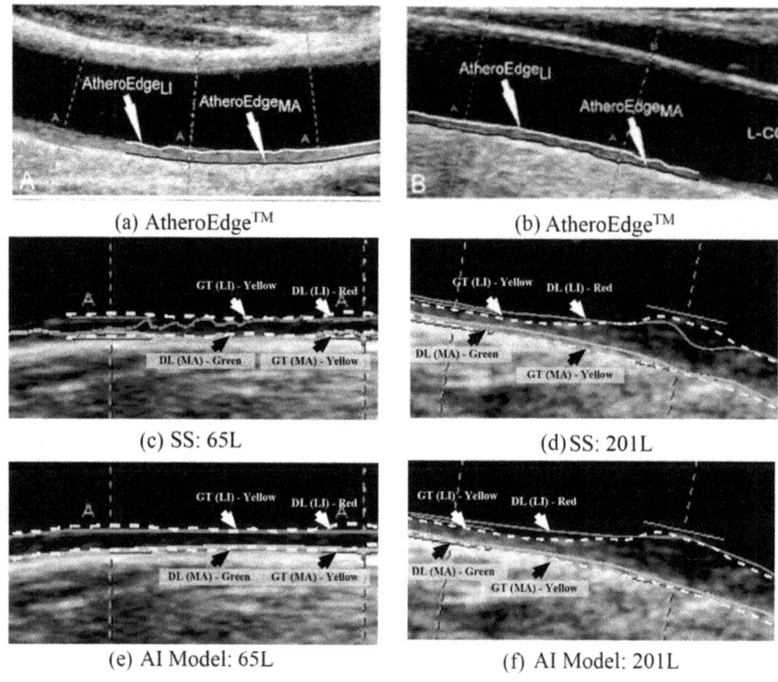


<table>
<tr><td>(a) AtheroEdge™</td><td>(b) AtheroEdge™</td></tr>
<tr><td>(c) SS: 65L</td><td>(d) SS: 201L</td></tr>
<tr><td>(e) AI Model: 65L</td><td>(f) AI Model: 201L</td></tr>
</table>

**Figure 9.9.** Comparison of AtheroEdge™ ((a) and (b) [121]), SS ((c) and (d) [46]) and AI ((e) and (f) [111]) delineation of the LI and MA borders. ((a) and (b) Reproduced with permission from [121]. (c) and (d) Reproduced with permission of Springer from [162]. Copyright 2021 Society for Imaging Informatics in Medicine. (e) and (f) Reproduced with permission from [111]. Copyright 2018 Elsevier.)

surrounding the walls and get around them to delineate correct limits. Low-level segmentation methods, such as scale-space, have failed to account for noisy data in their calculations, resulting in less precise demarcation.

### 9.5.1 Benchmarking

Molinari *et al* [122] have already briefly detailed the first and second-generation approaches. Table 9.3 presents the benchmarking table employing first-, second-, and third-generation methodologies utilized throughout the last two decades in this area. The techniques of the first generation are addressed first. Liang *et al* [123] employed DP to quantify cIMT from 50 photos in the year 2000. Stein *et al* [124] and Faita *et al* [125] employed edge detection approaches to calculate cIMT from 50 and 150 CCA images with cIMT errors of $0.012 \pm 0.006$ mm and $0.010 \pm 0.038$ mm, respectively. Ikeda *et al* [48] employed a bulb edge point detection methodology to estimate cIMT from 649 images, with a bias of $0.0106 \pm 0.00310$ mm between the anticipated and ground truth. Following that, a combination of first- and second-generation approaches will be discussed. Molinari

**Table 9.3.** Benchmarking table showing the ML and DL methods for cIMT and PA measurements. (Reproduced with permission of Springer from [162]. Copyright 2021 Society for Imaging Informatics in Medicine.)

| SN | Generation | Author, reference | Method | Cross-validation protocol | Data size | cIMT/PA error | Performance (cIMT error and PA errors) |
|---|---|---|---|---|---|---|---|
| 1 | First | Liang et al [123] | Dynamic programming | NA | 50 | cIMT | 0.042 ± 0.02 mm |
| 2 | | Stein et al [124] | Edge detection | NA | 50 | cIMT | 0.012 ± 0.006 mm |
| 3 | | Faita et al [125] | Edge detection | NA | 150 | cIMT | 0.010 ± 0.038 mm |
| 4 | | Ikeda et al [48] | Bulb edge point detection | NA | 649 | cIMT | 0.0106 ± 0.00310 mm |
| 5 | First + second | Molinari et al [72] | Level set, morphological image processing | NA | 200 | cIMT | 0.144 ± 0.179 mm |
| 6 | | Molinari et al [126] | CARES | NA | 647 | cIMT | 0.172 ± 0.222 mm |
| 7 | | Molinari et al [24] | CAMES 1.0 | NA | 657 | cIMT | 0.154 ± 0.227 mm |
| 8 | | Molinari et al [127] | CAUDLES | NA | 630 | cIMT | 0.224 ± 0.252 mm |
| 9 | | Molinari et al [128] | FOAM | NA | 665 | cIMT | 0.150 ± 0.169 mm |
| 10 | Second | Gutierrez et al [129] | Active contours | NA | 180 | cIMT | 0.090 ± 0.060 mm |
| 11 | | Molinari et al [130] | Snakes | NA | 200 | cIMT | 0.01 ± 0.01 mm |
| 12 | | Molinari et al [131] | CALEX 1.0 | NA | 665 | cIMT | 0.191 ± 0.217 mm |
| 13 | Third | Rosa et al [92] | ANN | 60% training, 40% testing | 60 | cIMT | 0.03763 ± 0.02518 mm (mean) 0.03670 ± 0.02429 mm (polyline) 0.03683 ± 0.02450 mm (center) |
| 14 | | Rosa et al [93] | RBNN | Jack-knifing (leave one out) | 25 | cIMT | 0.065 ± 0.046 mm |
| 15 | | Molinari et al [23] | FKMC | K13 (92% training, 8% testing) | 200 | cIMT | 0.054 ± 0.035 mm (polyline) |

| 16 | Rosa *et al* [110] | ANN + autoencoder | K2 (50% training, 50% testing) | 55 | cIMT | $0.0499 \pm 0.0498$ mm |
| 17 | Biswas *et al* [111] | FCN | K10 (90% training, 10% testing) | 396 | cIMT | $0.126 \pm 0.134$ mm (DL1)<br>$0.124 \pm 0.10$ mm (DL2) |
| 18 | Elisa *et al* [103] | FCN | K10 (90% training, 10% testing) | 396 | TPA | $20.52$ mm$^2$ (DL1)<br>$19.44$ mm$^2$ (DL2) |
| 19 | Biswas *et al* [104] | CNN + FCN | K10 (90% training, 10% testing) | 250 | cIMT, TPA value, TPA | $0.0935 \pm 0.0637$ mm (cIMT error),<br>$21.5794 \pm 7.9975$ mm$^2$ (DL TPA),<br>$2.7939 \pm 2.3702$ mm$^2$ (PA error) |
| 20 | Del Mar *et al* [112] | FCN | K10 (90% training, 10% testing) | 331 | cIMT error | $0.02$ mm |

*et al* [72] estimated cIMT from 200 photos using a mix of level set and morphological image analysis. The cIMT error was found to be 0.144 ± 0.179 mm. Molinari *et al* developed CARES [126], CAMES [24], CAUDLES [127], and FOAM [128], which are fusion-based patented methodologies for cIMT estimation. From 647, 657, 630, and 665 photos, the cIMT error for each of them was 0.172 ± 0.222 mm, 0.154 ± 0.227 mm, 0.224 ± 0.252 mm, and 0.150 ± 0.169 mm, respectively. Some of the second-generation techniques are now discussed. Gutierrez *et al* [129] created an active contour model for measuring cIMT from 180 images in 2002. Molinari *et al* [130] created a snake-based model to estimate cIMT from 200 images. Molinari *et al* [131] created a new CALEX 1.0 model for calculating cIMT from 665 images in 2014. The active contour, snakes, and CALEX 1.0 had cIMT errors of 0.090 ± 0.060 mm, 0.01 ± 0.01 mm, and 0.191 ± 0.217 mm, respectively. Rosa *et al* [92] employed an ANN as the ML model for cIMT error from 60 ultrasonography CCA images using three metrics: mean, Suri's polyline, and centerline, with average values of 0.03763±0.02518 mm, 0.03670±0.02429 mm, and 0.03683±0.02450 mm, respectively. Rosa *et al* [93] also utilized RBNN to compute the cIMT error, which was 0.065 ± 0.046 mm, using a database of 25 CCA images. Molinari *et al* [23] computed cIMT from 200 ultrasound CCA scans using an unsupervised FKMC methodology. The obtained cIMT error was 0.054 ± 0.035 mm. Rosa *et al* [110] presented the computation of cIMT from 55 ultrasound pictures with a mean cIMT error of 0.0499 ± 0.0498 mm using a mixture of ML (ANN) and a DL autoencoder. Biswas *et al* [111] used a database of 396 carotid images to demonstrate the DL paradigm for cIMT calculation utilizing an FCN by combining the ground truth from two separate observers. When evaluating the two ground truths, the cIMT error was 0.126±0.134 mm and 0.124±0.100 mm, respectively. Utilizing the same FCN methodology, Elisa *et al* [103] obtained a TPA of 20.52 mm$^2$ and 19.44 mm$^2$, respectively, using the two sets of ground facts. Biswas *et al* [104] went on to show how to compute cIMT from 250 ultrasound CCA pictures using a two-stage (CNN + FCN) DL method. In the second stage, the authors employed a patching-based strategy to extract the ROI from the first stage and outline the LI and MA borders. The the cIMT error was 0.0935 ± 0.0637 mm. In the sense that both stages were DL phases, the patch-based method was a one-of-a-kind approach. FCN was used on 331 carotid ultrasound scans with a cumulative cIMT error of 0.02 mm in another investigation by Del Mar *et al* [112].

### 9.5.2 A short note on cardiovascular risk assessment

The use of intelligence paradigms such as ML and DL to measure or stratify cardiovascular risk can aid in both monitoring and reducing CVD risk. Many studies [20, 79, 89, 132] have found a substantial link between covariates such blood biomarkers and traditional risk variables including age, grayscale median values, and stroke risk. Behavioral patterns such as smoking, eating, and other image based phenotypes such as cIMT and PA values can be linked with blood biomarkers to improve stroke risk assessment.

### 9.5.3 A note on the clinical impact of AI methods on cIMT/PA techniques

It is worth noting that DL has just recently begun to permeate the vascular field, particularly vascular ultrasound. Prototypes have been tried lately, and our group has been at the forefront of this field. While we may go over the designs, they have not yet made it to the stage of clinical application, unlike our non-AI based models, which are now in use (see AtheroEdge™ 2.0, AtheroPoint™, Roseville, CA, USA [79, 89, 132–135]). However, we expect that as time goes on, more transfer learning (TL), DL, and reinforcement learning (RL) applications will reach the clinical world, and the diagnostic ultrasonography community will begin to use them. Some scientific validation can be performed by applying registration procedures to match the plaque regional information between cross modalities [136].

### 9.5.4 A note on inter- and intra-observer variability analysis on the evaluation of AI models

Inter- and intra-observer variability are obviously essential factors to consider when evaluating the efficacy of cIMT/PA systems. Using AtheroEdge™ 1.0 and 2.0 systems, our group has attempted variability analysis on cIMT and other applications [137–139]. Biswas *et al* [140] used deep learning systems to demonstrate the effect of inter-observer variability. The authors used three sets of hand tracings on 407 ultrasound scans to demonstrate the inter-operator variability of the DL system for lumen boundary detection and performance evaluation. Manual tracings were used to train each DL system, which were then subjected to cross-validation runs for the training and testing protocols. The precision of merit for three DL systems using 407 US images was 99.61%, 97.75%, and 99.89%, respectively, using Suri's polyline distance measure. The DL lumen segmented region's Jaccard index against three GT regions was 0.94, 0.94, and 0.93, respectively, while the Dice similarity was 0.97, 0.97, and 0.97. The AUCs for the three DL systems were 0.95, 0.91, and 0.93, respectively. The proposed DL system outperformed standard approaches in the literature. The variability analysis was performed using two hand tracings taken from two vascular radiologists in another recent DL method using patch-based methodology [141]. The absolute cIMT error of the authors' AI based system was $0.0935 \pm 0.0637$ mm (i.e. 20% less than the prior method [111]) when using GT1 (i.e. $0.1164 \pm 0.1122$ mm) and 18% less when using GT2 (i.e. $0.1146 \pm 0.0940$ mm), respectively. The authors in [48, 130] attempted to analyze the effect of ultrasound scanners on the performance of cIMT as part of the system's performance evaluation, such as dependability.

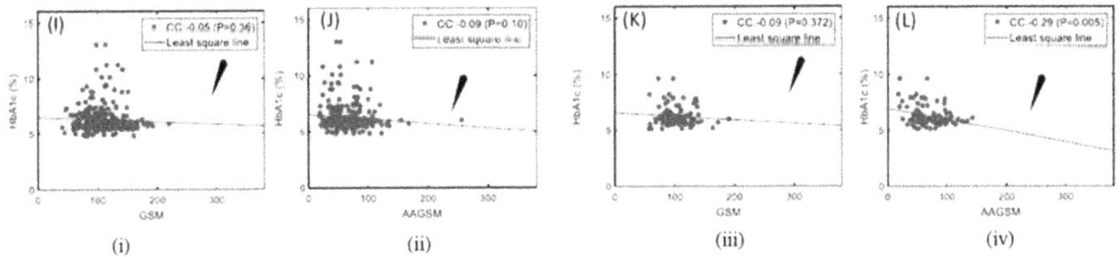

**Figure 9.10.** (i) GSM versus HbA1c for male; (ii) AAGSM versus HbA1c for male; (iii) GSM versus HbA1c for female; and (iv) AAGSM versus HbA1c for female. (Image courtesy of AtheroPoint™. Reproduced with permission of Springer from [162]. Copyright 2021 Society for Imaging Informatics in Medicine.)

### 9.5.5 A short note on 10 year risk estimation using cIMT and PA

CIMT and PA levels have been associated with diabetes and cardiovascular risk in numerous studies [79]. To calculate the 10 year risk, many risk metrics and calculators have been devised. As a function of age, PA, and traditional GSM, a novel measure called age-adjusted grayscale median (AAGSM) [28] has been devised. Figure 9.10 shows the relationship between HbA1c, AAGSM, and GSM, with a low CC of −0.05 ($p = 0.36$) between HbA1c (male) and GSM, a low CC of −0.09 ($p = 0.10$) between HbA1c (male) and AAGSM, a low CC of −0.09 ($p = 0.372$) between HbA1c (female) and GSM, and a high CC of −0.29 These connections were crucial in dividing diabetic patients into low-, medium-, and high-risk groups.

In another study, cIMT [133] has been used to calculate the risk of diabetic cohorts using logistic regression. Another risk score 'AtheroEdge Composite Risk Score 1.0' (AECRS $1.0_{10yr}$) [89] computed the 10 year risk of carotid image phenotypes by integrating conventional cardiovascular risk factors and reported the highest area-under-the-curve of 0.927.

### 9.5.6 Statistical power analysis and diagnostic-odds ratio

The sample size for training is validated using statistical power analysis. If $z*$ is the standardized value from the $z$-table, $p$ is the data proportion, and MoE is the margin of error, then $n = \left[ (z^*)^2 \times \left( \frac{\hat{p}(1-\hat{p})}{\text{MoE}^2} \right) \right]$ is the required number of samples. With a 95% confidence interval, a 5% margin of error, and a 0.5 data percentage, this is used. Power analysis was used for training data size validation in a few recent AI studies [142–144]. A meta-analysis of the diagnosis is presented in the diagnostic-odds ratio (DOR) [145]. The DOR is determined by

$$\text{DOR} = \left( \frac{\text{sens.}}{1 - \text{sens.}} \right) \Big/ \left( \frac{1 - \text{spec.}}{\text{spec.}} \right), \tag{9.11}$$

where 'sens.' and 'spec.' allude to the study's sensitivity and specificity, respectively. The reduction rate of type 2 error (false negative) is another useful measure for understanding the diagnostic instrument's power [146]. If $\beta$ refers to a false negative rate, then $1 - \beta$ denotes the study's power.

*A short note on graphical processing units*
The availability of GPUs is fueling the growth of AI. As our research has discovered, AI typically entails millions of complex computations that, when executed sequentially in multicore CPUs, require a long time (often even several hours) to converge. GPUs, on the other hand, divide big problems into thousands of little problems, which are then distributed among thousands of cores to solve in parallel and then they are aggregated.

As a result, convergence in GPUs is substantially faster than in CPUs. GPU-S converged faster than the CPU-S and CPU-L for both Omnivore and MXNet in a comparative examination [147] of ImageNet8 [84] implementation for several deep learning libraries such as MXNet [148], SINGA [149], and Omnivore [147] on CPU-S (nine cores), CPU-L (33 cores), and GPU-S (36 GPUs) cluster platforms. Omnivore outperformed the other libraries (2.3X-CPU-S, 4.8X-GPU-S, 3.2X-CPU-L), but SINGA outperformed the others and was not implemented in CPU-L or GPU-S.

## 9.6 Conclusions

The research discussed three generations of cIMT/PA measuring systems, from traditional methods to intelligence-based methods based on machine and deep learning. The ability to undertake deeper number crunching to derive sophisticated information, resulting in improved accuracy, reliability, and stability of systems, is the driving force behind this progress. The improved results also demonstrate that the systems have a lot of clinical potential. Medical imaging has been substantially affected by deep learning powered by a GPU, ushering in new horizons in automated diagnosis and treatment.

# Appendix A Mathematical representations of the ML and DL paradigms

The term learnability evolved from the general tendency of humans and machines to improve performance based on experience. In recent times, computational models such as AI have shown the ability to discover patterns from unstructured data and infer logic and make decisions. AI models such as ML and DL have been used in the area of medical imaging to make life saving diagnostic decisions [82, 150–152]. Here we give a brief mathematical discourse on the ML and DL models for better understanding.

*Machine learning: a mathematical representation*

The mathematical approach is given as follows [153, 154]: let us say that the instance space is denoted as $X = \{x_1, x_2, \ldots, x_i\}$ and label space as $L = \{l_1, l_2, \ldots, l_j\}$. The characterization job is therefore denoted as $\hat{f} : X \to L$, where $\hat{f}(x_i)$ denotes the estimate of the desired function $f(x_i)$, that represents the true label of $x_i$. In general, the difference between $f$ and $\hat{f}$ is used as a feedback to fine-tune the ML model. The least square model is the most common loss model to compute the difference between the desired and actual output and given by

$$\varepsilon = \frac{1}{n} \| f(X) - \hat{f}(X) \|^2, \tag{A.1}$$

where $f(X)$ is the input to the ML model and $\hat{f}(X)$ is the predicted output. Different loss models are applied by different ML models. Although label estimation forms a major part of supervised learning, there are other paradigms of learning such as regression, where function approximation is performed. If $R \in$ real numbers, $X = R$, and $Y = R$, then $\hat{f} : X \to Y$, is an approximation function mapping from $X$ to $Y$. In another form of learning, instead of label outputs, probability is generated for each label for a test instance, i.e. $\hat{f} : X \to [0,1]^{|Y|}$.

Any instance is represented by the feature or independent variables, where the instance automatically becomes the dependent variable. The instance space is therefore defined as $g_k : X \to I_k$, where, $g_k$ is defined as the mapping from instance space $X$ to feature space $I_k$. Therefore, the instance space $X$ can be defined as the Cartesian product of $k$ domains $X = I_1 \times I_2 \times \ldots \times I_k$. As stated previously, ML models are two-stage systems where the features are extracted which are generally statistical information. As per the nature of features, they are categorized into quantitative, ordinal, and categorical. For images, the features are generally quantitative in nature, which basically deals in orientation, direction, scale, texture etc. ML models on the other hand can be categorized into probabilistic, logical, or geometric with the latter most commonly used with respect to medical images. The geometric models use lines, distances, curves, arcs etc. These models work by drawing a hyper-plane between two labeled clusters. The most common geometric ML models are $K$-nearest neighbors [155], support vector machines (SVMs) [156], artificial neural networks (ANN) [157], etc. The probabilistic models consider the most probable class of an instance by computing the maximum posterior probability, i.e. using a Bayesian classifier [158]. Logical models follow rule based constructs to build models such as decision trees [159]. The training and testing paradigms of the ML and DL models are the same, although the feature extraction part is done earlier for the former. The training and testing data are different for both cases. The models are trained to learn the mapping from instances to labels from labeled instances in the training phase. After learning, the performance of training is judged from testing the trained mode on unknown test instances. The training and testing paradigm is shown in figure 9.3 (i).

*Deep learning: a mathematical representation*

DL models are inspired by the working of the visual cortex of the brain. DL models, unlike ML, are single-stage models and generate features directly from the instances, thus overriding the need of for an intermediate feature extraction phase. The feature space represents an abstract representation of the instance space and hence this is also called representation learning. The DL network is composed of multiple layers of kernels which often perform repeated functions. These kernels generally generate down-sampled versions of the instance space also called feature maps. This approach of learning feature map $F_i$ from instance $x_i$ is shown by:

$$F_i = A_n\Big(B_n\big(C_n\big(...(A_1(B_1(C_1(x_i))))...\big)\big)\Big), \tag{A.2}$$

where $x_i \in X$, $A$, $B$, and $C$ are different kernels applied repeatedly on the instance $x_i$. $F_i$ is the feature vector which is output after $n$ applications of $A$, $B$, and $C$. In the case of convolution neural networks (CNNs) [84], the kernel filter values are optimized using backpropagation rules [160] based on loss functions such as the least square model. Fully convolutional networks (FCNs) [109] are another supervised DL model most commonly used for semantic segmentation of images. Another form of learning is unsupervised learning where labeled instances are not available [161]. Unsupervised learning applies different methodologies to extract or find interesting patterns within unstructured data. $K$-means clustering [162] is an important ML unsupervised learning method. Autoencoders [107] on the other hand are an important DL model used generally for unsupervised learning.

- J S Suri, D Kumar, Medical image enhancement system, US Patent App. 11/ 609,743

# References

[1] Benjamin E J, Muntner P and Bittencourt M S 2019 Heart disease and stroke statistics-2019 update: a report from the American Heart Association *Circulation* **139** e56–528

[2] Benjamin E J, Virani S S, Callaway C W, Chamberlain A M, Chang A R, Cheng S, Chiuve S E, Cushman M, Delling F N and Deo R 2018 Heart disease and stroke statistics—2018 update: a report from the American Heart Association *Circulation* **137** e67–e492

[3] Ross R 1999 Atherosclerosis—an inflammatory disease *New Engl. J. Med.* **340** 115–26

[4] Suri J S, Kathuria C and Molinari F 2010 *Atherosclerosis Disease Management* (Berlin: Springer)

[5] Saba L, Sanches J M, Pedro L M and Suri J S 2014 *Multi-modality Atherosclerosis Imaging and Diagnosis* (Berlin: Springer)

[6] Libby P 2020 The heart in COVID19: primary target or secondary bystander? *JACC: Basic Transl. Sci.* **5** 537–42

[7] Basta G, Schmidt A M and De Caterina R 2004 Advanced glycation end products and vascular inflammation: implications for accelerated atherosclerosis in diabetes *Cardiovasc. Res.* **63** 582–92

[8] Colwell J A, Lopes-Virella M and Halushka P V 1981 Pathogenesis of atherosclerosis in diabetes mellitus *Diabetes Care* **4** 121–33

[9] Saam T, Yuan C, Chu B, Takaya N, Underhill H, Cai J, Tran N, Polissar N L, Neradilek B and Jarvik G P 2007 Predictors of carotid atherosclerotic plaque progression as measured by noninvasive magnetic resonance imaging *Atherosclerosis* **194** e34–42

[10] Hashimoto H, Tagaya M, Niki H and Etani H 2009 Computer-assisted analysis of heterogeneity on B-mode imaging predicts instability of asymptomatic carotid plaque *Cerebrovasc. Dis.* **28** 357–64

[11] Ross R and Glomset J A 1976 The pathogenesis of atherosclerosis *New Engl. J. Med.* **295** 369–77

[12] Ross R 1986 The pathogenesis of atherosclerosis—an update *New Engl. J. Med.* **314** 488–500

[13] Carr S, Farb A, Pearce W H, Virmani R and Yao J S 1996 Atherosclerotic plaque rupture in symptomatic carotid artery stenosis *J. Vasc. Surg.* **23** 755–66

[14] Iadecola C 2020 Revisiting atherosclerosis and dementia *Nat. Neurosci.* **23** 691–2

[15] Lucatelli P, Raz E, Saba L, Argiolas G M, Montisci R, Wintermark M, King K S, Molinari F, Ikeda N and Siotto P 2016 Relationship between leukoaraiosis, carotid intima-media thickness and intima-media thickness variability: preliminary results *Eur. Radiol.* **26** 4423–31

[16] Saba L, Sanfilippo R, Balestrieri A, Zaccagna F, Argiolas G M, Suri J S and Montisci R 2017 Relationship between carotid computed tomography dual-energy and brain leukoaraiosis *J. Stroke Cerebrovasc. Dis.* **26** 1824–30

[17] Wingo A P, Fan W, Duong D M, Gerasimov E S, Dammer E B, Liu Y, Harerimana N V, White B, Thambisetty M and Troncoso J C 2020 Shared proteomic effects of cerebral atherosclerosis and Alzheimer's disease on the human brain *Nat. Neurosci.* **23** 696–700

[18] Viswanathan V, Jamthikar A D, Gupta D, Puvvula A, Khanna N N, Saba L, Viskovic K, Mavrogeni S, Turk M and Laird J R 2020 Integration of eGFR biomarker in image-based CV/stroke risk calculator: a South Asian-Indian diabetes cohort with moderate chronic kidney disease *Int. Angiol.: A J. Int. Union Angiol.* **39** 290–306

[19] Puvvula A, Jamthikar A D, Gupta D, Khanna N N, Porcu M, Saba L, Viskovic K, Ajuluchukwu J N, Gupta A and Mavrogeni S 2020 Morphological carotid plaque area is associated with glomerular filtration rate: a study of South Asian Indian patients with diabetes and chronic kidney disease *Angiology* **71** 520–35

[20] Khanna N N, Jamthikar A D, Gupta D, Piga M, Saba L, Carcassi C, Giannopoulos A A, Nicolaides A, Laird J R and Suri H S 2019 Rheumatoid arthritis: atherosclerosis imaging and cardiovascular risk assessment using machine and deep learning-based tissue characterization *Curr. Atheroscler. Rep.* **21** 7

[21] Boi A, Jamthikar A D, Saba L, Gupta D, Sharma A, Loi B, Laird J R, Khanna N N and Suri J S 2018 A survey on coronary atherosclerotic plaque tissue characterization in intravascular optical coherence tomography *Curr. Atheroscler. Rep.* **20** 33

[22] Liu K and Suri J S 2005 Automatic vessel indentification for angiographic screening *US Patent* WO/2003/008989

[23] Molinari F, Zeng G and Suri J S 2010 Intima-media thickness: setting a standard for a completely automated method of ultrasound measurement *IEEE Trans. Ultrason. Ferroelectr. Freq. Control* **57** 1112–24

[24] Molinari F, Pattichis C S, Zeng G, Saba L, Acharya U R, Sanfilippo R, Nicolaides A and Suri J S 2011 Completely automated multiresolution edge snapper—a new technique for an accurate carotid ultrasound IMT measurement: clinical validation and benchmarking on a multi-institutional database *IEEE Trans. Image Process.* **21** 1211–22

[25] Herder M, Johnsen S H, Arntzen K A and Mathiesen E B 2012 Risk factors for progression of carotid intima-media thickness and total plaque area: a 13-year follow-up study: the Tromsø Study *Stroke* **43** 1818–23

[26] Moreno P R, Falk E, Palacios I F, Newell J B, Fuster V and Fallon J T 1994 Macrophage infiltration in acute coronary syndromes. Implications for plaque rupture *Circulation* **90** 775–8

[27] Cuadrado-Godia E, Maniruzzaman M, Araki T, Puvvula A, Rahman M J, Saba L, Suri H S, Gupta A, Banchhor S K and Teji J S 2018 Morphologic TPA (mTPA) and composite risk score for moderate carotid atherosclerotic plaque is strongly associated with HbA1c in diabetes cohort *Comput. Biol. Med.* **101** 128–45

[28] Kotsis V, Jamthikar A D, Araki T, Gupta D, Laird J R, Giannopoulos A A, Saba L, Suri H S, Mavrogeni S and Kitas G D 2018 Echolucency-based phenotype in carotid atherosclerosis disease for risk stratification of diabetes patients *Diabetes Res. Clin. Pract.* **143** 322–31

[29] Saba L, Banchhor S K, Londhe N D, Araki T, Laird J R, Gupta A, Nicolaides A and Suri J S 2017 Web-based accurate measurements of carotid lumen diameter and stenosis severity: an ultrasound-based clinical tool for stroke risk assessment during multicenter clinical trials *Comput. Biol. Med.* **91** 306–17

[30] Baradaran H, Ng C R, Gupta A, Noor N M, Al-Dasuqi K W, Mtui E E, Rijal O M, Giannopoulos A, Nicolaides A and Laird J R 2017 Extracranial internal carotid artery calcium volume measurement using computer tomography *Int. Angiol.* **36** 445–61

[31] Saba L, Than J C, Noor N M, Rijal O M, Kassim R M, Yunus A, Ng C R and Suri J S 2016 Inter-observer variability analysis of automatic lung delineation in normal and disease patients *J. Med. Syst.* **40** 142

[32] Saba L, Bhavsar A, Gupta A, Mtui E, Giambrone A, Baradaran H, Lavra F, Laird J, Nicolaides A and Suri J 2015 Automated calcium burden measurement in internal carotid artery plaque with CT: a hierarchical adaptive approach *Int. Angiol.* **34** 290–305

[33] Lloyd-Jones D M, Wilson P W, Larson M G, Beiser A, Leip E P, D'Agostino R B and Levy D 2004 Framingham risk score and prediction of lifetime risk for coronary heart disease *Am. J. Cardiol.* **94** 20–4

[34] Preiss D and Kristensen S L 2015 The new pooled cohort equations risk calculator *Can. J. Cardiol.* **31** 613–9

[35] Ridker P M, Buring J E, Rifai N and Cook N R 2007 Development and validation of improved algorithms for the assessment of global cardiovascular risk in women: the Reynolds Risk Score *JAMA* **297** 611–9

[36] Stevens R J, Kothari V, Adler A I, Stratton I M and Holman R R 2001 Group UKPDS: the UKPDS risk engine: a model for the risk of coronary heart disease in Type II diabetes (UKPDS 56) *Clin. Sci.* **101** 671–9

[37] Kothari V, Stevens R J, Adler A I, Stratton I M, Manley S E, Neil H A and Holman R R 2002 UKPDS 60: risk of stroke in type 2 diabetes estimated by the UK Prospective Diabetes Study risk engine *Stroke* **33** 1776–81

[38] Tillin T, Hughes A D, Whincup P, Mayet J, Sattar N, McKeigue P M, Chaturvedi N and Group S S 2014 Ethnicity and prediction of cardiovascular disease: performance of QRISK2 and Framingham scores in a UK tri-ethnic prospective cohort study (SABRE—Southall And Brent REvisited) *Heart* **100** 60–7

[39] Board J B S 2014 Joint British Societies' consensus recommendations for the prevention of cardiovascular disease (JBS3) *Heart* **100** ii1–67

[40] Seabra J and Sanches J 2012 *Ultrasound Imaging: Advances and Applications* (New York: Springer)

[41] Suri J S and Laxminarayan S 2003 *Angiography and Plaque Imaging: Advanced Segmentation Techniques* (Boca Raton, FL: CRC Press)

[42] Molinari F M, Meiburger K, Zeng G, Acharya U R, Liboni W, Nicolaides A and Suri J S 2012 Carotid artery recognition system: a comparison of three automated paradigms for ultrasound images *Med. Phys.* **39** 378–91

[43] Molinari F, Meiburger K M, Acharya U R, Zeng G, Rodrigues P S, Saba L, Nicolaides A and Suri J S 2011 CARES 3.0: a two stage system combining feature-based recognition and edge-based segmentation for CIMT measurement on a multi-institutional ultrasound database of 300 images *2011 Annual Int. Conf. of the IEEE Engineering in Medicine and Biology Society* (Piscataway, NJ: IEEE) pp 5149–52

[44] Wendelhag I, Liang Q, Gustavsson T and Wikstrand J 1997 A new automated computerized analyzing system simplifies readings and reduces the variability in ultrasound measurement of intima-media thickness *Stroke* **28** 2195–200

[45] Cheng D-c, Schmidt-Trucksäss A, Cheng K-s and Burkhardt H 2002 Using snakes to detect the intimal and adventitial layers of the common carotid artery wall in sonographic images *Comput. Methods Programs Biomed.* **67** 27–37

[46] Kumar P K, Araki T, Rajan J, Saba L, Lavra F, Ikeda N, Sharma A M, Shafique S, Nicolaides A and Laird J R 2017 Accurate lumen diameter measurement in curved vessels in carotid ultrasound: an iterative scale-space and spatial transformation approach *Med. Biol. Eng. Comput.* **55** 1415–34

[47] Ikeda N, Gupta A, Dey N, Bose S, Shafique S, Arak T, Godia E C, Saba L, Laird J R and Nicolaides A J U 2015 Improved correlation between carotid and coronary atherosclerosis SYNTAX score using automated ultrasound carotid bulb plaque IMT measurement *Ultrasound Med. Biol.* **41** 1247–62

[48] Ikeda N, Dey N, Sharma A, Gupta A, Bose S, Acharjee S, Shafique S, Cuadrado-Godia E, Araki T and Saba L 2017 Automated segmental-IMT measurement in thin/thick plaque with bulb presence in carotid ultrasound from multiple scanners: stroke risk assessment *Comput. Methods Programs Biomed.* **141** 73–81

[49] Han J, Pei J and Kamber M 2011 *Data Mining: Concepts and Techniques* (Amsterdam: Elsevier)

[50] LeCun Y, Bengio Y and Hinton G 2015 Deep learning *Nature* **521** 436–44

[51] Eggleston H G 1950 A property of Hausdorff measure *Duke Math. J.* **17** 491–8

[52] Saba L, Molinari F, Meiburger K, Piga M, Zeng G, Rajendra U A, Nicolaides A and Suri J 2012 What is the correct distance measurement metric when measuring carotid ultrasound intima-media thickness automatically? *Int. Angiol.* **31** 483–9

[53] Bartels S, Franco A R and Rundek T 2012 Carotid intima-media thickness (cIMT) and plaque from risk assessment and clinical use to genetic discoveries *Persp. Med.* **1** 139–45

[54] Gustavsson T, Liang Q, Wendelhag I and Wikstrand J 1994 A dynamic programming procedure for automated ultrasonic measurement of the carotid artery *Computers in Cardiology 1994* (Piscataway, NJ: IEEE) pp 297–300

[55] Ballard D H 1987 Generalizing the Hough transform to detect arbitrary shapes *Readings in Computer Vision* (Amsterdam: Elsevier) pp 714–25

[56] Destrempes F, Meunier J, Giroux M-F, Soulez G and Cloutier G 2011 Segmentation of plaques in sequences of ultrasonic B-mode images of carotid arteries based on motion estimation and a Bayesian model *IEEE Trans. Biomed. Eng.* **58** 2202–11

[57] Giraldi G, Rodrigues P, Suri J and Singh S 2011 Dual active contour models for medical image segmentation *Image Segment.* **129** 129–52

[58] Suri J S, Haralick R M, Sheehan F and Jamin V 1997 Effect of edge detection, pixel classification, and classification-edge fusion over LV calibration: a two stage automatic system *SCIA'97: 10th Scandinavian Conf. on Image Analysis (Lappeenranta, 9–11 June )* pp 197–204

[59] Hill P R, Canagarajah C N and Bull D R 2003 Image segmentation using a texture gradient based watershed transform *IEEE Trans. Image Process.* **12** 1618–33

[60] El-Baz A and Suri J S 2011 *Lung Imaging and Computer Aided Diagnosis* (Boca Raton, FL: CRC Press)

[61] Radeva P and Suri J S 2019 *Vascular and Intravascular Imaging Trends, Analysis, and Challenges: Plaque Characterization* vol 2 (Bristol: IOP Publishing)

[62] Molinari F, Zeng G and Suri J S 2010 Carotid wall segmentation and IMT measurement in longitudinal ultrasound images using morphological approach *Int. Symp. on Biomedical Imaging (Rotterdam)*

[63] Molinari F, Meiburger K M, Saba L, Acharya U R, Ledda G, Zeng G, Ho S Y S, Ahuja A T, Ho S C and Nicolaides A 2012 Ultrasound IMT measurement on a multi-ethnic and multi-institutional database: our review and experience using four fully automated and one semi-automated methods *Comp. Methods Prog. Biomed.* **108** 946–60

[64] Patel A K, Suri H S, Singh J, Kumar D, Shafique S, Nicolaides A, Jain S K, Saba L, Gupta A and Laird J R 2016 A review on atherosclerotic biology, wall stiffness, physics of elasticity, and its ultrasound-based measurement *Curr. Atheroscler. Rep.* **18** 83

[65] Kumar P K, Araki T, Rajan J, Laird J R, Nicolaides A and Suri J S 2018 State-of-the-art review on automated lumen and adventitial border delineation and its measurements in carotid ultrasound *Comput. Methods Programs Biomed.* **163** 155–68

[66] Molinari F, Meiburger K M and Suri J 2011 Automated high-performance cIMT measurement techniques using patented AtheroEdge™: a screening and home monitoring system *Annual Int. Conf. of the IEEE Engineering in Medicine and Biology Society* (Piscataway, NJ: IEEE) pp 6651–4

[67] Golemati S, Stoitsis J, Sifakis E G, Balkizas T and Nikita K S 2007 Using the Hough transform to segment ultrasound images of longitudinal and transverse sections of the carotid artery *Ultrasound Med. Biol.* **33** 1918–32

[68] Stoitsis J, Golemati S, Kendros S and Nikita K 2008 Automated detection of the carotid artery wall in B-mode ultrasound images using active contours initialized by the Hough transform *2008 30th Annual Int. Conf. of the IEEE Engineering in Medicine and Biology Society* (Piscataway, NJ: IEEE) pp 3146–9

[69] Petroudi S, Loizou C P and Pattichis C S 2010 Atherosclerotic carotid wall segmentation in ultrasound images using Markov random fields *Proc. of the 10th IEEE Int. Conf. on Information Technology and Applications in Biomedicine* (Piscataway, NJ: IEEE) pp 1–5

[70] Suri J S, Liu K, Reden L and Laxminarayan S 2002 A review on MR vascular image processing algorithms: acquisition and prefiltering: part I *IEEE Trans. Inform. Technol. Biomed.* **6** 324

[71] Delsanto S, Molinari F, Giustetto P, Liboni W and Badalamenti S 2006 CULEX-completely user-independent layers extraction: ultrasonic carotid artery images segmentation *2005 IEEE Engineering in Medicine and Biology 27th Annual Conf.* (Piscataway, NJ: IEEE) pp 6468–71

[72] Molinari F, Krishnamurthi G, Acharya U R, Sree S V, Zeng G, Saba L, Nicolaides A and Suri J S 2012 Hypothesis validation of far-wall brightness in carotid-artery ultrasound for feature-based IMT measurement using a combination of level-set segmentation and registration *IEEE Trans. Instrum. Meas.* **61** 1054–63

[73] Molinari F, Liboni W, Pantziaris M and Suri J 2011 CALSFOAM-completed automated local statistics based first order absolute moment' for carotid wall recognition, segmentation and IMT measurement: validation and benchmarking on a 300 patient database *Int. Angiol.* **30** 227–41

[74] Kass M, Witkin A and Terzopoulos D 1988 Snakes: active contour models *Int. J. Comput. Vision* **1** 321–31

[75] Molinari F, Zeng G and Suri J S 2010 An integrated approach to computer-based automated tracing and its validation for 200 common carotid arterial wall ultrasound images: a new technique *J. Ultrasound Med.* **29** 399–418

[76] Suri J S, Liu K, Singh S, Laxminarayan S N, Zeng X and Reden L 2002 Shape recovery algorithms using level sets in 2-D/3-D medical imagery: a state-of-the-art review *IEEE Trans. Inf. Technol. Biomed.* **6** 8–28

[77] Liguori C, Paolillo A and Pietrosanto A 2001 An automatic measurement system for the evaluation of carotid intima-media thickness *IEEE Trans. Instrum. Meas.* **50** 1684–91

[78] Molinari F, Zeng G and Suri J S 2011 Inter-greedy technique for fusion of different segmentation strategies leading to high-performance carotid IMT measurement in ultrasound images *Atherosclerosis Disease Management* (Berlin: Springer) pp 253–79

[79] Khanna N N, Jamthikar A D, Gupta D, Araki T, Piga M, Saba L, Carcassi C, Nicolaides A, Laird J R and Suri H S 2019 Effect of carotid image-based phenotypes on cardiovascular risk calculator: AECRS1. 0 *Med. Biol. Eng. Comput.* **57** 1553–66

[80] Than M P, Pickering J W, Sandoval Y, Shah A S, Tsanas A, Apple F S, Blankenberg S, Cullen L, Mueller C and Neumann J T 2019 Machine learning to predict the likelihood of acute myocardial infarction *Circulation* **140** 899–909

[81] Ambale-Venkatesh B, Yang X, Wu C O, Liu K, Hundley W G, McClelland R, Gomes A S, Folsom A R, Shea S and Guallar E 2017 Cardiovascular event prediction by machine learning: the multi-ethnic study of atherosclerosis *Circ. Res.* **121** 1092–101

[82] Biswas M, Kuppili V, Saba L, Edla D R, Suri H S, Cuadrado-Godia E, Laird J, Marinhoe R, Sanches J and Nicolaides A 2019 State-of-the-art review on deep learning in medical imaging *Front. Biosci.* **24** 392–426

[83] Bishop C M 2006 *Pattern Recognition and Machine Learning* (Berlin: Springer)

[84] Krizhevsky A, Sutskever I and Hinton G E 2012 ImageNet classification with deep convolutional neural networks *Adv. Neural Inform. Process. Syst.* **60** 84–90

[85] Naik V, Gamad R and Bansod P 2013 Carotid artery segmentation in ultrasound images and measurement of intima-media thickness *BioMed Res. Int.* **2013** 1–10

[86] Spence J D, Eliasziw M, DiCicco M, Hackam D G, Galil R and Lohmann T 2002 Carotid plaque area: a tool for targeting and evaluating vascular preventive therapy *Stroke* **33** 2916–22

[87] Spence J D 2002 Ultrasound measurement of carotid plaque as a surrogate outcome for coronary artery disease *Am. J. Cardiol.* **89** 10–5

[88] Mathiesen E B, Johnsen S H, Wilsgaard T, Bønaa K H, Løchen M-L and Njølstad I 2011 Carotid plaque area and intima-media thickness in prediction of first-ever ischemic stroke: a 10-year follow-up of 6584 men and women: the Tromsø Study *Stroke* **42** 972–8

[89] Khanna N N, Jamthikar A D, Gupta D, Nicolaides A, Araki T, Saba L, Cuadrado-Godia E, Sharma A, Omerzu T and Suri H S 2019 Performance evaluation of 10-year ultrasound image-based stroke/cardiovascular (CV) risk calculator by comparing against ten conventional CV risk calculators: a diabetic study *Comput. Biol. Med.* **105** 125–43

[90] Martis R J, Acharya U R, Prasad H, Chua C K, Lim C M and Suri J S 2013 Application of higher order statistics for atrial arrhythmia classification *Biomed. Signal Process. Control* **8** 888–900

[91] Maniruzzaman M, Kumar N, Menhazul Abedin M, Shaykhul Islam M, Suri H S, El-Baz A S and Suri J S 2017 Comparative approaches for classification of diabetes mellitus data: machine learning paradigm *Comput. Methods Programs Biomed.* **152** 23–34

[92] Menchón-Lara R-M, Bastida-Jumilla M-C, Morales-Sánchez J and Sancho-Gómez J-L 2014 Automatic detection of the intima-media thickness in ultrasound images of the common carotid artery using neural networks *Med. Biol. Eng. Comput.* **52** 169–81

[93] Menchón-Lara R-M and Sancho-Gómez J-L 2014 Ultrasound image processing based on machine learning for the fully automatic evaluation of the carotid intima-media thickness *12th Int. Workshop on Content-Based Multimedia Indexing (CBMI)* (Piscataway, NJ: IEEE) pp 1–4

[94] Meyer F 1994 Topographic distance and watershed lines *Signal Process.* **38** 113–25

[95] Elnakib A, Gimel'farb G, Suri J S and El-Baz A 2011 Medical image segmentation: a brief survey *Multi Modality State-of-the-Art Medical Image Segmentation and Registration Methodologies* (Berlin: Springer) p 1–39

[96] Suri J S 2000 Computer vision, pattern recognition and image processing in left ventricle segmentation: the last 50 years *Pattern Anal. Appl.* **3** 209–42

[97] Yegnanarayana B 1994 Artificial neural networks for pattern recognition *Sadhana* **19** 189–238

[98] Schwenker F, Kestler H A and Palm G 2001 Three learning phases for radial-basis-function networks *Neural Netw.* **14** 439–58

[99] Huang G-B, Zhu Q-Y and Siew C-K 2006 Extreme learning machine: theory and applications *Neurocomputing* **70** 489–501

[100] Miche Y, Sorjamaa A, Bas P, Simula O, Jutten C and Lendasse A 2009 OP-ELM: optimally pruned extreme learning machine *IEEE Trans. Neural Networks* **21** 158–62

[101] Dunn J C 1973 A Fuzzy Relative of the ISODATA Process and Its Use in Detecting Compact Well-Separated Clusters *Journal of Cybernetics* **3** 32–57

[102] Hartigan J A and Wong M A 1979 Algorithm AS 136: a $k$-means clustering algorithm *J. R. Stat. Soc.* C **28** 100–8

[103] Cuadrado-Godia E, Srivastava S K, Saba L, Araki T, Suri H S, Giannopolulos A, Omerzu T, Laird J, Khanna N N and Mavrogeni S 2018 Geometric total plaque area is an equally powerful phenotype compared with carotid intima-media thickness for stroke risk assessment: a deep learning approach *J. Vasc. Ultrasound* **42** 162–88

[104] Biswas M *et al* 2020 Two-stage artificial intelligence model for jointly measurement of atherosclerotic wall thickness and plaque burden in carotid ultrasound: a screening tool for cardiovascular/stroke risk assessment *Comp. Biol. Med.* **123** 103847

[105] Saba L *et al* 2019 The present and future of deep learning in radiology *Eur. J. Radiol.* **114** 14–24

[106] Hinton G E 2009 Deep belief networks *Scholarpedia* **4** 5947

[107] Baldi P 2012 Autoencoders, unsupervised learning, and deep architectures *Proc. of ICML Workshop on Unsupervised and Transfer Learning* pp 37–49

[108] He K, Zhang X, Ren S and Sun J 2016 Deep residual learning for image recognition *Proc. of the IEEE Conf. on Computer Vision and Pattern Recognition* **2016** 770–8

[109] Long J, Shelhamer E and Darrell T 2015 Fully convolutional networks for semantic segmentation *Proc. of the IEEE Conf. on Computer Vision and Pattern Recognition* (Piscataway, NJ: IEEE) pp 3431–40

[110] Menchón-Lara R-M and Sancho-Gómez J-L 2015 Fully automatic segmentation of ultrasound common carotid artery images based on machine learning *Neurocomputing* **151** 161–7

[111] Biswas M, Kuppili V, Araki T, Edla D R, Godia E C, Saba L, Suri H S, Omerzu T, Laird J R and Khanna N N 2018 Deep learning strategy for accurate carotid intima-media thickness measurement: an ultrasound study on Japanese diabetic cohort *Comput. Biol. Med.* **98** 100–17

[112] del Mar Vila M, Remeseiro B, Grau M, Elosua R, Betriu À, Fernandez-Giraldez E and Igual L 2020 Semantic segmentation with DenseNets for carotid artery ultrasound plaque segmentation and CIMT estimation *Artif. Intell. Med.* **103** 101784

[113] Molinari F, Meiburger K M, Saba L, Acharya U R, Famiglietti L, Georgiou N, Nicolaides A, Mamidi R S, Kuper II and Suri J S 2014 Automated carotid IMT measurement and its validation in low contrast ultrasound database of 885 patient Indian population epidemiological study: results of AtheroEdge® software *Multi-modality Atherosclerosis Imaging and Diagnosis* (Berlin: Springer) pp 209–19

[114] Saba L, Molinari F, Meiburger K M, Acharya U R, Nicolaides A and Suri J S 2013 Inter- and intra-observer variability analysis of completely automated cIMT measurement software (AtheroEdge™) and its benchmarking against commercial ultrasound scanner and expert readers *Comput. Biol. Med.* **43** 1261–72

[115] Suri J S, Haralick R M and Sheehan F H 2000 Greedy algorithm for error correction in automatically produced boundaries from low contrast ventriculograms *Pattern Anal. Appl.* **3** 39–60

[116] Biswas M, Kuppili V, Edla D R, Suri H S, Saba L, Marinhoe R T, Sanches J M and Suri J S 2018 Symtosis: a liver ultrasound tissue characterization and risk stratification in optimized deep learning paradigm *Comput. Methods Programs Biomed.* **155** 165–77

[117] Carvalho D D, Akkus Z, van den Oord S C, Schinkel A F, van der Steen A F, Niessen W J, Bosch J G and Klein S 2014 Lumen segmentation and motion estimation in B-mode and contrast-enhanced ultrasound images of the carotid artery in patients with atherosclerotic plaque *IEEE Trans. Med. Imaging* **34** 983–93

[118] Gastounioti A, Golemati S, Stoitsis J and Nikita K 2010 Kalman-filter-based block matching for arterial wall motion estimation from B-mode ultrasound *IEEE Int. Conf. on Imaging Systems and Techniques* (Piscataway, NJ: IEEE) pp 234–9

[119] Ilea D E, Duffy C, Kavanagh L, Stanton A and Whelan P F 2012 Fully automated segmentation and tracking of the intima media thickness in ultrasound video sequences of the common carotid artery *IEEE Trans. Ultrason. Ferroelectr. Freq. Control* **60** 158–77

[120] Araki T, Ikeda N, Molinari F, Dey N, Acharjee S M, Saba L, Nicolaides A and Suri J S 2014 Effect of geometric-based coronary calcium volume as a feature along with its shape-based attributes for cardiological risk prediction from low contrast intravascular ultrasound *J. Med. Imaging Informatics* **4** 255–61

[121] Molinari F, Zeng G and Suri J S 2010 A state of the art review on intima–media thickness (IMT) measurement and wall segmentation techniques for carotid ultrasound *Comput. Methods Prog. Biomed.* **100** 201–21

[122] Liang Q, Wendelhag I, Wikstrand J and Gustavsson T 2000 A multiscale dynamic programming procedure for boundary detection in ultrasonic artery images *IEEE Trans. Med. Imaging* **19** 127–42

[123] Stein J H, Korcarz C E, Mays M E, Douglas P S, Palta M, Zhang H, LeCaire T, Paine D, Gustafson D and Fan L 2005 A semiautomated ultrasound border detection program that facilitates clinical measurement of ultrasound carotid intima-media thickness *J. Am. Soc. Echocardiogr.* **18** 244–51

[124] Faita F, Gemignani V, Bianchini E, Giannarelli C, Ghiadoni L and Demi M 2008 Real-time measurement system for evaluation of the carotid intima-media thickness with a robust edge operator *J. Ultrasound Med.* **27** 1353–61

[125] Molinari F, Acharya U R, Zeng G, Meiburger K M and Suri J S 2011 Completely automated robust edge snapper for carotid ultrasound IMT measurement on a multi-institutional database of 300 images *Med. Biol. Eng. Comput.* **49** 935–45

[126] Molinari F, Meiburger K M, Zeng G, Nicolaides A and Suri J S 2011 CAUDLES-EF: carotid automated ultrasound double line extraction system using edge flow *J. Digital Imaging* **24** 1059–77

[127] Molinari F and Suri J S 2011 Automated measurement of carotid artery intima-media thickness *Ultrasound and Carotid Bifurcation Atherosclerosis* (Berlin: Springer) pp 177–92

[128] Gutierrez M A, Pilon P E, Lage S, Kopel L, Carvalho R and Furuie S 2002 Automatic measurement of carotid diameter and wall thickness in ultrasound images *Computers in Cardiology* (Piscataway, NJ: IEEE) pp 359–62

[129] Molinari F, Liboni W, Giustetto P, Badalamenti S and Suri J S 2009 Automatic computer-based tracings (ACT) in longitudinal 2-D ultrasound images using different scanners *J. Mech. Med. Biol.* **9** 481–505

[130] Molinari F, Acharya U R, Saba L, Nicolaides A and Suri J S 2014 Hypothesis validation of far wall brightness in carotid artery ultrasound for feature-based IMT measurement using a combination of level set segmentation and registration *Multi-Modality Atherosclerosis Imaging and Diagnosis* (Berlin: Springer) pp 255–67

[131] Khanna N N, Jamthikar A D, Araki T, Gupta D, Piga M, Saba L, Carcassi C, Nicolaides A, Laird J R and Suri H S 2019 Nonlinear model for the carotid artery disease 10-year risk prediction by fusing conventional cardiovascular factors to carotid ultrasound image phenotypes: a Japanese diabetes cohort study *Echocardiography* **36** 345–61

[132] Cuadrado-Godia E, Jamthikar A D, Gupta D, Khanna N N, Araki T, Maniruzzaman M, Saba L, Nicolaides A, Sharma A and Omerzu T 2019 Ranking of stroke and cardiovascular risk factors for an optimal risk calculator design: logistic regression approach *Comput. Biol. Med.* **108** 182–95

[133] Jamthikar A *et al* 2020 Artificial intelligence framework for predictive cardiovascular and stroke risk assessment models: a narrative review of integrated approaches using carotid ultrasound 104043

[134] Jamthikar A D, Puvvula A, Gupta D, Johri A M, Nambi V, Khanna N N, Saba L, Mavrogeni S, Laird J R and Pareek G 2020 Cardiovascular disease and stroke risk assessment in patients with chronic kidney disease using integration of estimated glomerular filtration rate, ultrasonic image phenotypes, and artificial intelligence: a narrative review *Int. Angiol.* **2** 150–64

[135] Narayanan R, Kurhanewicz J, Shinohara K, Crawford E D, Simoneau A and Suri J S 2009 Enhanced measurement MRI-ultrasound registration for targeted prostate biopsy *IEEE Int. Symp. on Biomedical Imaging: From Nano to Macro* (Piscataway, NJ: IEEE) pp 991–4

[136] Saba L, Banchhor S K, Araki T, Viskovic K, Londhe N D, Laird J R, Suri H S and Suri J S 2018 Intra-and inter-operator reproducibility of automated cloud-based carotid lumen diameter ultrasound measurement *Indian Heart J.* **70** 649–64

[137] Saba L, Than J C M, Noor N M, Rijal O M, Kassim R M, Yunus A, Ng C R and Suri J S 2016 Inter-observer variability analysis of automatic lung delineation in normal and disease patients *J. Med. Syst.* **40** 142

[138] Zhang S, Suri J S, Salvado O, Chen Y, Wacker F K, Wilson D L, Duerk J L and Lewin J S 2005 Inter-and intra-observer variability assessment of in vivo carotid plaque burden quantification using multi-contrast dark blood MR images *Stud. Health Technol. Inform.* **113** 384–93

[139] Biswas M *et al* 2019 Deep learning fully convolution network for lumen characterization in diabetic patients using carotid ultrasound: a tool for stroke risk *Med. Biol. Eng. Comput.* **57** 543–64

[140] Biswas M, Saba L, Chakrabartty S, Khanna N N, Song H, Suri H S, Sfikakis P P, Mavrogeni S, Viskovic K and Laird J R 2020 Two-stage artificial intelligence model for jointly measurement of atherosclerotic wall thickness and plaque burden in carotid ultrasound: a screening tool for cardiovascular/stroke risk assessment *Comput. Biol. Med.* **123** 103847

[141] Jamthikar A, Gupta D, Saba L, Khanna N N, Araki T, Viskovic K, Mavrogeni S, Laird J R, Pareek G and Miner M 2020 Cardiovascular/stroke risk predictive calculators: a comparison between statistical and machine learning models *Cardiovasc. Diag. Ther.* **10** 919

[142] Jamthikar A, Gupta D, Khanna N N, Saba L, Laird J R and Suri J S 2020 Cardiovascular/ stroke risk prevention: a new machine learning framework integrating carotid ultrasound image-based phenotypes and its harmonics with conventional risk factors *Indian Heart J.* **72** 258–64

[143] Viswanathan V *et al* 2020 Low-cost preventive screening using carotid ultrasound in patients with diabetes *Front. Biosci.* **25** 1132–71

[144] Glas A S, Lijmer J G, Prins M H, Bonsel G J and Bossuyt P M 2003 The diagnostic odds ratio: a single indicator of test performance *J. Clin. Epidemiol.* **56** 1129–35

[145] Kadam P and Bhalerao S 2010 Sample size calculation *Int. J. Ayurveda Res.* **1** 55

[146] Hadjis S, Zhang C, Mitliagkas I, Iter D and Ré C 2016 Omnivore: an optimizer for multi-device deep learning on CPUS and GPUS, arXiv:1606.04487

[147] Chen T, Li M, Li Y, Lin M, Wang N, Wang M, Xiao T, Xu B, Zhang C and Zhang Z 2015 Mxnet: a flexible and efficient machine learning library for heterogeneous distributed systems, arXiv:1512.01274

[148] Ooi B C, Tan K-L, Wang S, Wang W, Cai Q, Chen G, Gao J, Luo Z, Tung A K and Wang Y 2015 SINGA: a distributed deep learning platform *Proc. of the 23rd ACM Int. Conf. on Multimedia* pp 685–8

[149] Saba L, Tiwari A, Biswas M, Gupta S K, Godia-Cuadrado E, Chaturvedi A, Turk M, Suri H S, Orru S and Sanches J M 2019 Wilson's disease: a new perspective review on its genetics, diagnosis and treatment *Front. Biosci.* **11** 166–85

[150] Tandel G S, Biswas M, Kakde O G, Tiwari A, Suri H S, Turk M, Laird J R, Asare C K, Ankrah A A and Khanna N 2019 A review on a deep learning perspective in brain cancer classification *Cancers* **11** 111

[151] Sharma A M, Gupta A, Kumar P K, Rajan J, Saba L, Nobutaka I, Laird J R, Nicolades A and Suri J S 2015 A review on carotid ultrasound atherosclerotic tissue characterization and stroke risk stratification in machine learning framework *Curr. Atheroscler. Rep.* **17** 55

[152] Flach P 2004 The many faces of ROC analysis in machine learning *ICML Tutor.* **20** 538–46

[153] Flach P A 2001 On the state of the art in machine learning: a personal review *Artif. Intell.* **131** 199–222

[154] Cover T and Hart P 1967 Nearest neighbor pattern classification *IEEE Trans. Inf. Theory* **13** 21–7

[155] Cortes C and Vapnik V 1995 Support-vector networks *Mach. Learn.* **20** 273–97

[156] Majumder M 2015 Artificial neural network *Impact of Urbanization on Water Shortage in Face of Climatic Aberrations* (Berlin: Springer) pp 49–54

[157] Resnik P and Hardisty E 2010 Gibbs sampling for the uninitiated *Maryland University College Park Institute for Advanced Computer Studies*

[158] Quinlan J R 1986 Induction of decision trees *Mach. Learn.* **1** 81–106

[159] LeCun Y, Touresky D, Hinton G and Sejnowski T 1988 A theoretical framework for back-propagation *Proc. of the 1988 Connectionist Models Summer School: 1988 CMU* (Pittsburgh, PA: Morgan Kaufmann) pp 21–8

[160] Barlow H B 1989 Unsupervised learning *Neural Comput.* **1** 295–311

[161] Alsabti K, Ranka S and Singh V 1997 An efficient k-means clustering algorithm *Proc. First Workshop High Performance Data Mining*

[162] Biswas M *et al* 2021 A review on joint carotid intima-media thickness and plaque area measurement in ultrasound for cardiovascular/stroke risk monitoring: artificial intelligence framework *J. Digit. Imaging* **34** 581–604

# Part IV

Machine and deep learning in liver imaging

**IOP** Publishing

# Multimodality Imaging, Volume 1
### Deep learning applications
**Mainak Biswas and Jasjit S Suri**

# Chapter 10

# Ultrasound fatty liver disease risk stratification using an extreme learning machine framework

**Mainak Biswas and Jasjit S Suri**

The build-up of fat in liver cells causes fatty liver disease (FLD), which can lead to fatal conditions including liver cancer. In this regard, numerous FLD detection and characterization models based on machine learning (ML) have been used. These ML algorithms use a huge number of ultrasonic grayscale features, a feature selection approach for picking the best features, and a variety of training/testing combinations. As a result of the disparity between the grayscale feature type and the classifier used, they are computationally costly, slow, and do not guarantee excellent performance.

Here, a new tissue classification method (a type of Symtosis system) based on an extreme learning machine (ELM) for risk characterization of ultrasound liver imaging is proposed. The ELM is a single hidden layer neural network. The parameters from the input to hidden layer are generated randomly. The parameters from the hidden to output layers are only trained, and this is done in one pass (without iteration), making ELM more rapid than traditional ML methods. Using four different $K$-fold cross-validation methods ($K = 2$, 3, 5, and 10) on three different data sizes, i.e. S0 representing the original, S4 representing four splits, S8 representing eight splits (a sum of 12 instances), and 46 different grayscale features, we use ELM to stratify FLD ultrasound (US) pictures and compare them to support vector machine (SVM) results.

Our findings show that the ELM outperforms the SVM in terms of sensitivity, specificity, accuracy, and area-under-the-curve (AUC) for all cross-validation (CV) protocols (K2, K3, K5, and K10) and all types of US datasets (S0, S4, and S8) using an US liver database of 63 patients (27 normal/36 abnormal). The ELM showed a performance of 96.75% accuracy compared to 89.01% for the SVM for the S8 dataset, with AUCs of 0.97 and 0.91, respectively, using the K10 CV methodology. Further tests revealed that the ELM classifier had a average speed improvement of 40% when compared to the SVM. We used two biometric classes to validate the

Symtosis system. We tested the Symtosis method using two classes of biometric face public data and found it to be 100% accurate.

## 10.1 Introduction

In the past two decades, liver-related mortality has consistently been among the top 12 causes of death, and is the fourth largest cause of death in individuals from 45–54 years old. [1]. Fatty liver diseases (FLDs) are caused by the presence of too much fat in the liver cells. Steatosis is the accumulation of fat in the liver cells, which can be induced by metabolic syndromes, alcohol intake, obesity due to insulin resistance, and a range of other reasons [1, 2]. FLDs are divided into two categories: alcoholic and non-alcoholic. Non-alcoholic FLD (NAFLD) affects the majority of the people in Western countries who have the disease [3]. FLD can cause significant conditions, e.g. steatohepatitis, cirrhosis, and liver cancer. This disease is treatable in its early stages, and patients who are diagnosed with FLD early on have a higher chance of living a long life. Furthermore, when compared to the treatment of more severe liver illnesses, FLD detection is less expensive. For the time being, liver biopsy is the gold standard for detecting FLD. The biopsy procedure is inconvenient, prone to sample mistakes, and is invasive [4]. FLD can be detected using a variety of non-invasive imaging techniques, i.e. computed tomography (CT) and magnetic resonance imaging (MRI). Radiation concerns are a problem with CT [5] and MRI can only detect a tiny amount of fat [6]. When it comes to detecting fatty infiltration, MRI performs well [1, 7]. Ultrasound (US) pictures, which are widely adopted for FLD imaging, are an alternative to these modalities [8]. Machine learning (ML) was used to classify US liver pictures, and the specificity and sensitivity were both above 80% [9]. As a result, one of the most often used scanning techniques for FLD detection is US [10].

In the literature, two strategies for the characterization of liver disease have been proposed: (a) ML and (b) signal processing. Suri and his team developed a tissue characterization system [11] using the discrete wavelet transform (DWT) [12], high order spectra (HOS) [13], and texture features [14], which were computed using US liver images and then fed to a decision tree (DT) based classifier, resulting in a 93.3% accuracy. Acharya *et al* suggested a fuzzy classifier for identification of Hashimoto thyroiditis from thyroid pictures [15–17] using the wavelet transform [18] in the same class in 2014. The performance of the system was 84.6 % accuracy. Subramanya *et al* [19] used the support vector machine (SVM) classifier to achieve a performance of 84.9% accuracy on US liver scans in 2014. Ma *et al* established a kurtosis-based [20] scanning method for the identification and grading of FLD in US liver images using a signal processing approach in 2015, signifying an accuracy of 81.2%. Suri and his colleagues (Saba *et al* [21]) employed a back propagation neural network (BPNN) with ten hidden layers to mine 128 characteristics from US liver images using six different feature mining algorithms in 2016. The accuracy of the BPNN was 97.6%.

The SVM [22] is a popular supervised machine learning approach. For stratification, SVMs use two major strategies. First, it uses kernel methods to convert the problem from its original input space to a high-dimensional space termed the feature space, which allows for linear separation of training samples belonging to various classes. Second, it seeks the optimal hyper-plane for splitting the two classes. These ML algorithms make use of a huge number of ultrasonic grayscale characteristics, a pooling approach for picking the best features, and several training and testing combinations. As a result of the disparity between the grayscale features and classifier type, they are computationally costly, slow, and do not guarantee excellent performance. We offer the extreme learning machine (ELM) [23, 24] paradigm, which considers computational speed and performance in relation to data size. The ELM is a single hidden layer feed-forward neural network (SLFFNN) with random initialization of the input-to-hidden layer parameters. The ELM uses only the least square loss model, which utilizes a closed form solution generated by the Moore–Penrose pseudo-inverse [25], to train the hidden-to-output layer parameters. The error is minimized in a least squares sense, and it is likely to be more accurate or at least similar to iterative neural network models [21]. Furthermore, ELM provides a one-stop shop for training parameters, leading us to predict that the ELM is likely to be a faster technique than the SVM, achieving minimum least squares error in one pass regardless of the quantity of training data. The ELM, unlike other neural network architectures [21], is a single-layered neural network architecture that requires little resource management and is likely to perform better. As a result, we believe ELM will provide a better approach for FLD risk classification than traditional ML systems.

In this work, we use a US liver dataset to conduct a complete analysis of the two approaches. As shown in figure 10.1, we created a computer-assisted technique for detecting and stratifying FLD-affected (diseased) and FLD-unaffected (controls or normal) liver pictures. Before being sent into the tissue characterization module, the input US pictures are processed and partitioned. Before feeding the dataset, four types of cross-validation ($K = 2, 3, 5,$ and 10) are performed. We also sub-sample the

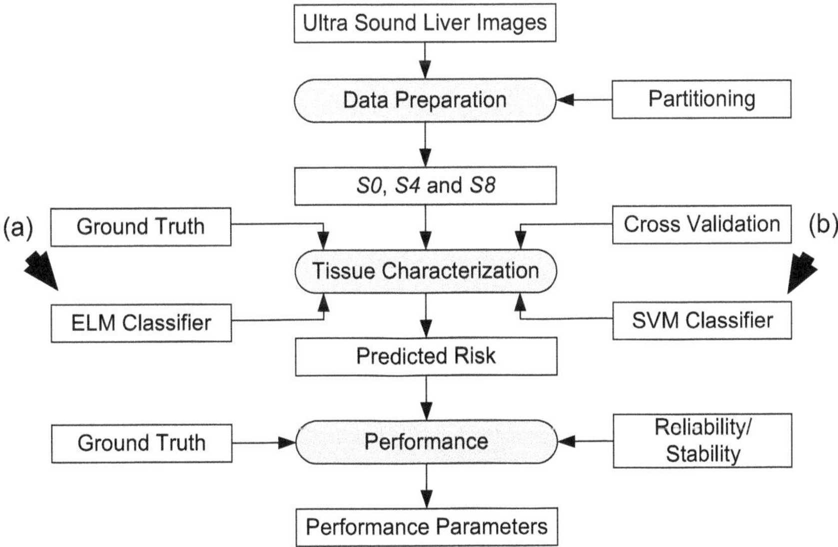


**Figure 10.1.** The overall Symtosis system using the ELM based risk assessment system. (Reproduced with permission from [54]. Copyright 2017 Springer.)

original photos (S0) into four parts (S4) and eight parts (S8) because our data are restricted. The tissue characterization module then generates a risk prediction based on the ground truth (i.e. biopsy reports) and cross-validation type. Finally, the anticipated risk is compared to the actual risk, yielding the performance parameters. Figure 10.1 depicts the complete system in its entirety. Using the Gabor filter, gray level co-occurrence matrix (GLCM), and gray level run length matrix (GRLM), the system produces 46 features. We have limited ourselves to only limited feature mining without feature selection paradigms because the aim of this chapter is to merely understand and harness the ELM, as well as to benchmark against the SVM and BPNN. The accuracy of the system using the K10 protocol on three kinds of datasets (S0, S4, and S8) is 92.4%, 94.8%, and 96.7%, respectively, whereas the accuracy of the SVM based system is 86%, 87.9%, and 89%, respectively. It is also worth noting that as $K$ (cross-validation folds) and sub-sampling go up, the accuracy values go up. Compared to the SVM, we obtain a 40% increase in ELM speed. We also compare our architecture to BPNN [21], which is likewise based on a neural network (NN) based system and has a similar accuracy as well as an efficient architecture and good speed.

The ELM based tissue characterization system is further tested using a biometric facial dataset, where it achieves a performance of 100% accuracy, demonstrating a higher degree of generalization than current machine learning algorithms such as the SVM. The demographics of the data and the picture acquisition procedure in the US are discussed in section 10.2. Section 10.3 introduces several feature mining strategies and develops the ELM paradigm's mathematical underpinning. In section 10.4 the experimental technique is provided and the results are reported in section 10.5. In section 10.6 we compare our results to those of traditional SVM based classification and in section 10.7 we analyze our findings.

## 10.2 Data demographics, collection, and preparation

*Demographics, ethics clearance, and gold standard.* The Instituto Superior Tecnico (IST), University of Lisbon, Portugal, had written consent from all informed patients, a total of 63 patients (37 abnormal and 26 normal) and their data were gathered after IRB approval. Retrospective analysis was done on the images. It was decided to choose a patient with a normal BMI. Figure 10.2 depicts the normal and

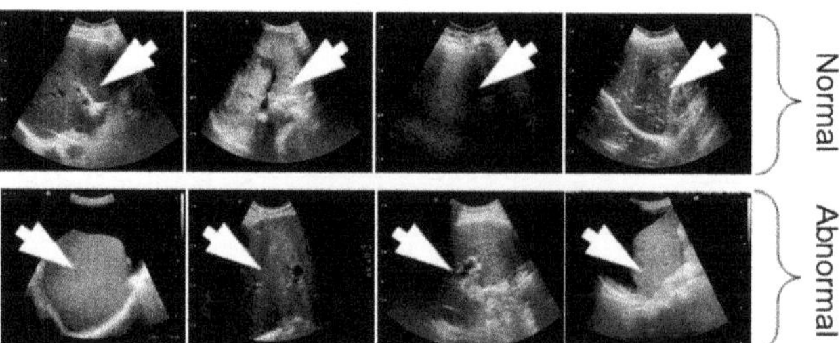


**Figure 10.2.** Top row: ordinary liver images; bottom row: abnormal liver images. (Reproduced with permission from [54]. Copyright 2017 Springer.)

anomalous US scanned pictures. By collecting a liver biopsy and examining it in a tissue pathological laboratory, the gold standard or ground truth label for each patient (normal or abnormal) was determined [11].

*Liver ultrasound scanning, data collection, and preparation.* Medical specialists assisted in the US scanning and analysis of the patients. The scanned photos were captured with a Philips CX 50 US machine. The US scanner had a frequency range of 1–5 MHz and 160 curved piezoelectric components. Grayscale photos at a resolution of 1024 × 1024 pixels were taken. Each grayscale image was saved with a resolution of 8 bits per pixel. For input, the manufacturers were given a standard computer interface. We used this interface to obtain the image data from the patient. Before receiving input photographs, we double-checked the settings and calibration of the US machine. Standardization was carried out according to the method proposed by Qayyum *et al* [27]. Specifically, US scanning was performed on 20 patients who had a normal liver and a normal BMI (18.5–24.9). The image outcome was then investigated. Standardization was carried out based on the findings. A US device with a 15 cm picture depth and a frequency of 3.5 MHz was used. The image featured two focal zones, with the center one measuring 7.5 cm. The dynamic range was fixed at 70 dB for this experiment, but the gain was adjusted dependent on the patient biotype. To eliminate this changeable parameter, time gain compensation (TGC) was established at the center point for all scanned image examinations. The protocol's standardization is aided by the fixed central position. To obtain the anatomical landmarks of the liver, several transducer angles and orientations were used based on patient biotype. To avoid major patient motion, patients were kept in a supine, comfortable position during scanning. The liver is branched into two lobes, one in the epigastric area and the other in the right hypochondrium [26]. The effects of FLD illness can be observed in both the liver and the spleen. We utilized a scanned image of the right lobe liver as it is the most important region of the liver. Each image was divided into a 128 × 128 pixel region-of-interest (ROI) along the medial axis.

### 10.2.1 Sub-sampling of US datasets (S4 and S8)

The learning of the ELM based Symtosis generalizes faster if the training samples are increased, therefore, the training samples should be increased. Thus, the original DICOM pictures are subsampled into two datasets: S4 and S8. We divided the original datasets (S0) into four and eight sections, dubbed the S4 and S8 US liver datasets, respectively. Figures 10.3 and 10.4 illustrate examples of S4 and S8 pictures, respectively.

## 10.3 Methodology

The functionality of classifiers in the Symtosis system (see figure 10.1) has been examined. The computational expense of obtaining support vectors in the dataset is the main obstacle in using the SVM. The use of kernel functions to obtain linear solutions in high-dimensional space for non-linearly separable data adds to the mathematical strain [28]. The ELM solves the classification problem in one iteration, eliminating the need for an iterative technique. Figure 10.5 shows the internal

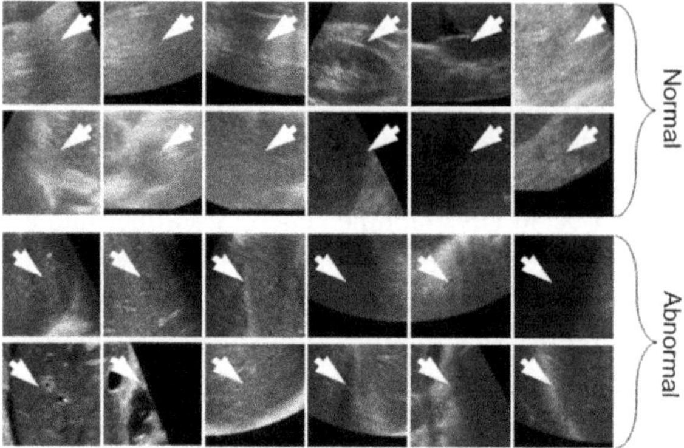


**Figure 10.3.** S4 datasets. Top two rows: ordinary liver images; bottom two rows: abnormal liver images. (Reproduced with permission from [54]. Copyright 2017 Springer.)

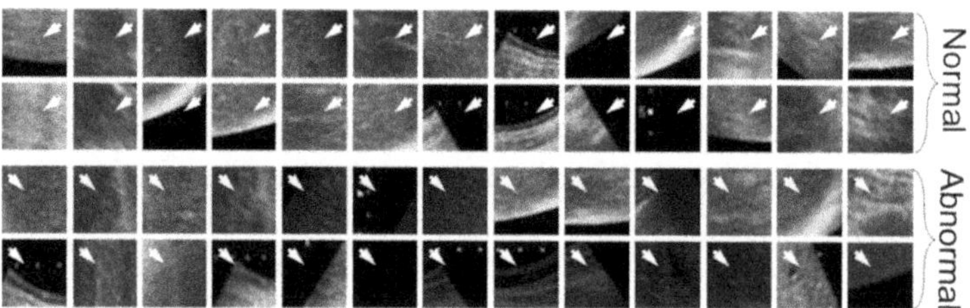


**Figure 10.4.** S8 datasets. Top two rows: ordinary liver images; bottom two rows: abnormal liver images. (Reproduced with permission from [54]. Copyright 2017 Springer.)

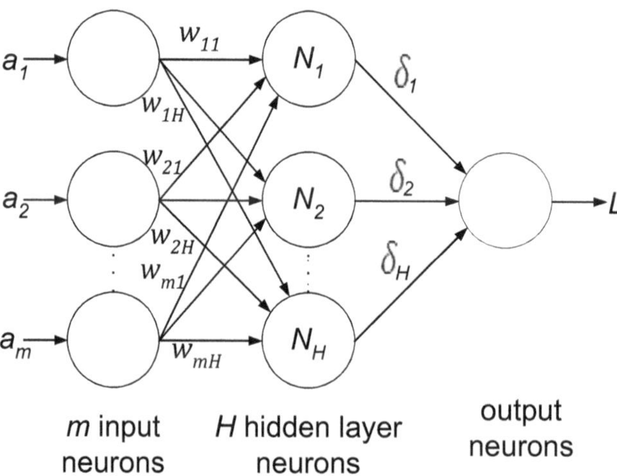


**Figure 10.5.** The design of the ELM paradigm using SLFFNN in the Symtosis class for liver tissue stratification. (Reproduced with permission from [54]. Copyright 2017 Springer.)

architecture of the Symtosis system's ELM based tissue classification, which allows the SLFFNN to be trained and optimized in a single pass. It can be seen that the ELM uses the generalized pseudo-inverse matrix activation function in conjunction with the well-known targets to obtain the optimal hidden-to-output parameters in a single iteration, lowering the system's computational cost. The activation function is made up of a amalgamation of input features and randomized input-to-hidden layer parameters, as shown below. In this work, US liver images were used to mine features and an ELM based CADx system is proposed for FLD disease identification and risk characterization [23, 24, 29]. The Gabor, GLCM, and GRLM features [30–34] are the three feature mining algorithms based on texture that we apply. The Gabor feature mining is based on the scale and direction of the pixel distribution in a picture. Finally, the GRLM matrix determines the neighboring pixel of a reference pixel and texture features are generated after the GLCM mines statistical second-order features. The ELM architecture and mathematical underpinning are explored in length in section 10.3.1, while the tissue characterization algorithm is presented in section 10.3.2. In section 10.3.3, the details of feature mining algorithms are described.

### 10.3.1 The three-layered ELM architecture

As described earlier, the ELM is an SLFFNN that can be taught in one pass, which makes it more rapid than current ML techniques. In the SLFFNN there are three layers of neurons (or nodes), with parameters from the input to hidden nodes being randomly initiated and subsequently fixed without iteration. The parameters from the hidden to output layer are the only ones that need to be learned. ELM tends to reach a global optimum quickly because it learns the parameters in a single pass. Figure 10.5 depicts the ELM architecture. It is a three-tiered structure. The input is accepted by the first layer, which then passes it on to the concealed layer.

Let the number of training images be represented by the vector $P_{\text{trg}}$ and testing images be represented by $P_{\text{tst}}$. Let the target vectors analogous to the training and testing images be represented by $L_{\text{trg}}$ and $L_{\text{tst}}$. Let the weight vector be $W$ from the input to hidden layer. Let $Q$ be the output of the application of the activation function to the input data. Let $\delta$ be the hidden-to-output layer vector of training parameters. Then, the least square solution is given by $\left\| Q\hat{\delta} - L_{\text{trg}} \right\| = \min_{\delta} \left\| Q\delta - L_{\text{trg}} \right\|$, where $\delta$ is the least squares solution of $Q\delta = L_{\text{trg}}$. For a larger training dataset the minimum norm of the least squares solution of the linear system is represented by $\hat{\delta} = Q^{\dagger}L_{\text{trg}}$, where $Q^{\dagger}$ is the Moore–Penrose [35] generalized inverse of matrix $Q$. The complete set of symbols and their significance are given in table B.2.

### 10.3.2 Tissue characterization and risk stratification using ELM and SVM frameworks

The tissue and risk characterization system is based on a traditional machine learning (ML) system design, in which the given data are divided into training and testing for cross-validation protocol creation. Figure 10.6 illustrates this. This is

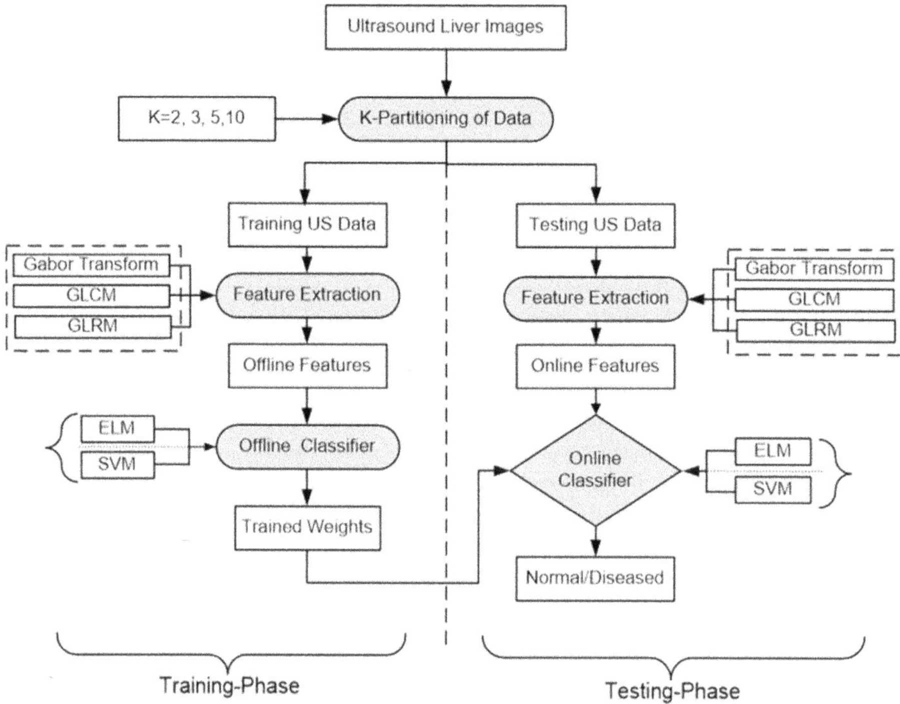


**Figure 10.6.** Tissue characterization and risk prediction system in ELM and SVM frameworks. (Reproduced with permission from [54]. Copyright 2017 Springer.)

made up of two parts: the training phase and the testing phase, which are depicted as the left and right halves of figure 10.6. The training phase produces training parameters or coefficients, whereas the testing phase predicts the label class.

The testing phase is largely identical to the training phase, with the exception that the training phase generates training parameters using ground truth labels and grayscale features calculated from training US images. The testing phase then predicts the class of the test images, which is derived using the transformation of the testing features through the training parameters. It is worth noting that both ELM and SVM use the same feature computation protocol. The ELM based tissue characterization system employs the SLFFNN, which is made up of three sets of neurons connected by weighted lines. The parameters from the input to hidden layers are chosen at random, whereas the parameters from the hidden to output layers must be learned. The number of input neurons equals the number of picture features mined. The number of buried layer neurons is estimated to be two hundred. The hidden neurons are connected to all of the input neurons.

Among the activation functions employed, the sigmoid, sine hard limit, triangle basis function, and radial basis function are the most prevalent forms of in the ELM. The sigmoid function was chosen as the activation function in our experiment. The number of output neurons is determined by the classification issue type. The output layer neuron is coupled to each of the hidden neurons. The least squares solution can be transformed into algorithmic steps using the notations as illustrated in the pseudo-code block below. Note that the maximum margin hyper-plane between two classes is found out from the calculated support vectors generated from the SVM during the training phase if the SVM is modified in figure 10.6.

### 10.3.3 Feature mining

The goal of feature mining was to compute a small number of features in order to grasp the power of the ELM while comparing it to the SVM. The feature mining methods used in our experiment, Gabor, GLCM, and GRLM, are discussed here. The directions and scales, as well as texture repeatability, were used to choose these traits [36].

#### 10.3.3.1 Gabor based directional features

The Gabor filter is a type of edge detection filter that combines Gaussian and complex-plane waves. It seeks to reduce uncertainty in both the spatial and frequency domains by using this combination. This function's dilations and rotations yield Gabor filters that are similar. It aids in the alignment and identification of scale-tunable edges and lines. It aids in the expansion of an image and its containment in spatial frequency representation. A sinusoidal wave can be used to depict the Gabor transform's impulse response (a plane wave for distinct frequency and aligned 2D Gabor). This is given as

$$f(p, q) = \exp\left\{-\frac{1}{2}\left[\left(\frac{p}{\sigma_p}\right)^2 + \left(\frac{q}{\sigma_q}\right)^2\right]\right\} \exp\left[j2\pi(Up + Vq)\right], \tag{10.1}$$

where $(p, q)$ symbolize the rectilinear spatial-domain coordinates, $(U, V)$ represent the points that are the specific 2D frequency of the complex sinusoid, and $(\sigma_p, \sigma_q)$ depict the spatial extent and bandwidth of $f$. Figure 10.7 shows the Gabor filters used for feature mining.

#### 10.3.3.2 Gray level co-occurrence matrix

The gray level co-occurrence matrix (GLCM) is a well-known texture mining technique [37–41]. The GLCM depicts the spatial relationship between adjacent pixels. It calculates how often a pixel with a certain gray level or intensity appears in relation to its neighbors in a range of directions. The statistical distribution of pixel

| $0^\circ$ | $30^\circ$ | $60^\circ$ | $90^\circ$ | $120^\circ$ | $150^\circ$ |
|---|---|---|---|---|---|

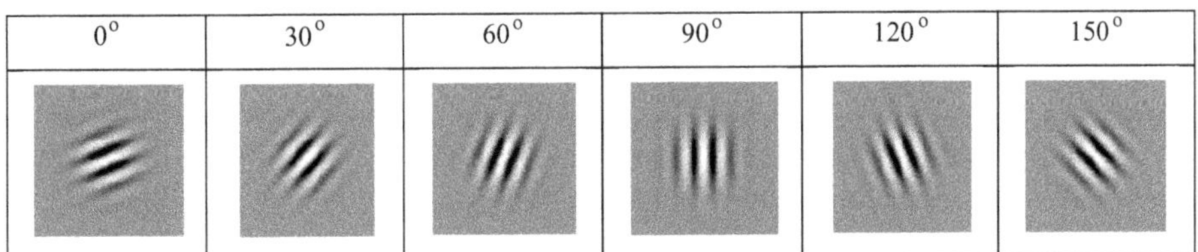

**Figure 10.7.** Gabor filter representation at different orientations. (Reproduced with permission from [54]. Copyright 2017 Springer.)

intensities is used to compute features. Second-order statistics are used in GLCM based feature mining. The texture feature derived from the co-occurrence matrix is never used for analysis directly. The probability of two pixels with gray levels $i, j$ that are positioned in the inter-distance $d$ direction $\theta$ is calculated using the gray level co-occurrence matrix.

The probability is represented by $p(i, j|d, \theta)$. Angle $\theta$ and distance $d$ are used to illustrate the spatial relationship. We calculate characteristics based on the calculated likelihood.

### 10.3.3.3 Gray level run length matrix

The gray level run length matrix (GRLM) is a group of collinear pixels with the same gray level. The GRLM's primary role is to mine texture features and images of gray intensity pixels in a given orientation from which reference pixels are computed. The run length represented as $S(i, j|d, \theta)$ is the number of pixels in a given direction that have the same gray intensity.

## 10.4 Experimental protocol

We use a CV experimental approach to assess the generalization strength of each methodology. The effect of four cross-validation techniques on stratification accuracy utilizing all three types of datasets is discussed in section 10.4.1. Using various cross-validation protocols, the influence of the percentage of dataset size on sub-sampling data on the system's accuracy is investigated in section 10.4.2. We compare the time required by the ELM and SVM in this experiment because the ELM is a single pass method. The comparative time analysis for the ELM and SVM algorithms is presented in section 10.4.3.

### 10.4.1 Experiment 1: the effect of the size of the training data on accuracy using four CV protocols

The goal of this experiment is to determine the effect of the size of the training data on risk stratification performance. We can adjust the number of patients in the training datasets using the cross-validation process. We used four different cross-validation protocols: K2, K3, K5, and K10, which stand for two-fold, three-fold, five-fold, and ten-fold, respectively. The dataset includes each fold. The dataset is equally partitioned into two parts in K2 cross-validation, with one portion utilized for training and the other for testing.

### 10.4.2 Experiment 2: the effect of the training set size using the sub-sampling strategy

It is critical to comprehend how the size of the training data affects the ELM architecture. Because no iterations are required, as with NN or BPNN, the scale of the training data can have a greater impact on the ELM system's performance. As a result, we subsampled the original database (S0) into two types of datasets: the S4 and S8 datasets. The images were separated into four parts and eight parts, respectively, as illustrated in figures 10.3 and 10.4. All CV protocols (i.e. K2, K3, K5, and K10 cross-validations) must be run for all S0, S4, and S8.

### 10.4.3 Experiment 3: ELM and SVM time complexity

As the ability to learn incredibly quickly lies at the heart of the ELM, the temporal complexity of the ELM system must be computed for both the training and testing phases. This necessitates computing the times for all CV protocols (i.e. K2, K3, K5, and K10) as well as all three types of datasets (i.e. S0, S4, and S8), resulting in a total of 12 time comparisons.

## 10.5 Results

The findings of the three experiments performed in the ELM framework on the US liver dataset are presented in this section. Using the four CV protocols, section 10.5.1 illustrates the effect of training data size. Section 10.5.2 shows the findings of employing a sub-sampling technique to examine the effect of training set size. The results of the time analysis are reported in section 10.5.3.

### 10.5.1 Experiment 1: the effect of training data size on accuracy using the four CV protocols

If $\eta_{\mathrm{sys}}$ is the system accuracy, $k$ represents the cross-validation method, i.e. K2, K3, K5, and K10, $t$ represents the index of trial numbers, $T$ represents the total number of trials, $i$ represents the index of data size, and $N_L$ represents the total size of the liver dataset, then the average accuracy for each cross-validation protocol, $k$, of the system can be mathematically expressed as

$$\eta_{\mathrm{sys}}(k) = \frac{\sum\limits_{t-1}^{T}\sum\limits_{i-1}^{N_L}\eta(k, i, t)}{T \times N_L}. \tag{10.2}$$

$T = 20$ trials are carried out in total. Table 10.1 shows the average accuracy, sensitivity, specificity, and timing for all methods. It is worth noting that the same formula applies to the Symtosis systems based on the SVM and ELM. For all cross-validations, it is evident that the ELM outperforms the SVM. With K10 cross-validation, the ELM achieves 92.4% accuracy, whereas the SVM achieves just 86.42%. When compared to the SVM, the ELM has a greater average specificity and

Table 10.1. The ELM and SVM performance comparison for the S0 dataset. (Reproduced with permission from [54]. Copyright 2017 Springer.)

| CV* | Accuracy (%) | | Sensitivity (%) | | Specificity (%) | | AUC | |
|---|---|---|---|---|---|---|---|---|
| Classifier | ELM | SVM | ELM | SVM | ELM | SVM | ELM | SVM |
| K2 | 81.70 | 76.14 | 85.10 | 84.90 | 78.52 | 74.52 | 0.82 | 0.76 |
| K3 | 82.70 | 75.40 | 85.70 | 76.80 | 84.66 | 75.40 | 0.81 | 0.74 |
| K5 | 89.00 | 83.50 | 87.16 | 80.16 | 87.42 | 85.40 | 0.89 | 0.83 |
| K10 | 92.40 | 86.42 | 91.30 | 88.20 | 92.10 | 86.30 | 0.92 | 0.86 |

*CV: Cross-validation protocol.

sensitivity. Tables B.1 and B.2 in appendix B provide the results for the S4 and S8 datasets, respectively.

## 10.5.2 Experiment 2: the effect of the training data size in parts during the CV protocols

We perform the experiment with different data sizes to see how the training dataset size affects the ELM architecture. In this investigation the S0, S4, and S8 datasets were used. As the quantity of the training dataset grew larger, so did the accuracy. For all training data sizes the S4 outperforms the S0, whereas the S8 surpasses both the S4 and the S0. Figure 10.8 shows the accuracy gained for the SVM and ELM with various dataset sizes for each cross-validation.

## 10.5.3 Experiment 3: time comparison between ELM and SVM

In terms of training and testing time, the ELM performs better. Tables 10.2, 10.3, and 10.4 show the time evaluation between the ELM and SVM classifiers for S0, S4,

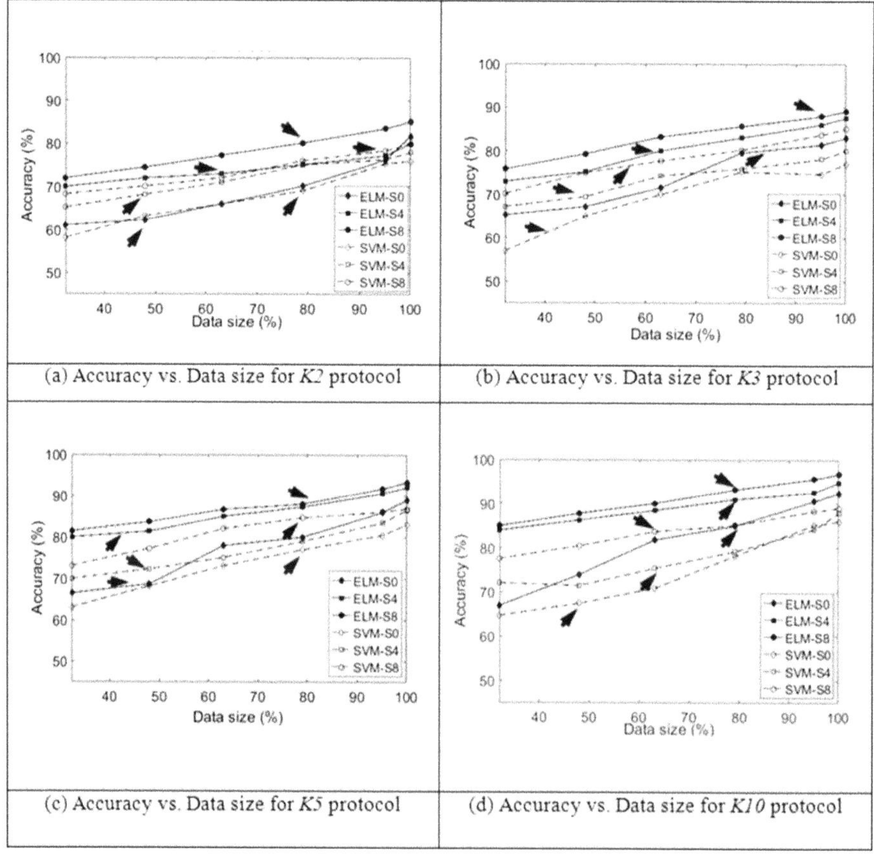



**Figure 10.8.** Accuracy analysis for (a) K2, (b) K3, (c) K5, and (d) K10 cross-validations for different data sizes. (Reproduced with permission from [54]. Copyright 2017 Springer.)

**Table 10.2.** ELM and SVM time comparison for S0. (Reproduced with permission from [54]. Copyright 2017 Springer.)

| Training and testing time | CV[*] | ELM | SVM | Speed increase (%)[***] |
|---|---|---|---|---|
| **Average training time (ms[**])** | K2 | 2.1 | 7.2 | 70.8 |
| | K3 | 4.3 | 7.9 | 45.6 |
| | K5 | 7.4 | 9.6 | 22.9 |
| | K10 | 9.3 | 13.3 | 30.1 |
| **Average testing time (ms[**])** | K2 | 2.7 | 3.6 | 25.0 |
| | K3 | 2.3 | 3.1 | 25.8 |
| | K5 | 2.2 | 2.9 | 24.1 |
| | K10 | 2.1 | 2.2 | 4.5 |

[*]CV: cross-validation protocol; [**]ms: milliseconds; [***]speed increase $= \dfrac{\mid \mathrm{Time_{SVM}} - \mathrm{Time_{ELM}} \mid}{\mathrm{Time_{SVM}}} \times 100$.

**Table 10.3.** ELM and SVM time comparison for S4. (Reproduced with permission from [54]. Copyright 2017 Springer.)

| Training and testing time | CV[*] | ELM | SVM | Speed increase (%)[***] |
|---|---|---|---|---|
| **Average training time (ms[**])** | K2 | 7.1 | 14.5 | 51.0 |
| | K3 | 8.0 | 14.8 | 46.0 |
| | K5 | 9.7 | 15.1 | 35.8 |
| | K10 | 10.3 | 16.0 | 35.6 |
| **Average testing time (ms[**])** | K2 | 4.6 | 8.4 | 45.2 |
| | K3 | 4.1 | 8.0 | 48.8 |
| | K5 | 3.6 | 7.5 | 52.00 |
| | K10 | 3.0 | 7.1 | 57.8 |

CV: cross-validation protocol; ms: milliseconds; speed increase $= \dfrac{\mid \mathrm{Time_{SVM}} - \mathrm{Time_{ELM}} \mid}{\mathrm{Time_{SVM}}} \times 100$.

**Table 10.4.** ELM and SVM time comparison for S8. (Reproduced with permission from [54]. Copyright 2017 Springer.)

| Training and testing time | CV[*] | ELM | SVM | Speed increase (%)[***] |
|---|---|---|---|---|
| **Average training time (ms[**])** | K2 | 8.7 | 17.5 | 50.3 |
| | K3 | 10.1 | 17.9 | 43.6 |
| | K5 | 13.2 | 18.1 | 27.1 |
| | K10 | 15.3 | 19.5 | 21.5 |
| **Average testing time (ms[**])** | K2 | 5.3 | 9.8 | 45.9 |
| | K3 | 5.0 | 9.3 | 46.2 |
| | K5 | 4.8 | 8.6 | 44.2 |
| | K10 | 4.1 | 7.8 | 47.4 |

[*]CV: cross-validation protocol; [**]ms: milliseconds; [***]speed increase $= \dfrac{\mid \mathrm{Time_{SVM}} - \mathrm{Time_{ELM}} \mid}{\mathrm{Time_{SVM}}} \times 100$.

and S8. The training period for K2 is shorter than for K10, however, the testing duration is the opposite. This occurs when the data size for training grows from K2 to K10, whereas the data size for testing reduces from K2 to K10. In all three datasets, the SVM takes longer for training and testing for all forms of cross-validations. For K10 cross-validation for S0, the ELM design uses 2.1 milliseconds (ms) for testing and 9.3 ms for training. The amount of time spent testing is almost non-existent. For S0, the average speed improvement for the ELM is more than the SVM by 31%. For K10 cross-validation with the ELM the testing time for S4 is 3.0 ms and the training time is 10.3 ms. The maximum testing time for K2 is 4.6 ms, whereas the ELM classifier takes 4.6 ms. For ELM and SVM the time increased from S0 to S4, although it was still minimal. In K10 cross-validation the SVM training takes 16.0 ms. For S4 the increase in speed for the ELM compared to the SVM is approximately 47%. When using the S8 dataset the SVM takes up to 19.5 ms to train, while the ELM takes only 15.3 ms. The performance of the ELM over the SVM for S8 is increased by 41%. Overall, the ELM outperforms the SVM by about 40% on average. Our claim is further validated using biometric face data (appendix A).

## 10.6 Performance evaluations

The ELM system's performance is determined by plotting the area-under-the-curve (AUC) and receiver operating characteristic (ROC) curve for each set of CV protocols. The performance characteristics, including accuracy, sensitivity, and specificity, which are detailed in section 10.6.1, are also recorded. In section 10.6.2 the stability and dependability analyses are assessed.

### 10.6.1 ROC curves

The ELM's performance was calculated using ROC curves for all three types of datasets: S0, S4, and S8. We used four different cross-validation techniques for each dataset: K2, K3, K5, and K10. We show the 12 ROC curves in figures 10.9(A)–(C). The ROC curves were generated using the ELM and SVM for each permutation of K and S. Tables 10.1, B.1, and B.2 exhibit the AUCs for the S0, S4, and S8 datasets. For each dataset (S0, S4 and S8), the K10 performs better of all four CV protocols and better performance was shown by the ELM when compared to the SVM (0.97 versus 0.91).

### 10.6.2 Reliability and stability analysis

The dependability and stability of the ELM are examined in this section. This evaluation is critical because it shows how the system operates under varied settings and under recurring conditions. This demonstrates that the outcomes are regular and reproducible. The basis of the reliability index (RI) depends on the divergence of the accuracy with respect to its average as the data size increases [33]. The reliability index $\zeta_{N_L}$ (%) is specified as

$$\zeta_{N_L}\,(\%) = \left(1 - \frac{\mu_{N_L}}{\sigma_{N_L}}\right) \times 100, \tag{10.3}$$

where $\mu_{N_L}$ is the average accuracy and $\sigma_{N_L}$ represents the standard deviation for all US liver images, $N_L$.

The stability evaluation examines how the system responds to repeated situations. We accomplish this by employing a control theory approach to dynamics [34]. To begin, a 5 % variation stability requirement is established, i.e. the system is deemed to be unstable if it varies by more than 5%. The standard deviation (SD) is then calculated for each calculation using varying data sizes. We can say that the system is stable if the SD < 5%. The RIs for all four $K$-fold protocols in table 10.5 are greater than 0.95, showing that the ELM classification system is quite reliable. We used second-class biometric facial data to further confirm our ELM and SVM categorization (appendix A).

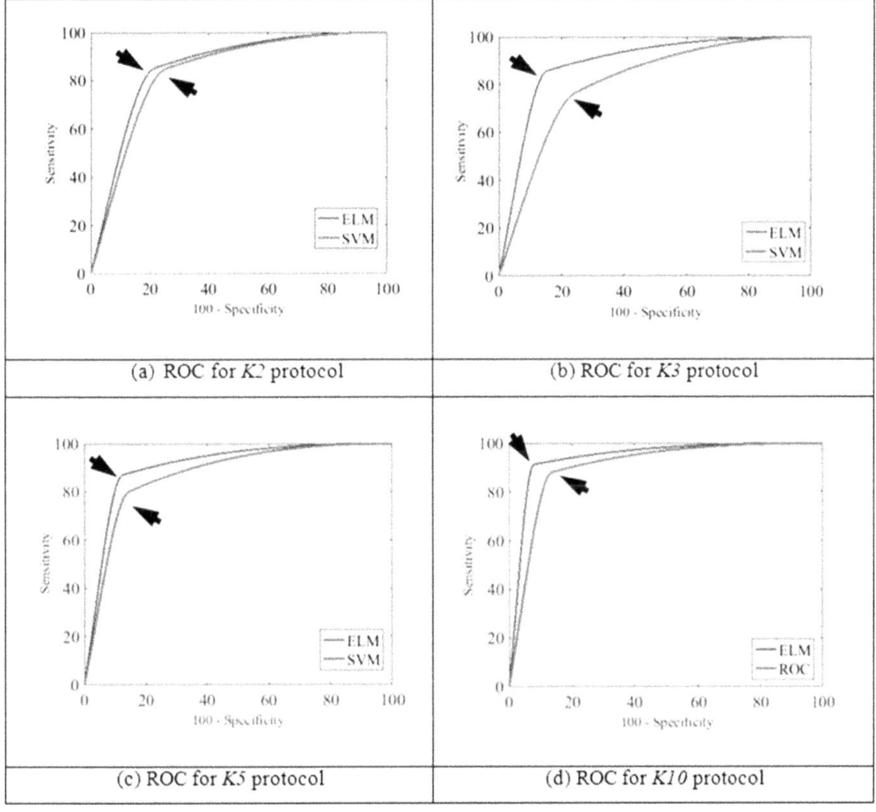

Figure 10.9. ROC curves for (a) K2, (b) K3, (c) K5, and (d) K10 cross-validations using the S0 dataset. (Reproduced with permission from [54]. Copyright 2017 Springer.)

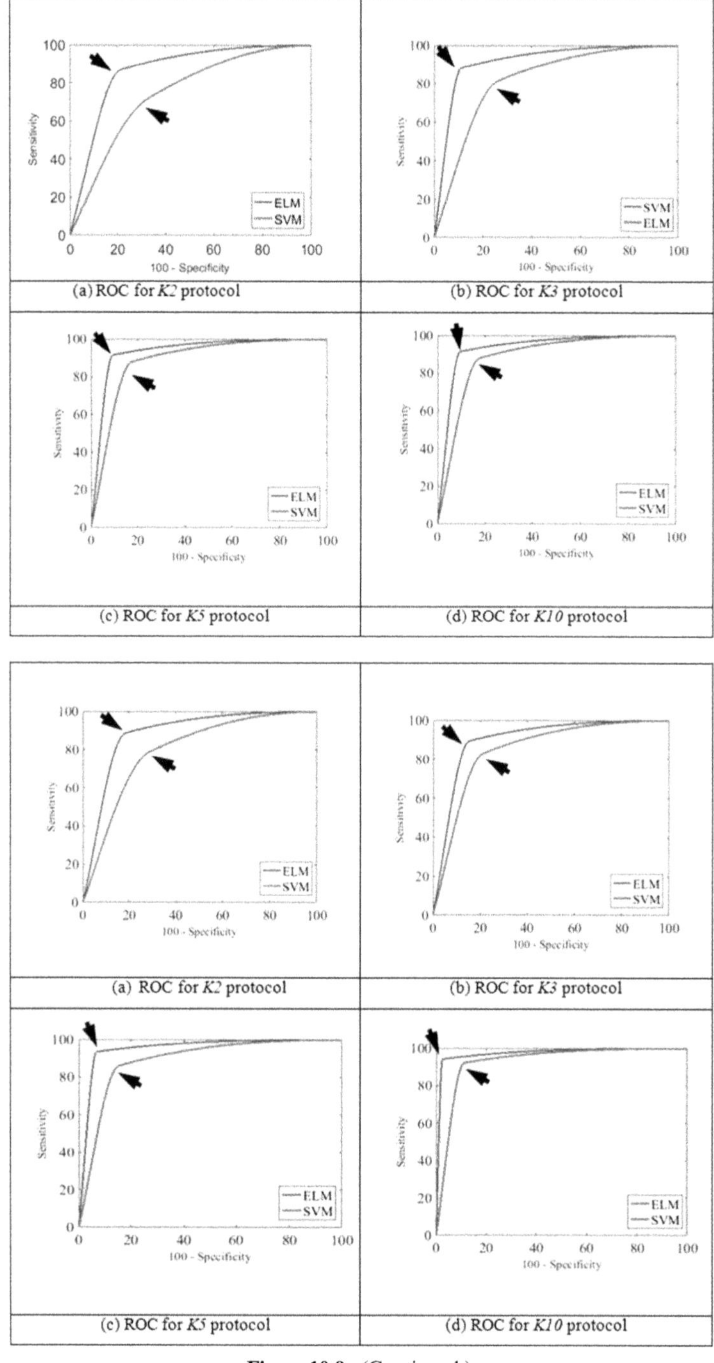

**Figure 10.9.** (Continued.)

## 10.7 Discussion

This study presented a reliable and rapid tissue characterization system (a type of Symtosis system) based on the ELM for FLD disease stratification in US liver imaging. The SLFFNN was taught using the ELM. The parameters from the input to hidden layers were initiated at random to save time and money. The parameters from the hidden to output layers were the only ones to be trained, and this was done in one pass (without iteration), making the ELM quicker than the SVM model. A previously established, the SVM based characterization system was compared to an ELM based characterization system. It is worth noting that the ELM and SVM systems have the same set of features. GRLM, GLCM, and Gabor were the three most popular grayscale features. The study's major goal was to contrast the ELM with the SVM. We compared the ELM to the BPNN since the ELM is an NN based system. The ELM was shown to have equal accuracy and speed when the number of features was reduced by one third and the number of hidden layers was reduced by one third (as proved by the BPNN [21]). The scientific validation was carried out by utilizing the biometric facial datasets listed in appendix A.

### 10.7.1 Benchmarking

There is not a huge amount of research on the use of a CADx based system for liver diagnostics and risk assessment. Suri and his colleagues used a decision tree (DT) to classify a US liver dataset [11] and fuzzy classifiers to diagnose Hashimoto thyroiditis [25] (shown in table 10.6). In this work, three sets of features were generated, and the DT based classifier was used to apply them. With the idea that the pixels are distributed non-linearly, the higher order spectra (HOS) texture and discrete wavelet transform (DWT) were created. The texture was excellent for capturing the varied granular structures in the US photos. Following feature reduction, DT based classification was used, providing 93.3% accuracy. Acharya *et al* [29] proposed a fuzzy classifier (FC) for Hashimoto thyroiditis identification from US thyroid imaging in 2014. The wavelet transform was used to mine the features. The algorithm was tested on 526 US pictures and obtained 84.6% accuracy. Subramanya *et al* [19] employed 53 US liver pictures in 2014, which were divided into four categories: 12 were normal, 14 were mild, 14 were moderate, and 13 were severe. First order statistics (FOS), gradient based (Gr), mutual information based (MI), GRLM, GLCM, and Laws' texture were among the six types of features generated. The use of the SVM resulted in 84.9% accuracy. Suri and his team [21] recently used a BPNN on US liver scans and achieved 97.6% accuracy. The next subsection discusses a brief evaluation of the BPNN and the ELM and their comparison. Liu *et al* [42] used an amalgamation of a liver capsule detection technique and a trained convolution neural network [43] model for feature mining, as well as the SVM as a classifier, to obtain 89.2% accuracy.

The ELM was employed to classify diseased patients from the S0, S4, and S8 datasets, with accuracies of 92.4%, 94.8%, and 96.7%, respectively. Among all the

**Table 10.5.** ELM reliability index for varying data sizes using different *K*-fold cross-validations. (Reproduced with permission from [54]. Copyright 2017 Springer.)

| Data size (Images) | Normal/abnormal | Reliability index (%) | | | |
| --- | --- | --- | --- | --- | --- |
| | | K2 | K3 | K5 | K10 |
| 64 | 27/37 | 99.7 | 98.9 | 100.0 | 99.9 |
| 252 | 104/148 | 99.6 | 98.9 | 99.4 | 99.8 |
| 4032 | 1664/2368 | 99.8 | 99.9 | 100.0 | 99.9 |

cross-validations, K10 outperformed K2, K3, and K5 in terms of performance. Table 10.1 shows that for all cross-validation methods (and K10), the ELM accuracy is better than for the SVM (81.70%, 82.70%, 89.00%, and 92.40%, against 76.14%, 75.40%, 83.50%, and 86.42,% respectively). Tables 10.2, 10.3, and 10.4 further show that the average speed increase for the ELM over the SVM is around 40%, confirming the premise that the ELM is quicker than the SVM. The stability study in table 10.5 reveals that the ELM is a very dependable and stable system. It is seen that the ELM accuracy increases as the data size increases.

### 10.7.2 ELM and BPNN comparison

Both the ELM and BPNN are NN based techniques. Suri's adaptation of BPNN yielded 97.6% accuracy, whereas the ELM yielded 96.7% accuracy. These findings (also given in table 10.6) suggest that the BPNN outperformed the ELM in terms of accuracy, implying that the BPNN is a stronger classifier. The benefits of the ELM, on the other hand, considerably surpass those of the BPNN. As the ELM is an SLFNN, it has a significantly smaller network complexity than the BPNN, which can include a greater number of hidden layers and neurons (up to ten layers in [21]). Second, the BPNN convergence is much slower than for the ELM since each parameter in the BPNN design is updated iteratively. Due to its simple matrix multiplications and single hidden layer, the ELM achieves equivalent accuracy in a single pass. Furthermore, unlike the BPNN, which requires three times the number of features, the ELM achieves an accuracy difference of less than 1% (0.9%) with only 46 conventional features (128). Overall, these advantages justify the use of the ELM over the BPNN and SVM, both of which are classified as ML approaches.

### 10.7.3 A special note on the ELM and SVM

There are two stages to SVM training. The features are mapped from their original dimension to a higher-dimensional space using a nonlinear mapping function (or kernel functions) in stage one. The optimization method is used in the next stage to find the maximum separating hyper-plane of two different categories of data in this feature space while minimizing training errors. As the optimization issue is quadratic and convex, it is simple to solve. An SLFFNN is trained in two steps using the ELM: feature mapping and linear parameter solving. The hidden layer parameters are set

**Table 10.6.** Benchmarking. (Reproduced with permission from [54]. Copyright 2017 Springer.)

| Authors | Data type | Classifier type | Types of features | No. of features | Data size | Accuracy (%) |
|---|---|---|---|---|---|---|
| Acharya *et al* (2012) [11] | Liver | Decision tree | HOS, texture, DWT | | 100 | 93.3 |
| Acharya *et al* (2014) [29] | Thyroid | FC | Wavelet transform | 40 | 526 | 84.6 |
| Subramanya *et al* (2014) [19] | Liver | SVM | FOS, Gr, MI, GRLM, GLCM, Laws | 636 | 53 | 84.9 |
| Saba *et al* (2016) [21] | Liver | BPNN | BG, Fourier, DCT, Harlick, Gupta, Gabor | 128 | 62 | 97.6 |
| Liu *et al* (2017) [42] | Liver | SVM | CNN based features | 500 | 91 | 89.2 |
| Proposed work | | ELM | | 46 | 64 | 92.4 |
| Proposed work | | | | 46 | 252 | 94.8 |
| Proposed work | | | | 46 | 4032 | 96.7 |

FC: Fuzzy Classifier; FOS: First Order Statistics; Gr: Gradient based features; Laws: Laws texture features; BG: Basic geometric

at random in the first step in order to translate the input data into a feature space using sigmoid nonlinear functions. The parameters of the hidden-to-output layer are solved in the next stage of ELM learning by minimizing the error. To minimize the loss of projected mistakes, the Moore–Penrose pseudo-inverse is used [25, 34]. In a single step, the parameters of the hidden-to-output neurons are learned. For datasets which are smaller than the size of the hidden neurons, the ELM requires a time proportional to the number of hidden neurons [27]. The ELM differs from the SVM in that the input-to-hidden layer parameters do not need to be modified. The least squares solution for regression, binary, and multi-class classification in the ELM is only reliant on the input data and the amount of training samples [44]. As it only takes one pass to find the hidden-to-output layer parameters by multiplying the Moore–Penrose inverse of the activation function output with the target, the ELM has a lower computational complexity than the SVM. Therefore, the time complexity of the ELM for smaller datasets is relative to the quantity of hidden nodes, and it takes a time proportionate to the quantity of hidden neurons, which is substantially larger than the size of tiny datasets with only one feed-forward pass. When dealing with nonlinearly separable data, the SVM's computing cost increases because it must be solved in a high-dimensional space using kernel functions. Furthermore, the kernel functions of the SVM differ depending on the application, whereas the ELM provides a more universal solution to the classification problem. The SVM computation becomes more complex as the size of the training dataset grows as it involves finding more support vectors from the entire dataset. Although the ELM and SVM both use the same loss function, the ELM has a looser optimization constraint than the SVM, which optimizes using the least squares model while the latter uses the biggest separating margin technique. The ELM is faster than other classifiers because the parameters between the input and hidden layers are constant; the model only learns the hidden-output parameters, which is the same as learning a linear model [44, 45]. Experiments have shown that the ELM with random hidden nodes can run up to ten times quicker than the SVM. As a result of this analysis, it can be appropriately concluded and justified that the ELM produces more accurate findings than the SVM.

For error minimization, the ELM uses a constrained least squares model. It uses the gradient descent derivative of error in a single pass over the whole feature space [46–48], to achieve minimal training error. The use of all features in the SVM allows for the existence of noisy data, which prevents it from converging to a single optimal separation hyper-plane. As a result, the SVM must identify feature categories and use feature selection methods to reduce noisy features in order to achieve improved accuracy. As a result, in the absence of feature selection techniques, we can confidently state that the ELM outperforms the SVM in terms of accuracy.

### 10.7.4 Strengths, weaknesses, and future work

It is evident that the ELM is quicker and more accurate than the SVM; nonetheless, evaluating the ELM for Big Data applications [49] is required to determine its true strength. The ELM must also be tested on the abdomen [50] as well as other medical imaging applications [51]. A comparison with recent deep learning [52] techniques is also required. Only simple feature mining methods are used in the experiments. We propose using improved feature selection techniques in the future, based on principal component analysis, discriminant analysis, or mutual information. To improve classifier accuracy, it is proposed that a larger training dataset is necessary for efficient classification. The high categorization accuracy of the basic ELM in this study warrants additional research into multiple ELM versions for classification of US liver imaging.

## 10.8 Conclusions

This study used the ELM to propose a better strategy for FLD categorization that was compared to the SVM. The ELM is an SLFFNN, with input-to-hidden layer parameters generated at random to save time and hidden-to-output parameters trained only once. In comparison to the SVM, the ELM was faster due to its simplified architecture and single pass. Furthermore, since the least squares paradigm has been altered, the results are better with a smaller number of features. On three different data sizes, the Symtosis system used four different $K$-fold cross-validation protocols ($K = 2$, 3, 5, and 10), S0—original, S4—four splits, and S8—eight splits (a total of 12 cases) were created with 46 different grayscale features mined from Gabor, GLCM, and GRLM feature sets. On the S8 dataset, using the K10 cross-validation methodology, the ELM showed an accuracy of 96.75% compared to 89.01% for the SVM, with AUCs of 0.97 and 0.91, respectively. Further tests revealed that the ELM classifier had a mean dependability of 99% and a mean speed improvement of 40% when compared to the SVM. We used two classes of biometric face public data to validate the Symtosis system, demonstrating 100% accuracy.

# Appendix A  Scientific validation

Scientific validation is always a part of the system design process. Another set of liver datasets, the results of which are known *a priori*, must be run for validation. We employ a facial biometric dataset to test the classification accuracy because such clinical data is difficult to come by. Dr Libor Spacek of the University of Essex's Department of Computer Science provided data on the biometric facial dataset Face94 [53]. The ELM and the SVM were used to validate this dataset, which included both male and female faces.

*Face94 dataset*

Using the Face94 dataset, we conducted experiments to validate our findings. The Face94 data collection contains 153 unique photos of people expressing various emotions and positions while sitting at a fixed distance from the camera. There are two classes: male and female, with a total of 2660 photos, 2260 of which are male and 400 of which are female. Figure A.1 shows a subset of the photos. To ensure generality, a cross-validation protocol (K2, K3, K5, and K10) is used. Table A.1 shows the validation findings. Across all cross-validation processes, the ELM provides 100% accuracy.

**Figure A. 1.** Male images (top row) and female images (bottom row) from the Face94 dataset.

**Table A.1.** Comparison between the ELM and SVM for the Face94 dataset.

| CV* | Accuracy (%) | | Sensitivity (%) | | Specificity (%) | | AUC | |
|---|---|---|---|---|---|---|---|---|
| Classifier | ELM | SVM | ELM | SVM | ELM | SVM | ELM | SVM |
| K2 | 100 | 98.46 | 100 | 97.50 | 100 | 99.29 | 1 | 0.98 |
| K3 | 100 | 98.85 | 100 | 98.06 | 100 | 99.52 | 1 | 0.99 |
| K5 | 100 | 98.69 | 100 | 98.67 | 100 | 98.71 | 1 | 0.99 |
| K10 | 100 | 98.88 | 100 | 98.33 | 100 | 99.36 | 1 | 0.99 |

*CV: Cross-validation protocol

## Appendix B Results of the ELM/SVM classifier for the S4 and S8 datasets

**Table B.1.** Comparison between the ELM based and SVM based learning methods for the S4 dataset.

| CV* | Accuracy (%) | | Sensitivity (%) | | Specificity (%) | | AUC | |
|---|---|---|---|---|---|---|---|---|
| Classifier | ELM | SVM | ELM | SVM | ELM | SVM | ELM | SVM |
| K2 | 83.05 | 71.14 | 86.91 | 71.65 | 78.52 | 67.42 | 0.83 | 0.73 |
| K3 | 88.05 | 78.83 | 88.25 | 81.45 | 88.77 | 73.72 | 0.88 | 0.77 |
| K5 | 92.18 | 84.65 | 91.27 | 87.77 | 90.42 | 82.34 | 0.91 | 0.86 |
| K10 | 94.78 | 87.91 | 92.32 | 89.37 | 95.12 | 84.50 | 0.94 | 0.88 |

*CV: Cross-validation protocol

**Table B.2.** Comparison between the ELM based and SVM based learning methods for the S8 dataset.

| CV* | Accuracy (%) | | Sensitivity (%) | | Specificity (%) | | AUC | |
|---|---|---|---|---|---|---|---|---|
| Classifier | ELM | SVM | ELM | SVM | ELM | SVM | ELM | SVM |
| K2 | 85.19 | 74.45 | 88.71 | 78.81 | 81.41 | 71.23 | 0.85 | 0.73 |
| K3 | 90.03 | 80.16 | 89.41 | 82.96 | 84.65 | 78.84 | 0.87 | 0.81 |
| K5 | 93.27 | 85.14 | 93.46 | 85.95 | 93.14 | 84.49 | 0.93 | 0.86 |
| K10 | 96.75 | 89.01 | 94.23 | 92.23 | 97.59 | 88.87 | 0.97 | 0.91 |

*CV: Cross-validation protocol

## References


[1] Saverymuttu S H, Joseph A E and Maxwell J D 1986 Ultrasound scanning in the detection of hepatic fibrosis and steatosis *Br. Med. J.* **292** 13–5

[2] Mohamed W S, Mostafa A M, Mohamed K M and Serwah A H 2015 Effects of fenugreek, nigella, and termis seeds in nonalcoholic fatty liver in obese diabetic albino rats *Arab J. Gastroenterol.* **16** 1–9

[3] Wieckowska A and Feldstein A E 2005 Nonalcoholic fatty liver disease in the pediatric population: a review *Curr. Opin. Pediatr..* **17** 636–41

[4] Ratziu V, Charlotte F, Heurtier A, Gombert S, Giral P, Bruckert E, Grimaldi A, Capron F and Poynard Tand LIDO Study Group 2005 Sampling variability of liver biopsy in nonalcoholic fatty liver disease *Gastroenterology* **128** 1898–906

[5] Cheung R C 2009 *Complications of Liver Biopsy, Gastrointestinal Emergencies, Gastrointestinal Emergencies* ed T C K Tham, J S A Collins and R Soetikno (Oxford : Blackwell) pp 72–9

[6] Saadeh S, Younossi Z M, Remer E M, Gramlich T, Ong J P, Hurley M, Mullen K D, Cooper J N and Sheridan M J 2009 The utility of radiological imaging in nonalcoholic fatty liver disease *Gastroenterology* **123** 745–50

[7] Wang D, Fang Y, Hu B and Cao H 2012 B-scan image feature mining of fatty liver *Sixth Int. Conf. Internet Computing for Science and Engineering* pp 188–92

[8]   Yajima Y, Ohta K, Narui T, Abe R, Suzuki H and Ohtsuki M 1983 Ultrasonographical diagnosis of fatty liver: significance of the liver–kidney contrast *Tohoku J. Exp. Med.* **139** 43–50

[9]   Mathiesen U L, Franzen L E, Aselius H, Resjö M, Jacobsson L, Foberg U, Frydén A and Bodemar G 2002 Increased liver echogenicity at ultrasound examination reflects degree of steatosis but not of fibrosis in asymptomatic patients with mild/moderate abnormalities of liver transaminases *Dig. Liver Dis.* **34** 516–22

[10]  Mendler M H, Bouillet P, Le Sidaner A, Lavoine E, Labrousse F, Sautereau D and Pillegand B 1998 Dual-energy CT in the diagnosis and quantification of fatty liver: limited clinical value in comparison to ultrasound scan and single-energy CT, with special reference to iron overload *J. Hepatol.* **28** 785–94

[11]  Acharya U R, Sree S V, Ribeiro R, Krishnamurthi G, Marinho R T, Sanches J and Suri J S 2012 Data mining framework for fatty liver disease classification in ultrasound: a hybrid feature mining paradigm *Med. Phys.* **39** 4255–64

[12]  Shensa M J 1992 The discrete wavelet transform: the discrete wavelet transform: wedding the atrous and Mallat algorithms *IEEE Trans. Signal Process.* **40** 2464–82

[13]  Bruce L M, Koger C H and Li J 2002 Dimensionality reduction of hyperspectral data using discrete wavelet transform feature mining *IEEE Trans. Geosci. Remote Sens.* **40** 2331–8

[14]  Manjunath B S and Ma W Y 1996 Texture features for browsing and retrieval of image data *IEEE Trans. Pattern Anal. Mach. Intell.* **18** 837–42

[15]  Ishibuchi H, Nakashima T and Murata T 1999 Performance evaluation of fuzzy classifier systems for multidimensional pattern classification problems *IEEE Trans. Syst. Man Cybern.* B **29** 601–18

[16]  Kuncheva L 2000 *Fuzzy Classifier Design* (Berlin: Springer)

[17]  Uebele V, Abe S and Lan M-S 1995 A neural-network-based fuzzy classifier *IEEE Trans. Syst., Man, Cybernet.* **25** 353–61

[18]  Antonini M, Barlaud M, Mathieu P and Daubechies I 1992 Image coding using wavelet transform *IEEE Trans. Image Process.* **1** 205 20

[19]  Subramanya M B, Kumar V, Mukherjee S and Saini M A 2015 CAD system for B-mode fatty liver ultrasound images using texture features *J. Med. Eng. Technol.* **39** 123–30

[20]  Ma H Y, Zhou Z, Wu S, Wan Y L and Tsui P H 2016 A computer-aided diagnosis scheme for detection of fatty liver *in vivo* based on ultrasound kurtosis imaging *J. Med. Syst.* **40** 33

[21]  Saba L, Dey N, Ashour A S, Samanta S, Nath S S, Chakraborty S, Sanches J, Kumar D, Marinho R and Suri J S 2016 Automated stratification of liver disease in ultrasound: an online accurate feature classification paradigm *Comput. Methods Programs Biomed.* **130** 118–34

[22]  Vapnik V N 1999 An overview of statistical learning theory *IEEE Trans. Neural Networks* **10** 988–99

[23]  Huang G-B, Zhu Q-Y and Sie C-K 2006 ELM: theory and applications *Neurocomputing* **70** 489–501

[24]  Huang G-B, Zhu Q-Y and Siew C-K 2004 ELM: a new learning scheme of feedforward neural networks *Proc. 2004 IEEE Int. Joint Conf. Neural Networks* **2** 985–90

[25]  Rao C R and Mitra S K 1971 *Generalized Inverse of Matrices and its Applications* (New York: Wiley)

[26]  Kadah Y M, Farag A A, Zurada J M, Badawi A M and Youssef A M 1996 Classification algorithms for quantitative tissue characterization of diffuse liver disease from ultrasound images *IEEE Trans. Med. Imaging* **15** 466–78

[27] Qayyum A 2009 MR spectroscopy of the liver: principles and clinical applications *Radiographics* **29** 1653–64

[28] Chorowski J, Wang J and Zurada J M 2014 Review and performance comparison of SVM- and ELM-based classifiers *Neurocomputing* **128** 507–16

[29] Acharya U R, Sree S V, Krishnan M M, Molinari F, ZieleŸnik W, Bardales R H, Witkowska A and Suri J S 2014 Computer-aided diagnostic system for detection of Hashimoto thyroiditis on ultrasound images from a Polish population *J. Ultrasound Med.* **33** 245–53

[30] Mohanty A K, Beberta S and Lenka S K 2011 Classifying benign and malignant mass using GLCM and GLRLM based texture features from mammogram *Int. J. Eng. Res. Appl.* **1** 687–93

[31] Mohanaiah P S, Sathyanarayana P and GuruKumar L 2013 Image texture feature mining using GLCM approach *Int. J. Sci. Res. Publ.* **3** 1

[32] Herman P 2008 Comparative analysis of spectral approaches to feature mining for EEG-based motor imagery classification *IEEE Trans. Neural Syst. Rehabil. Eng.* **16** 317–26

[33] Anuradha K 2013 Statistical feature mining to classify oral cancers *J. Global Res. Comp. Sci.* **4** 8–12

[34] Barbu T 2010 Gabor filter-based face recognition technique *Proc. Roman. Acad.* **11** 277–83

[35] MacAusland R 2014 The Moore–Penrose inverse and least squares *Advanced Topics in Linear Algebra* Math 420 University of Puget Sound, Tacoma, WA

[36] Mirmehdi M 2008 *Handbook of Texture Analysis* (London: Imperial College Press)

[37] Acharya U R, Mookiah M R, Sree S V, Afonso D, Sanches J, Shafique S, Nicolaides A, Pedro L M, Fernandes J F and Suri J S 2013 Atherosclerotic plaque tissue characterization in 2D ultrasound longitudinal carotid scans for automated classification: a paradigm for stroke risk assessment *Med. Biol. Eng. Comput.* **51** 513–23

[38] Acharya R U, Faust O, Alvin A P, Sree S V, Molinari F, Saba L, Nicolaides A and Suri J S 2012 Symptomatic vs asymptomatic plaque classification in carotid ultrasound *J. Med. Syst.* **36** 1861–71

[39] Acharya U R, Faust O, Sree S V, Molinari F, Saba L, Nicolaides A and Suri J S 2012 An accurate and generalized approach to plaque characterization in 346 carotid ultrasound scans *IEEE Trans. Instrum. Meas.* **4** 1045–53

[40] Shrivastava V K 2016 Computer-aided diagnosis of psoriasis skin images with HOS, texture and color features: a first comparative study of its kind *Comput. Methods Programs Biomed.* **126** 98–109

[41] Shrivastava V K, Londhe N D, Sonawane R S and Suri J S 2015 Reliable and accurate psoriasis disease classification in dermatology images using comprehensive feature space in machine learning paradigm *Expert Syst. Appl.* **42** 6184–95

[42] Liu X, Song J L, Wang S H, Zhao J W and Chen Y Q 2017 Learning to diagnose cirrhosis with liver capsule guided ultrasound image classification *Sensors* **17** 149

[43] LeCun Y, Bengio Y and Hinton G 2015 Deep learning *Nature* **521** 436–44

[44] Huang G-B 2012 ELM for regression and multiclass classification *IEEE Trans. Syst. Man Cybern. B* **42** 513–29

[45] Zhu Q-Y 2005 Evolutionary ELM *Pattern Recognit.* **38** 1759–63

[46] Mao L, Zhang L, Liu X, Li C and Yang H 2014 Improved extreme learning machine and its application in image quality assessment *Math. Probl. Eng.* **2014** 426152

[47] Tang J, Deng C and Huang G-B 2016 ELM for multilayer perceptron *IEEE Trans Neural Netw. Learn. Syst.* **27** 809–21

[48] Demuth H B, Beale M H, De Jess O and Hagan M T 2014 *Neural Network Design* (Stillwater, OK: Martin Hagan)

[49] El-Baz A and Suri J S 2018 *Big Data in Medical Imaging* (Boca Raton, FL: CRC Press)

[50] El-Baz A S, Saba L and Suri J S 2014 *Abdomen and Thoracic Imaging* (Berlin: Springer)

[51] El-Baz A, Gimel'farb G and Suri J S 2015 *Stochastic Modeling for Medical Image Analysis* (Boca Raton, FL: CRC Press)

[52] Esses S J, Lu X, Zhao T, Shanbhogue K, Dane B, Bruno M and Chandarana H 2017 Automated image quality evaluation of T2-weighted liver MRI utilizing deep learning architecture *J. Magn. Reson. Imaging* **47** 723–28

[53] Spacek L Face94 Database http:// cswww.essex.ac.uk /mv/allfaces /faces94.html

[54] Kuppili V 2017 Extreme learning machine framework for risk stratification of fatty liver disease using ultrasound tissue characterization *J. Med. Syst.* **41** 1–20

**IOP** Publishing

# Multimodality Imaging, Volume 1
### Deep learning applications
**Mainak Biswas and Jasjit S Suri**

# Chapter 11

# Symtosis: deep learning based liver ultrasound tissue characterisation and risk stratification

**Mainak Biswas and Jasjit S Suri**

Fatty liver disease (FLD) is caused by build-up of fat within liver cells which, if left untreated, may lead to life-threatening diseases such as cancer. Machine learning (ML) modules implicitly depend on features extracted manually from ultrasound images, therefore their accuracy is essential for the integrity of healthcare systems.

In this work a deep learning convolutional neural network (CNN) model for FLD detection from ultrasound images is proposed under the class Symtosis. The proposed CNN model is 22 layers deep and consists of four types of layers: convolution, pooling, rectified linear unit, and dropout. The DL model uses a special operation called inception which is an amalgamation of multiple convolution filters in one layer which leads to better performance without increasing complexity. The DL based system is compared to the conventional ML models of the support vector machine (SVM) and extreme learning machine (ELM).

The dataset consists of ultrasound images from 63 patients of which 27 were normal and 36 diseased. The CV protocol for learning was K10 cross-validation, i.e. 90% training and 10% testing. The performance of the DL model was better than that of the ELM and SVM, the performance was 100%, 92%, and 82%, respectively. In the same order the AUC were 1.0, 0.92 and 0.79, respectively.

The DL system shows a better performance for FLD diagnosis and risk characterisation compared to the contemporary ML models of ELM and SVM.

## 11.1 Introduction

Fatty liver disease (FLD) has been identified as one of the top reasons for liver-related deaths among adults in the age group of 45–54 years in the United States in the last two decades [1]. Steatosis is the build-up of fat in liver cells. FLD can be caused by a multiplicity of factors, including metabolic syndrome, alcohol intake, and obesity owing to insulin resistance [1, 2]. Non-alcoholic FLD (NAFLD) is a

doi:10.1088/978-0-7503-2244-7ch11

frequent cause of chronic liver disease in the Western world, accounting for 19%–46% of all liver illnesses [2]. The classification of FLD tissue has been studied in a number of ways. A dual energy CT (DECT) based technique was used by Lamb *et al* [3] for categorisation of computed tomography (CT) images. Artificial neural networks (ANNs) were employed for the categorisation of magnetic resonance imaging (MRI) liver images [4]. In 2012 Suri and his team [5] created a computer-assisted liver data classification model for ultrasound (US). For the classification, they used a logical machine learning (ML) model, i.e. decision tree [6], with 93.3% accuracy. In 2016 Suri and his team [7] proposed the Levenberg–Marquardt back propagation neural network (BPNN) [8] under the class of Symtosis$^{TM}$. Ultrasound liver images with FLD are frequently hypoechoic, preventing statistical classifiers from achieving 100% accuracy. There is a strong demand for appropriate tissue feature classification by ML algorithms. Furthermore, during training, CV protocols necessitate several trials per combination, which is time consuming. Convolutional neural networks (CNNs) have recently become popular in classification [9]. Figure 11.1 shows an instance of a CNN model. The CNN uses a series of rectilinear units (ReLus) and pooling operations for feature mining. A convolution kernel is used on an input image to perform a multiplication of the image pixel by the kernel weights block-by-block, and then adding the results to generate feature maps. The ReLu is frequently used in conjunction with convolution. The ReLu performs rectification on feature maps to diminish the issue of vanishing gradient. The down-sampling of input data is accomplished through pooling. CNN now plays a role in a variety of real-world applications that have been buffered with optimisation techniques [10].

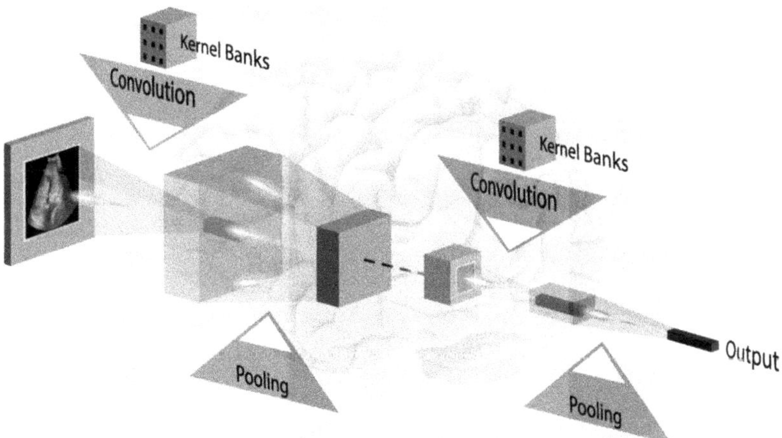


**Figure 11.1.** The basic deep learning CNN model. (Reproduced with permission from [40]. Copyright 2018 Elsevier.)

In this work a CNN deep learning (DL) model is used to detect hypoechoic FLD and categorise normal and diseased/abnormal ultrasound liver images automatically under the Symtosis$^{TM}$ system. This is one of the initial studies of its kind to compare three ML based classification methodologies: support vector machines (SVMs) [11], extreme learning machines (ELMs) [12], and the CNN model. In addition, we look into a specialised operation known as inception. The inception operation combines various magnitudes of convolution filters using the dimensionality reduction principle, then concatenates these various convolution pipelines to decrease the computational cost of the DL liver Symtosis$^{TM}$ system. The deepness (multiple hidden layers) of DL increases as a result of inception. The increased depth allows DL to extract better higher-level features, which improves the system's accuracy. Finally, in our deep learning CNN based model for tissue categorisation and risk assessment, we investigate a GPU based framework. The SVM, ELM, and DL have areas-under-the-curve (AUCs) of 0.79, 0.92, and 1.0, respectively, resulting in perfect 100% accuracy when using the K10 cross-validation (CV) protocol for optimised ultrasound liver images. Figure 11.2 depicts the DL model's overall architecture. The border-stripped ultrasound liver images were first preprocessed for the best DL stratification accuracy, which serves as the foundation for all data analysis. Furthermore, biometric facial data were used to validate our system.

The following is the layout of this chapter. The patient parameters and acquisition are covered in section 11.2, while the methodology for the DL based Symtosis$^{TM}$ system and two other traditional classification algorithms are covered in section 11.3. In sections 11.4, 11.5, and 11.6, the results, performance evaluation, and discussion are presented, respectively. Finally, the conclusion is in section 11.7.

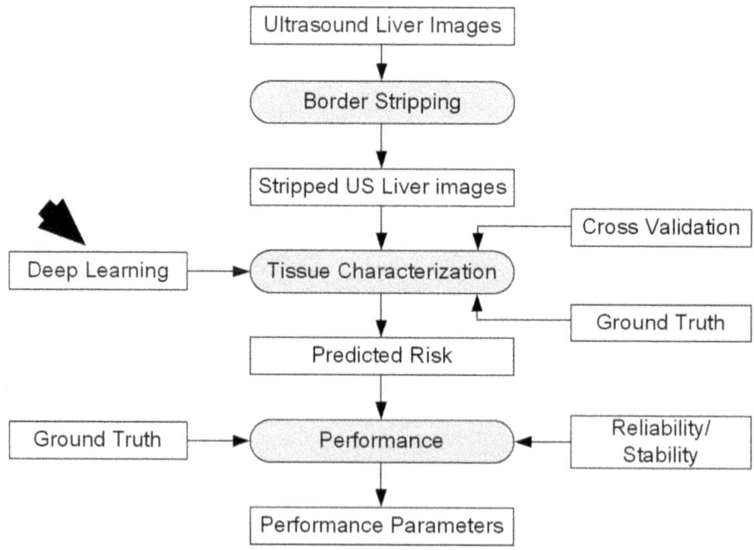


**Figure 11.2.** Symtosis$^{TM}$ system using deep learning. (Reproduced with permission from [40]. Copyright 2018 Elsevier.)

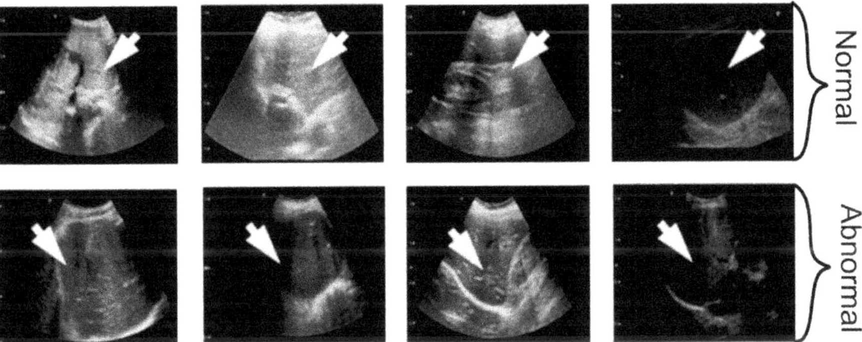


**Figure 11.3.** Normal liver images (top row) and abnormal liver images (bottom row). (Reproduced with permission from [40]. Copyright 2018 Elsevier.)

## 11.2 Patient demographics and acquisition

For this purpose, 63 patients were chosen from the Santa Maria Hospital's Gastroenterology Department (ethical approval given) in Lisbon, Portugal [13], which had previously been employed in our earlier study [5, 7]. CX c 50 (Philips Medical Systems) in the DICOM (Digital Imaging and Communications in Medicine) format was used to obtain the images in the United States. The scanner images had an 8 bit pixel resolution and were 1024 × 1024 pixels in size. The curved array transducer C5–1 was installed in the ultrasound scanner. The number of piezoelectric elements was 160 with a resonance frequency of 1–5 MHz in this device. Amongst the 63 patients, 36 were found to have FLD and the remaining 27 were found to be normal. The expert ultrasound equipment operators within the hospital had acquired the input images of normal and fatty liver. Depending on the indicators received from the laboratory biopsy reports, the results were categorised as normal and abnormal. These categorised photographs established our experiment's ground truth (GT). In the liver there is a small left lobe and a large right lobe. The right lobe livers were scanned to make the right liver the principal liver component. The parenchyma echogenic intensity of the liver and cortex are similar in an ultrasound, whereas the echogenic intensity of liver parenchyma is significantly higher than the renal cortex in the ultrasound image of a steatotic liver [14, 15]. A 128 × 28 pixel intermediate area of the region of interest (ROI) was removed from each image along the medial axis. Figure 11.3 shows the initial liver pictures, where the top row shows normal images and bottom row shows abnormal images.

## 11.3 Methodology

During the training process of the CNN [16], as shown in figure 11.1, the low level features are turned into high level features. To learn these sets of features derived from the initial image, the CNNs apply convolution and pooling repeatedly. In a simple CNN model, the first challenge is the overfitting, which leads to decreased classification test precision and accuracy. The second challenge stems from the need

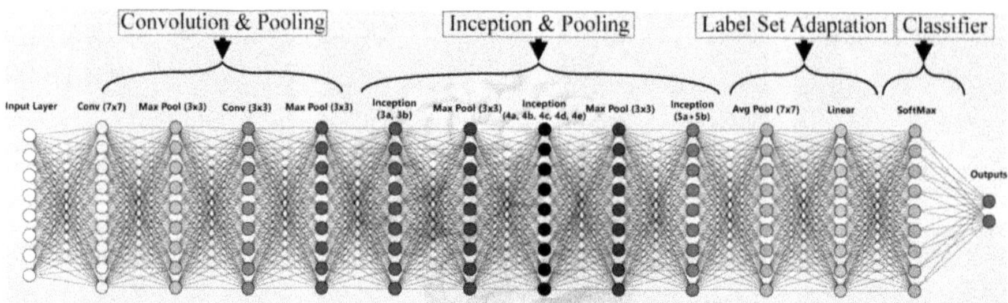


**Figure 11.4.** The proposed CNN model. (Reproduced with permission from [40]. Copyright 2018 Elsevier.)

for more layers as these layers are capable of generating better feature maps. This, however, increases the DL system's computational complexity. A 'dropout' strategy tries to resolve the first challenge by which it randomly limits the quantity of hidden layer weights. An 'inception model' whose task is to cut the computation costs is the resolution of the second issue. These two architectural integrations into CNN as a basic model are the novelty of the system for the stratification of liver disease. In figure 11.4 you can see this architecture. This DL architecture has two sets, nine inception points, five pooling layer, a dropout layer, a linear layer, and a softmax layer.

### 11.3.1 Risk stratification model

In terms of lung [17], hepatitis [5], prostrate [18], and ovarian [19] diseases, the basic model of risk can be characterised through the conventional literature approaches created by Suri and his team. The main difference between the present approach and the previous ML applications is the changes to the DL mining, the dimensionality reduction of features, and the training of NN weights in contrast to the conventional method. The new system process is shown in figure 11.5. As the DL generates a huge number of features because of convolutions, it is essential to carry out the embedded feature mining process in the ROI, which only includes tissue data but does not contain redundant background tissue or noise. We therefore use a cropping protocol to eliminate the background data. Following this cropping method, this concept employs the same cross-validation (CV) protocol as the standard machine learning systems ($K$-fold CVs: $K = 2$, 3, 5, and 10). The protocol for cross-validation K2 separates the dataset into two equal parts. There are two learning rounds, one for training and the other for testing the learning effectiveness, for each combination. In the same way, K3 splits the data into say A, B, and C, i.e. three parts. One part is chosen for testing, the two other sections for training are mixed. Likewise in K2, each combination has three training/testing rounds.

For K5 and K10 cross-validation, a similar concept is used. Suri has extended this concept to several organs such as the carotid [20], heart [21], liver [7, 22], breast [23, 24], and prostate [25], with a high degree of success. Recently, Suri's group with its

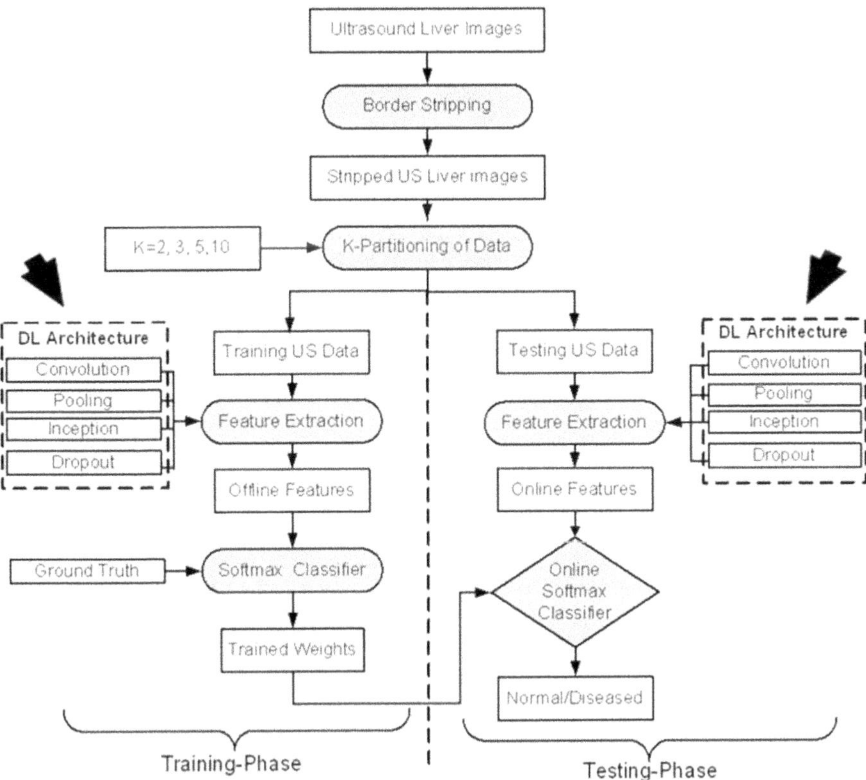


**Figure 11.5.** DL tissue characterisation system using Symtosis™. (Reproduced with permission from [40]. Copyright 2018 Elsevier.)

Bayesian classifier for the characterization of skin diseases has recently proved this in four different kinds of categories [26].

## 11.3.2 CNN architecture

In convolution, the feature maps are generated using various convolution filters (also known as kernels). The system's data flow is shown in figure 11.6. This DL architecture, with 22 layers (see table 11.1, column 'Depth'), includes convolution, bonding, starting, dropout, linear, and softmax layers for the classification of US liver images. The DL architecture is used to calculate the weight parameters of the training in ultrasound liver images. These trained parameters are then used for testing the unlabelled dataset to obtain the final precision using the softmax layer. Table 11.1 provides the details of DL layer data flow. In the table, '# 1 × 1' means the number of 1 × 1 convolutions applied to the images for reduction in dimensionality. 'Reduce # 3 × 3 CF by 1 × 1 CF' (convolution filter) refers to the use of 1 × 1 convolution filter/kernels before the use of the 3 × 3 convolution filter/ kernel number. Likewise for 'Reduce # 5 × 5 CF by 1 × 1 CF'. Also, the reduced

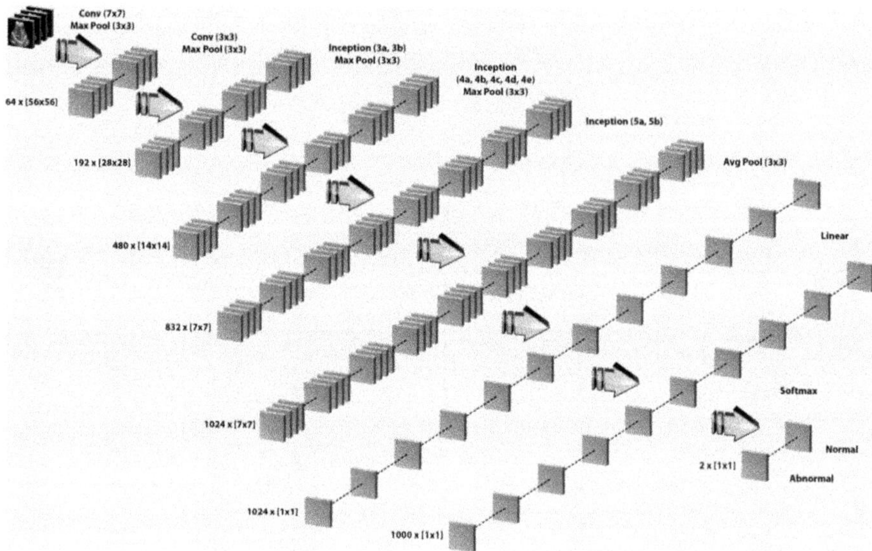

**Figure 11.6.** DL data flow. (Reproduced with permission from [40]. Copyright 2018 Elsevier.)

pooling number represents the number of $1 \times 1$ filters/kernels for dimensional reduction after pooling activities. The final column indicates the number of weights to be trained.

## 11.4 Results

The risk stratification outcomes may be summarised by analysing the deep learning system's ultimate performance and contrasting it with other standard ML systems (the SVMs and ELMs), as presented in figure 11.2. The findings are shown in two aspects: (i) image cropping to remove the background and (ii) evaluating the influence of the data size.

### 11.4.1 Image pre-processing for the DL, ELM, and SVM

It is essential, before assessing the correctness of the test data by utilising a CV methodology as shown in figure 11.5, that the background information is removed. Thus border stripping was used on the ultrasound liver images to eliminate the background to ensure that the only tissue region is in the region of interest. For the images that are fed into the Symtosis$^{\text{TM}}$ system and the architectures used by the SVM, ELM, and DL with the K10 CV protocol, the stratifying precisions are 82% for SVM, 92% for ELM, and 100% for DL. Figure 11.7 illustrates the curve between border stripping percentages versus exactness in 63 patients. It shows the maximum precision for 15% background removal in accordance with the K10 procedure of cross-validation. Figure 11.8 shows the analogue precision bar chart. Our study illustrated that the SVM and ELM had lower background sensitivity than DL. In

**Table 11.1.** DL architecture table showing different layers. (Reproduced with permission from [40]. Copyright 2018 Elsevier.)

| Type | Patch size/ stride | Output size | Depth | #** $1 \times 1$ $1 \times 1$ CF | Reduce # $3 \times 3$ CF* by # $3 \times 3$ conv. | | Reduce # $5 \times 5$ CF by $1 \times 1$ CF | #** $5 \times 5$ conv. | Reduce pool | Number of weights |
|---|---|---|---|---|---|---|---|---|---|---|
| Convolution | $7 \times 7/2$ | $112 \times 112 \times 64$ | 1 | | | | | | | $2.7 \times 10^3$ |
| Max pool | $3 \times 3/2$ | $56 \times 56 \times 64$ | 0 | | | | | | | |
| Convolution | $3 \times 3/1$ | $56 \times 56 \times 192$ | 2 | | 64 | 192 | | | | $112 \times 10^3$ |
| Max pool | $3 \times 3/2$ | $28 \times 28 \times 192$ | 0 | | | | | | | |
| Inception (3a) | | $28 \times 28 \times 256$ | 2 | 64 | 96 | 128 | 16 | 32 | 32 | $159 \times 10^3$ |
| Inception (3b) | | $28 \times 28 \times 480$ | 2 | 128 | 128 | 192 | 32 | 96 | 64 | $380 \times 10^3$ |
| Max pool | $3 \times 3/2$ | $14 \times 14 \times 480$ | 0 | | | | | | | |
| Inception (4a) | | $14 \times 14 \times 512$ | 2 | 192 | 96 | 208 | 16 | 48 | 64 | $364 \times 10^3$ |
| Inception (4b) | | $14 \times 14 \times 512$ | 2 | 160 | 112 | 224 | 24 | 64 | 64 | $437 \times 10^3$ |
| Inception (4c) | | $14 \times 14 \times 512$ | 2 | 128 | 128 | 256 | 24 | 64 | 64 | $463 \times 10^3$ |
| Inception (4d) | | $14 \times 14 \times 528$ | 2 | 112 | 144 | 288 | 32 | 64 | 64 | $580 \times 10^3$ |
| Inception (4e) | | $14 \times 14 \times 832$ | 2 | 256 | 160 | 320 | 32 | 128 | 128 | $840 \times 10^3$ |
| Max pool | $3 \times 3/2$ | $7 \times 7 \times 832$ | 0 | | | | | | | |
| Inception (5a) | | $7 \times 7 \times 832$ | 2 | 256 | 160 | 320 | 32 | 128 | 128 | $1072 \times 10^3$ |
| Inception (5b) | | $7 \times 7 \times 1024$ | 2 | 384 | 192 | 384 | 48 | 128 | 128 | $1388 \times 10^3$ |
| Avg. pool | $7 \times 7/1$ | $1 \times 1 \times 1024$ | 0 | | | | | | | |
| Dropout | 40% | $1 \times 1 \times 1024$ | 0 | | | | | | | |
| Linear | | $1 \times 1 \times 1000$ | 1 | | | | | | | $1000 \times 10^3$ |
| Softmax | | $1 \times 1 \times 2$ | 0 | | | | | | | |

*SN: Serial Number; CF: convolution filter; **#: Number of.

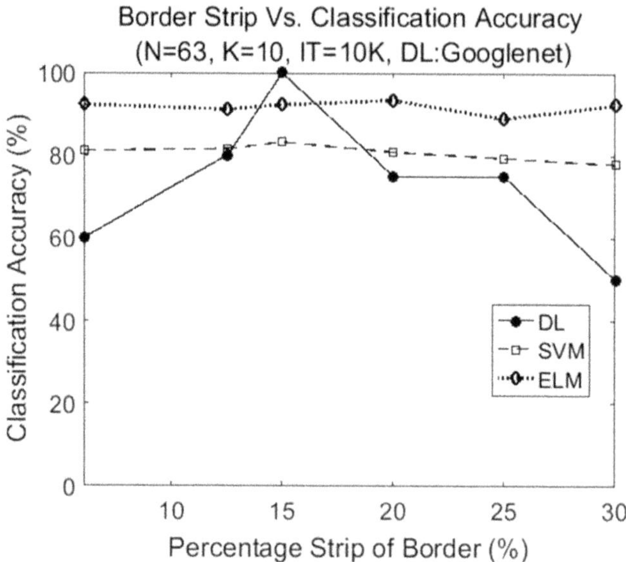



**Figure 11.7.** Accuracy versus cropping percentage of background borders. The accuracy was calculated using the K10 protocol in the (a) SVM, (b) ELM, and (c) DL frameworks. (Reproduced with permission from [40]. Copyright 2018 Elsevier.)

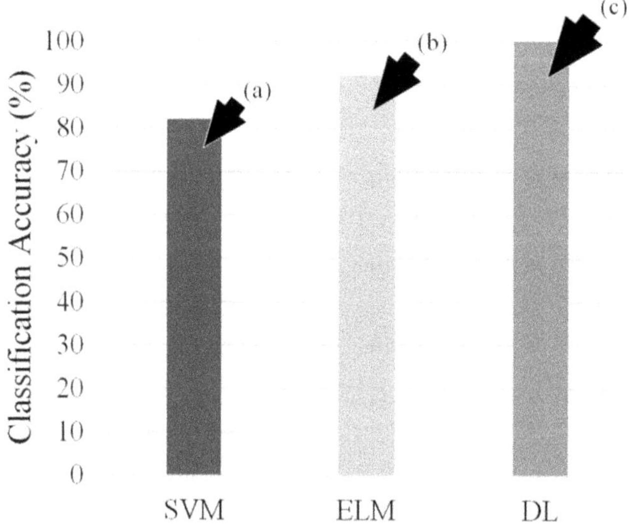



**Figure 11.8.** Accuracy bar chart. (Reproduced with permission from [40]. Copyright 2018 Elsevier.)

the DL scenario, one likely cause for this is because, in contrast to the SVM and ELM, it over-adapts in the initial round and does not include any iteration. In figure 11.8, with (a), (b), and (c) for the SVM, ELM, and DL study architectures, a

**Table 11.2.** Comparative performance of three models using the K10 cross-validation. protocol: SVM versus ELM versus DL. (Reproduced with permission from [40]. Copyright 2018 Elsevier.)

| Classifier type | Average ACC* (%) | Average sensitivity (%) | Average specificity (%) | Average PPV (%) | Average AUC |
|---|---|---|---|---|---|
| SVM | 82.08 | 64.21 | 93.56 | 86.31 | 0.79 |
| ELM | 92.22 | 93.33 | 90.83 | 84.58 | 0.92 |
| DL | 100.00 | 100.00 | 100.00 | 100.00 | 1.00 |

* Accuracy.

matching bar chart with 15% stripped images is shown for all three techniques. For the K10 cross-validation of 15% stretched ultrasound liver pictures the average accuracy, AUC, sensitivity, specificity, and positive predictive value (PPV) values are shown in table 11.2. For DL the result is 100% better than for the SVM and ELM for the accuracy, sensitivity, specificity, and PPV scores. The value for AUC is a perfect 1.0, higher than the value for the SVM and ELM. Deep learning delivers a solid performance for the cropping technique, as we have shown from our trials. The cropping function is used to remove information regarding background noise. Deep learning can be considered as 100% accurate at 15% border zone cutting. A further increase in cropping, however, removes FLD data from the liver, resulting in a decline in accuracy which leads in an optimised harvest value. This further indicates that the FLD tissue information is not always concentrated in images. The initial model gives effective outcomes when the regional border is stripped to 15%. The reason for this is that the liver tissue is represented best in this scenario and because of its low noise characteristics. Note that in a typical learning machine system, such as the SVM or ELM, these advanced capabilities do not exist.

### 11.4.2 The effect of data size on stratification accuracy

It is vital to understand the impact of the size of the data during training and testing in the DL model and compare the SVM and ELM techniques. The Symtosis$^{TM}$ system was implemented with the technique of figure 11.5 using four CV sets (K2, K3, K5, and K10). Figure 11.9 shows the results. The results show that DL model is better in terms of accuracy.

### 11.4.3 Stratification analysis using the 'liver segregation index'

In the three paradigms, i.e. SVM, ELM, and DL, it is crucial to understand how the Symtosis$^{TM}$ class of characteristics perform. The main idea is to calculate features, standardise them, and calculate the index that shows the difference between patients who have a normal and abnormal liver. We have defined the index as the liver segregation

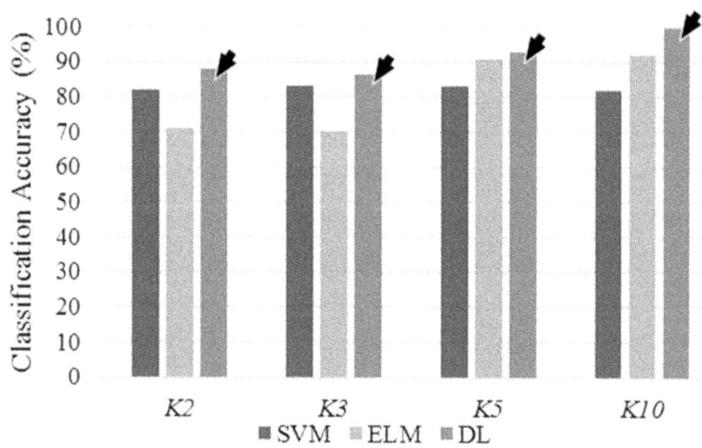



**Figure 11.9.** Accuracy bar chart for all CV protocols. (Reproduced with permission from [40]. Copyright 2018 Elsevier.)

**Table 11.3.** Average feature values for normal and abnormal patients. (Reproduced with permission from [40]. Copyright 2018 Elsevier.)

| | ELM/SVM | | | | DL | |
|---|---|---|---|---|---|---|
| Abnormal | Average abnormal liver values | SD* abnormal | Abnormal | Average abnormal liver values | | SD abnormal |
| Gabor | 0.500 | 0.506 | Deep | 0.259 | | 0.247 |
| GLRM | 0.295 | 0.258 | features | | | |
| GLCM | 0.486 | 0.287 | | | | |
| Normal | Average normal liver values | SD normal | Normal | Average normal liver values | | SD normal |
| Gabor | 0.462 | 0.380 | Deep | 0.285 | | 0.246 |
| GLRM | 0.321 | 0.396 | features | | | |
| GLCM | 0.504 | 0.284 | | | | |

* Standard deviation.

index (LSI) and it is provided mathematically as $\mathrm{LSI} = \dfrac{|\mu_{\mathrm{Nor}} - \mu_{\mathrm{Abnor}}|}{\mu_{\mathrm{Nor}}} \times 100$, where $\mu_{\mathrm{Nor}}$ and $\mu_{\mathrm{Abnor}}$ are the average of normal and abnormal features and are given as

$$\mu_{\mathrm{Nor}} = \frac{\sum_{j=1}^{P_{\mathrm{Nor}}} \sum_{i=1}^{N} x_i^{j}}{P_{\mathrm{Nor}} \times N}; \quad \mu_{\mathrm{Abnor}} = \frac{\sum_{j=1}^{P_{\mathrm{Abnor}}} \sum_{i=1}^{N} x_i^{j}}{P_{\mathrm{Abnor}} \times N}, \tag{11.1}$$

where $P_{\mathrm{Nor}}$ and $P_{\mathrm{Abnor}}$ indicate the number of normal and abnormal/diseased features, and $N$ signifies the total number of features. Table 11.3 shows the results for the average values of the features utilising a three normal class and abnormal

**Table 11.4.** Liver segregation index. (Reproduced with permission from [40]. Copyright 2018 Elsevier.)

| Classifier | Liver segregation index (LSI) (%) | | |
| --- | --- | --- | --- |
| ELM/SVM | Gabor | GLRM | GLCM |
| | 7.65 | 8.81 | 3.66 |
| DL | Deep learning features | | |
| | 9.86 | | |

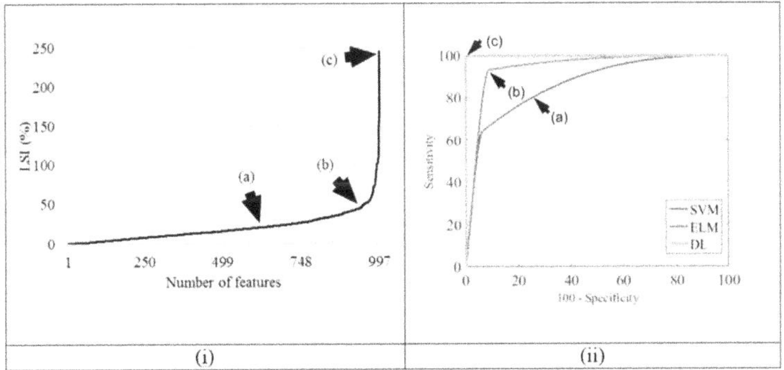

Figure 11.10. (i) LSI (%) versus the number of features using the DL framework. (ii) (a) SVM, (b) ELM, and (c) DL using the $K = 10$ ROC curve. (Reproduced with permission from [40]. Copyright 2018 Elsevier.)

class strategy. Table 11.4 shows the respective LSI values. Note that between the DL and ELM/SVM architectures there is an important distinction. The employment of the Gabor, GLRM, and GLCM [5] feature maps is part of the SVM/ELM. Another argument for greater precision in DL than in ELM and SVM is the division in the feature map.

Table 11.4 shows clearly that the LSI values for the DL features are higher than the LSI values for the ML features. The plot is provided in figure 11.10 (i) for LSI (%) versus deep characteristics. It is shown that LSI values increase gradually (arrow (a)) until the total number of features reaches 800 (arrow (b)) and thereafter increase sharply (arrow (c)).

## 11.5 Performance evaluation: ROC, reliability, and timing analysis

CV techniques are used to test the DL architecture. We mention the ROC analysis in section 11.5.1. The reliability analysis for various cross-validated $K$-folds is included in section 11.5.2 and the time analysis is presented in section 11.5.3.

### 11.5.1 ROC analysis

Figure 11.10(ii) shows the ROC curves for the SVM, ELM, and DL. The AUC for DL was improved more in comparison to that for the SVM and ELM.

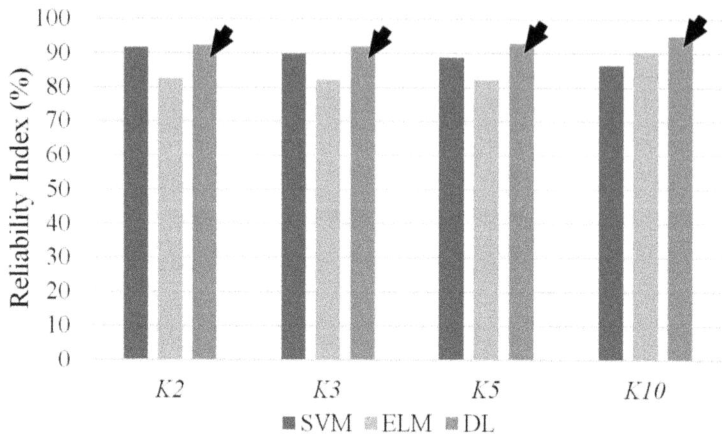


**Figure 11.11.** Bar chart of the reliability index. (Reproduced with permission from [40]. Copyright 2018 Elsevier.)

**Table 11.5.** Timing analysis. (Reproduced with permission from [40]. Copyright 2018 Elsevier.)

| Type of classifier | Avg. training time (ms) | Avg. testing time (ms) | PE index test |
| --- | --- | --- | --- |
| SVM | 0.029 016 | 0.002 262 | 0.188 269 |
| ELM | 0.007 020 | 0.001 560 | 0.143 868 |
| DL | 1.300 937 | 0.002 476 | 0.247 600 |

ms: milliseconds.

### 11.5.2 Reliability analysis

Reliability analysis is performed for all $K$-fold cross validations. The reliability index (RI) $\zeta_{N_L}(\%)$ is formulated as

$$\zeta_{N_L}(\%) = \left(1 - \frac{\mu_{N_L}}{\sigma_{N_L}}\right) \times 100, \tag{11.2}$$

where $\mu_{N_L}$ is the average accuracy and $\sigma_{N_L}$ is the standard deviation of accuracy for $N_L$ liver images. Figure 11.11 shows the bar chart for the reliability analysis. The chart demonstrates clearly the better reliability of the DL architecture than for the SVM and ELM.

### 11.5.3 Timing analysis

The DL architecture time analysis compared to the SVM and ELM is provided in table 11.5. It is shown that the DL architecture's average training and testing time is longer than for the SVM and ELM. In addition, for measuring composite performance, we calculate the classification precision product and test duration as the PE index (table 11.5).

## 11.6 Discussion

The objective of this research was the classification of FLD, also known as hepatic steatosis, using a Symtosis$^{TM}$ class system, a DL approach, along with the ML models SVM and ELM as benchmarks. FLD characterisation has been done previously with standard ML methods including SVM [11], BPNN [8], PNN [27], fuzzy classifier (FC) [26, 28], and decision tree (DT) [6]. The hypoechoic nature and the stratification in liver images in standard ML based methods have long been limitations. The ML approaches used for FLD identification and risk stratification in US limit the precision of the computed feature maps. Furthermore, smaller datasets with dynamically different fits [5, 29–33] can be used with ML methods. In this chapter, DL was used to diagnose and stratify FLD. The Symtosis$^{TM}$ system is validated using biometric face data and has been shown to be accurate [34, 35, 38, 39].

*Benchmarking*

For hepatic detection and risk classification, there are few DL applications. We consequently opted to use the newest ML models which were applied to US imagery to benchmark our DL method. Table 11.6 presents the outcomes of our benchmarking. The ML methods for US, such as liver, thyroid, and ovarian US, come mostly from a number of large organisations. It is crucial to see. The core of table 11.6 is column 5, which shows the feature maps modified for ML. These

**Table 11.6.** Benchmarking. (Reproduced with permission from [40]. Copyright 2018 Elsevier.)

| Reference | Disease | Dataset | No. of samples | Feature mining | Type of classifier | Accuracy (%) |
|---|---|---|---|---|---|---|
| Acharya *et al* (2012) [5] | FLD | Liver ultrasound images | 100 | HOS, texture, DWT | DT | 93.3 |
| Acharya *et al* (2014) [36] | HT | Thyroid ultrasound images | 526 | Wavelet transform | FC | 84.6 |
| Acharya *et al* (2014) [19] | OC | Transvaginal ultrasound images | 2600 | HIM, Gabor, entropies | PNN | 99.8 |
| Saba *et al* (2016) [7] | FLD | Liver ultrasound images | 62 | Harlick, Fourier, Gabor | BPNN | 97.6 |
| Liu *et al* (2017) [37] | LC | Liver ultrasound images | 91 | DET + T-CNN | SVM | 89.2 |
| Proposed method | FLD | Liver ultrasound images | 63 | CNN | Softmax | 100.0 |

FLD: fatty liver disease; LC: liver cirrhosis; OC: ovarian cancer; HT: Hashimoto thyroiditis; CNN: convolution neural networks; T-CNN: trained CNN; DET: detection of liver capsule; HIM: Hu's invariant moments; HOS: high order spectra; DWT: wavelet packet decomposition; SVM: support vector machine; BPNN: back propagation neural network; PNN: probabilistic neural network; FC: fuzzy classifier; DT: decision tree.

attributes largely consisted of texture, wavelength, Gabor, Fourier, and higher order spectra (HOS) features.

In 2012 Suri and his team [5] reached a precision of 93.3% in the liver stratification with a combination of high quality spectra, texture, and the breakdown of wavelets with a decision tree (DT) classification. In 2014 Suri and his team again tried a similar strategy using fuzzy classification and the use of the wavelet transform function [36], giving 84.6% precision in ultrasound thyroid photos. Suri and his team work were applied in [19] a neural network approach to ovarian images utilising a mixture of Hu's invariant times, Gabor features, and entropy based features, and a probabilistic 99.8% neural network classifier. In recent years, Suri and his team [7] have published a ultrasound liver neural network propagation classifier with characteristics such as Harlick, Fourier, and Gabor, resulting to 97.6% accuracy. Liu *et al* [37] have recently employed a combination of liver capsule detection and a trained CNN model for extracting features and have used the SVM for an accurate classification of 89.2%. With stronger and optimistic findings about 100%, CNN gives better results, albeit with a fewer than 63 individuals. As the classification accuracy approached 100% the CNN exhibited a high AUC value of 1.0 for K10.

*Strengths and weaknesses*
The power to decompose images into small patches and then extract a bigger deck of maps allows neural networks to learn, generalize, and classify FLD properly. This DL approach can span a network from one layer to 22 layers, ensuring that refined functions are selected with the dynamic calculation of millions of weights. The classification paradigm provides a model with inherent strength. The use of the starting layer in figure 11.12 significantly strengthens the CNN model by combining a few convolution operations with the use of a concept of dimensional reduction in a single layer. This leads to an increase in the depth of the suggested DL model

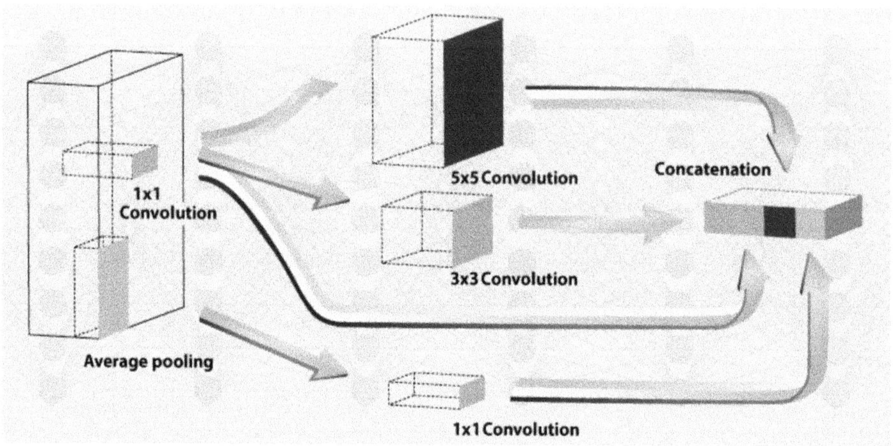


**Figure 11.12.** Inception layer showing the dimensionality reduction and concatenation process utilised in the DL architecture. (Reproduced with permission from [40]. Copyright 2018 Elsevier.)

without increasing calculation costs. In addition, the DL based technique offers various applications and expandability because of the reduced dimensionality and concatenation power of the inception model, when applied in GPU frames.

Despite the above distinctive and unique advantages, several obstacles have to be tackled in the future. We have to generalize results on a large dataset using multiple deep learning layers. As it handles millions of common weights, unlike ELM which has only one layer, the system has longer convergence times. Although it achieved virtually 100% precision and intriguing results, our solution calls for additional work for the stabilisation and construction of robust, rapid, and universal systems.

## 11.7 Conclusion

We presented a Symtosis DL based approach for the characterisation of US images of liver tissues and for the risk classification of both hyper- and hypoechoic liver images. Without increasing the computing costs, our DL based Symtosis$^{TM}$ system uses a basic model to reduce the dimensionality and speed in our DL network. The images were optimised by removing the backgrounds from the original liver images using border removal. With 15% background removal our results indicate perfect accuracy. The liver datasets were cross-validated (protocols K1, K3, K5 and K10) in comparison to common ML algorithms such as SVM and ELM. In addition, the findings of our experiment reveal that our Symtosis$^{TM}$ DL based system is reliable and stable in all cross-validations compared to SVM and ELM. The characteristics retrieved from the ML and our Symtosis DL system revealed a considerable increase in the liver segregation index for the DL functionality in comparison to ML based functions. The technique has been validated using two freely available facial biometric datasets.

## References

[1] Lazo M, Hernaez R, Eberhardt M S, Bonekamp S, Kamel I, Guallar E, Koitesh A, Brancati F L and Clark J M 2013 Prevalence of nonalcoholic fatty liver disease in the United States: the Third National Health and Nutrition Examination Survey, 1988–1994 *Am. J. Epidemiol.* **178** 38–45

[2] Browning J D, Szczepaniak L S, Dobbins R, Horton J D, Cohen J C, Grundy S M and Hobbs H H 2004 Prevalence of hepatic steatosis in an urban population in the United States: impact of ethnicity *Hepatology* **40** 1387–95

[3] Lamb *et al* 2015 Stratification of patients with liver fibrosis using dual-energy CT *IEEE Trans. Med. Imaging* **34** 807–15

[4] Guo D, Qiu T, Bian J, Kang W and Zhang L 2009 A computer-aided diagnostic system to discriminate SPIO-enhanced magnetic resonance hepatocellular carcinoma by a neural network classifier *Comput. Med. Imaging Graph.* **33** 588–92

[5] Acharya U R, Sree S V, Ribeiro R, Krishnamurthi G, Marinho R T, Sanches J and Suri J S 2012 Data mining framework for fatty liver disease classification in ultrasound: a hybrid feature mining paradigm *Med. Phys.* **39** 4255–64

[6] Quinlan J 1987 Simplifying decision trees *Int. J. Man-Mach. Stud.* **27** 221–34

[7] Saba L, Dey N, Ashour A S, Samanta S, Nath S S, Chakraborty S, Sanches J, Kumar D, Marinho R and Suri J S 2016 Automated stratification of liver disease in ultrasound: an

online accurate feature classification paradigm *Comput. Methods Prog. Biomed.* **130** 118–34

[8] Hornik K, Stinchcombe M and White H 1989 Multilayer feedforward networks are universal approximators *Neural Netw.* **5** 359–66

[9] Szegedy C, Liu W, Jia Y, Sermanet P, Reed S, Anguelov D, Erhan D, Vanhoucke V and Rabinovich A 2015 Going deeper with convolutions *Proc. IEEE Conf. on Computer Vision and Pattern Recognition* p 1–9

[10] Lo S C, Chan H P, Lin J S, Li H, Freedman M T and Mun S K 1995 Artificial convolution neural network for medical image pattern recognition *Neural Netw.* **8** 1201–14

[11] Vapnik V N 1999 An overview of statistical learning theory *IEEE Trans. Neural Networks* **10** 988–99

[12] Huang G B, Zhu Q Y and Siew C K 2006 Extreme learning machine: theory and applications *Neurocomputing* **70** 489–501

[13] Ribeiro R, Marinho R T, Velosa J, Ramalho F, Sanches J M and Suri J S 2011 The usefulness of ultrasound in the classification of chronic liver disease *Ann. Int. Conf. IEEE Engineering in Medicine and Biology Society* pp 5132–5

[14] Gangadhar K, Chintapalli K N, Cortez G and Nair S V 2014 MRI evaluation of fatty liver in day to day practice: quantitative and qualitative methods *Egypt. J. Radiol. Nucl. Med.* **45** 619–26

[15] Kadah Y M, Farag A A, Zurada J M and Badawi A M 1996 Classification algorithms for quantitative tissue characterization of diffuse liver disease from ultrasound images *IEEE Trans. Med. Imaging* **15** 466–78

[16] LeCun Y, Bengio Y and Hinton G 2015 Deep learning *Nature* **521** 436–44

[17] Noor N M, Than J C, Rijal O M, Kassim R M, Yunus A, Zeki A A, Anzidei M, Saba L and Suri J S 2015 Automatic lung segmentation using control feedback system: morphology and texture paradigm *J. Med. Syst.* **39** 22

[18] McClure P, Elnakib A, El-Ghar M A, Khalifa F, Soliman A, El-Diasty T, Suri J S, Elmaghraby A and El-Baz A 2014 *In-vitro* and *in-vivo* diagnostic techniques for prostate cancer: a review *J. Biomed. Nanotechnol.* **10** 2747–77

[19] Acharya U R, Mookiah M R, Sree S V, Yanti R, Martis R J, Saba L, Molinari F, Guerriero S and Suri J S 2014 Evolutionary algorithm-based classifier parameter tuning for automatic ovarian cancer tissue characterization and classification *Ultraschall Med* **35** 237–45

[20] Saba L *et al* 2017 Plaque tissue morphology-based stroke risk stratification using carotid ultrasound: a polling-based PCA learning paradigm *J. Med. Syst.* **41** 98

[21] Araki T *et al* 2016 PCA-based polling strategy in machine learning framework for coronary artery disease risk assessment in intravascular ultrasound: a link between carotid and coronary grayscale plaque morphology *Comput. Methods Prog. Biomed* **128** 137–58

[22] Kuppili V, Biswas M, Sreekumar A, Suri H S, Saba L, Edla D R, Marinhoe R T, Sanches J M and Suri J S 2017 Extreme learning machine framework for risk stratification of fatty liver disease using ultrasound tissue characterization *J. Med. Syst.* **41** 152

[23] Huang S F, Chang R F, Moon W K, Lee Y H, Chen D R and Suri J S 2008 Analysis of tumor vascularity using three-dimensional power Doppler ultrasound images *IEEE Trans. Med. Imaging* **3** 320–30

[24] Bikesh K S, Verma K, Thoke A S and Suri J S 2017 Risk stratification of 2D ultrasound-based breast lesions using hybrid feature selection in machine learning paradigm *Measurement* **105** 146–57

[25] Pareek G *et al* 2013 Prostate tissue characterization/classification in 144 patient population using wavelet and higher order spectra features from transrectal ultrasound images *Technol. Cancer Res. Treat.* **12** 545–57

[26] Shrivastava V K, Londhe N D, Sonawane R S and Suri J S 2017 A novel and robust Bayesian approach for segmentation of psoriasis lesions and its risk stratification *Comput. Methods Prog. Biomed.* **150** 9–22

[27] Specht D F 1990 Probabilistic neural networks *Neural Netw.* **3** 109–18

[28] Carse B, Fogarty T C and Munro A 1996 Evolving fuzzy rule based controllers using genetic algorithms *Fuzzy Sets Syst.* **80** 273–93

[29] Saba L *et al* 2016 Carotid artery intra-plaque attenuation variability using computed tomography *Neurovasc. Imaging* **2** 12

[30] Banchhor S K, Londhe N D, Araki T, Saba L, Radeva P, Laird J R and Suri J S 2017 Well-balanced system for coronary calcium detection and volume measurement in a low resolution intravascular ultrasound videos *Comp. Biol. Med.* **84** 168–81

[31] Araki T *et al* 2016 PCA-based polling strategy in machine learning framework for coronary artery disease risk assessment in intravascular ultrasound: a link between carotid and coronary grayscale plaque morphology *Comp. Methods Prog. Biomed.* **128** 137–58

[32] Shrivastava V K, Londhe N D, Sonawane R S and Suri J S 2015 Reliable and accurate psoriasis disease classification in dermatology images using comprehensive feature space in machine learning paradigm *Exp. Syst. Appl.* **42** 6184–95

[33] Acharya U R, Sree S V, Krishnan M M, Krishnananda N, Ranjan S, Umesh P and Suri J S 2013 Automated classification of patients with coronary artery disease using grayscale features from left ventricle echocardiographic images *Comp. Methods Prog. Biomed.* **112** 624–32

[34] Spacek L Face94 Database http://cswww.essex.ac.uk/mv/allfaces/faces94.html

[35] Spacek L Face95 Database http://cswww.essex.ac.uk/mv/allfaces/faces95.html

[36] Acharya U R, Sree S V, Krishnan M M, Molinari F, ZieleŸnik W, Bardales R H, Witkowska A and Suri J S 2014 Computer-aided diagnostic system for detection of Hashimoto thyroiditis on ultrasound images from a Polish population *J. Ultrasound Med.* **33** 245–53

[37] Liu X, Song J L, Wang S H, Zhao J W and Chen Y Q 2017 Learning to diagnose cirrhosis with liver capsule guided ultrasound image classification *Sensors* **17** 149

[38] Pang Y-H, Teoh A B J and Ngo D C L 2006 A discriminant pseudo Zernike moments in face recognition *J. Res. Pract. Inform. Technol.* **38** 197

[39] Chitaliya N G and Trivedi A I 2010 An efficient method for face feature mining and recognition based on contourlet transform and principal component analysis using neural network *Int. J. Comput. Appl.* **6** 28–34

[40] Biswas M *et al* 2018 Symtosis: a liver ultrasound tissue characterization and risk stratification in optimized deep learning paradigm *Comput. Methods Progr. Biomed.* **155** 165–177

# Part V

Deep learning in COVID-19

**IOP** Publishing

# Multimodality Imaging, Volume 1
### Deep learning applications
**Mainak Biswas and Jasjit S Suri**

# Chapter 12

# Characterization of COVID-19 severity in infected lungs via artificial intelligence transfer learning

**Sanagala S Skandha, Luca Saba, Suneet K Gupta, Vijaya K Kumar, Amer M Johri, Narendra N Khanna, Sophie Mavrogeni, John R Laird, Gyan Pareek, Petros P Sfikakis, Athanasios Protogerou, Monica Turk, Aditya M Sharma, Andrew Nicolaides, George D Kitas, Klaudija Viskovic, Tomaz Omerzu and Jasjit S Suri**

COVID-19 is a serious pandemic causing significant death rates, fear, and economic problems in society. The timely detection of COVID-19 can reduce its severity, but this requires a feasible solution. Artificial intelligence (AI) gives such a mechanism for quick and efficient detection of COVID-19 tissue. Therefore, we hypothesized that the AI transfer learning (TL) technique provides an efficient detection mechanism using computed tomography (CT) scans.

We proposed eight AI models out of which one is based on deep learning (DL) (SuriNet inspired by UNet) and seven use TL mechanisms (AlexNet, DenseNet121, 169, XceptionNet, ResNet50, InceptionV3 (IV3), and VGG16), which were developed for classification between controls versus COVID-19. The cohort contained 60 patients (30 COVID-19, 30 control). Furthermore, we optimized the efficiency of the proposed AI models by varying the cohort from augmentation folds 1× to 6× using the K10 cross-validation (CV; 90% training, 10% testing) protocol for performance evaluation and bispectrum analysis.

Our proposed models show the highest accuracy for DenseNet121 with 89.75% $\pm$ 5.85% with an area-under-the-curve (AUC) of 0.902, $p < 0.0001$. In comparison to this model, IV3 exhibited 89.54% $\pm$ 4.56% accuracy with an AUC of 0.895, $p < 0.0001$. The DL based SuriNet exhibited the best accuracy of 85.75% $\pm$ 9.56% among all the combinations of K10 CV.

In this work, we showed that our hypothesis was validated, with TL models efficiently classifying and characterizing the COVID-19 infected tissue.

## 12.1 Introduction

The novel coronavirus disease 2019, COVID-19, shook the world's health and economy. SARS-CoV-2 is a high-grade (reproduction number, Ro = 3) infection. The world is still battling COVID-19 and its different mutated variants even after widespread vaccination. COVID-19 causes many complications such as respiratory problems, strokes, hyper-inflammation, and pneumonia, and sometimes it leads to multiorgan failure [1]. As of September 2022, nearly 622 million people have been infected by COVID-19, with nearly 6.5 million fatalities [2]. The majority of deaths occurred in eight countries: the United States, Brazil, India, Mexico, the United Kingdom, Italy, Russia, and France [2]. This novel coronavirus damages the respiratory system causing breathing problems. The symptoms include loss of taste, anosmia, a high fever, dry cough, vomiting, and diarrhea [3, 4]. The danger of COVID-19 is that asymptomatic patients can spread the disease to healthy people [3]. It mainly affects the lungs and pulmonary tissues and causes acute respiratory distress syndrome (ARDS) [5]. A significant number of patients with COVID-19 require ventilator support [6].

Many of the first COVID-19 victims in China were admitted to hospital because of the lower respiratory tract (LRT) abnormality [3, 7]. However, these symptoms varied greatly across individuals. Some individuals had no symptoms, whereas others had hypoxia as a result of ARDS. Within nine days [7], LRT turns into ARDS in some individuals [8, 9]. Effective diagnosis of infected lungs requires observation under computer tomography (CT) or ultrasound [10–12]. Generally, this observation was done using a real-time transcription-polymerase chain reaction (RT-PCR) test [13]. Researchers have shown that CT is a more sensitive COVID-19 screening approach than standard approaches [14]. Chest radiography was shown to be ineffective in detecting the opaque image characteristics of COVID-19 in a recent investigation [15]. Among all the existing imaging techniques, CT has proven to be the gold standard for the detection of COVID-19 infection in the lungs [16]. Many researchers are studying the diagnosis of COVID-19 using artificial intelligence techniques such as deep learning (DL), machine learning (ML), and transfer learning (TL). Initially, a group of researchers started studying the diagnosis of COVID-19 using lung CT images with weakly supervised DL [17]. Li *et al* [18] proposed an end-to-end computer-aided diagnosis system based on DL to predict COVID-19 severity using lung CT scans. Some researchers segmented the COVID-19 affected lungs from the CT scans using UNet++ architectures for fast diagnosis [19, 20]. In addition, new approaches for identifying COVID-19 utilizing TL on lung CT images have been proposed in several studies [21–24]. TL gives a promising result within a short period [25–28]. In this study, we used six TL architectures, namely AlexNet, ResNet50, DenseNet121, 169, XceptionNet, and VGG16. The performance of the TL framework is compared with the novel DL architecture SuriNet, inspired by the UNet architecture. We benchmarked our performance against existing works on COVID-19 detection in lung CTs for classification into diseased versus control. As part of the characterization, we used AI based mean feature strength (MFS), and validated it using the signal processing based higher-order spectrum (HOS), also called bispectrum. We optimized the performance of the seven models with varying augmentation folds.

In this chapter, section 12.2 described the methodology. It contains patient demographics, pre-processing, and a description of the models implemented in this study. Section 12.3 gives an overview of the 3D optimized results using the K10 cross-validation protocol and area-under-the-curve (AUC) comparison. Section 12.4 describes the characterization techniques implemented in this study, i.e. MFS and bispectrum analysis. It also visually describes the SuriNet features. Finally, section 12.5 discusses the benchmarking, strengths, weaknesses, extensions, and conclusion.

## 12.2 Methodology

### 12.2.1 Patient demographics

We considered CT scans from 130 patients, 100 patients with COVID-19 and 30 patients without COVID-19. In the 100 COVID-19 patients, there were 68 male and 32 female patients with an age range of 17–93 and an age of $61.49 \pm 16$ years. In the non-COVID-19 patient group there were nine male and 21 female patients with an average age of $51.4 \pm 2$ years.

### 12.2.2 Data acquisition

The patients were placed in a deep inspiration breath-hold (DIBH) supine posture and were scanned with a Philips Ingenuity Core CT Scanner. No oral contrast or intravenous medications were provided to the patients. The CT scan was 120 kV, 225 mAs, with 1.08, 0.5 s, and $65 \times 0.625$ chosen as the spiral pitch factor, gantry rotation time, and detector configuration, respectively. A soft tissue kernel was used to rebuild 1 mm thick images using a $768 \times 768$ lung window and a $512 \times 512$ mediastinal window. Figures 12.1 and 12.2 provide the full scans of the COVID-19 and control classes.

### 12.2.3 Baseline characteristics

Table 12.1 presents the baseline characteristics of the cohort. We used MedCalc for performing a $t$-test and obtained a significant value of $p < 0.05$. Table 12.1 row 3 to row 6 represents the visual characteristics of lung CT data. In the table, non-COVID-19 patients are represented as NCoP and COVID-19 patients are represented as CoP. The ground-glass opacities (GGO), lung consolidations (CONS), and pleural effusion (PLE) show a significant difference between NCop and CoP with significant values of $p = 0.000\ 01$, $p = 0.004\ 53$, and $p = 0.004\ 13$, respectively. The common physiological symptoms of fever co-related with body temperature has a value of $p = 0.003\ 13$.

### 12.2.4 Segmentation

We used a freely available tool called XMedCon for lung segmentation [29]. The original images were converted from DICOM format to png format using a Python script, and the images were then fed to XMedCon for contrast enhancement and segmentation of the lung. Figure 12.3 represents the segmentation of the lung. The segmented region of the lung in the full scan is represented in RGB colors in figure 12.4.

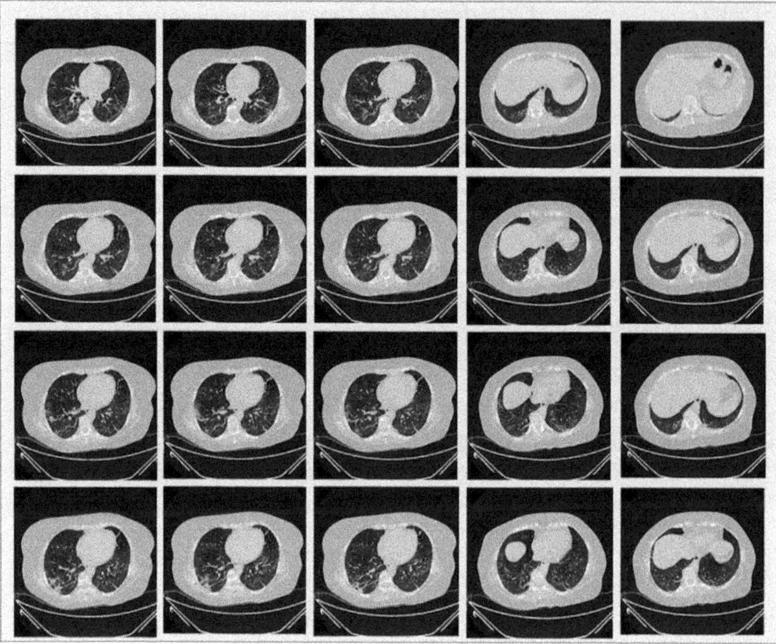

**Figure 12.1.** Full scans of the COVID-19 class.

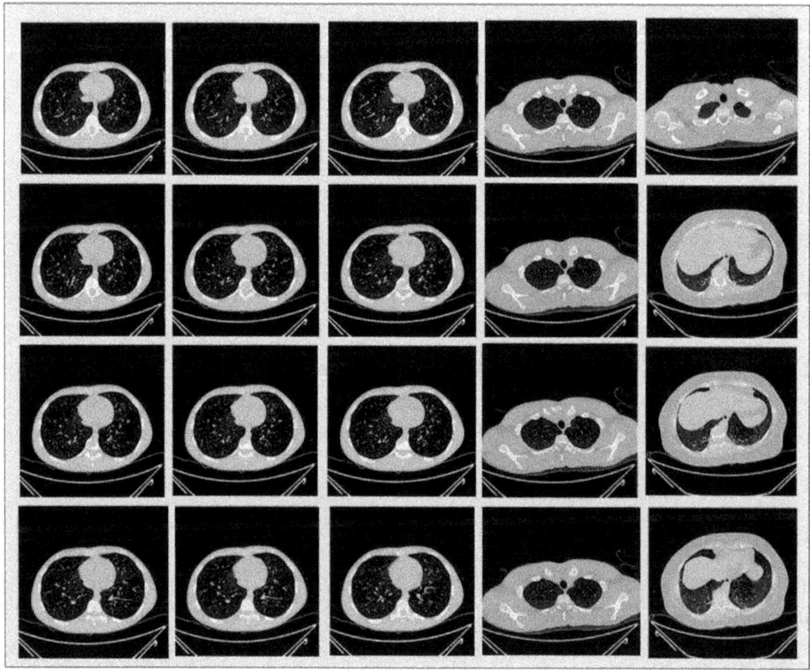

**Figure 12.2.** Full scans of the control class.

**Table 12.1.** Baseline characteristics of the patients divided into NCoP and CoP classes.

| Characteristic | Acronym | Description | NCoP ($N = 30$) | CoP ($N = 100$) | $p$-values |
|---|---|---|---|---|---|
| Age | | | 51.4 | 61.49 | 0.021 31 |
| Gender | | | 0.68 | 0.3 | 0.438 40 |
| GGO | Ground-glass opacities | An area of hazy increased lung opacity through which vessels and bronchial structures may still be seen. | 1.77 | 4.42 | 0.000 01 |
| CONS | Consolidations | A pulmonary consolidation is a region of normally compressible lung tissue that has filled with liquid instead of air. | 2.53 | 3.07 | 0.004 53 |
| PLE | Pleural effusion | The build-up of excess fluid between the layers of the pleura outside the lungs. | 0.63 | 0.12 | 0.004 13 |
| LNF | Lymph nodes | A lymph node or lymph gland is a kidney-shaped organ of the lymphatic system and the adaptive immune system. | 0.2 | 0.19 | 0.362 80 |
| Cough | | | 0.4 | 0.62 | 0.038 34 |
| Sore throat | | | 0.06 | 0.09 | 0.670 40 |
| Dyspnoea | | Shortness of breath. | 0.4 | 0.57 | 0.107 70 |
| Body temperature | | | 37.42 | 37.89 | 0.003 13 |

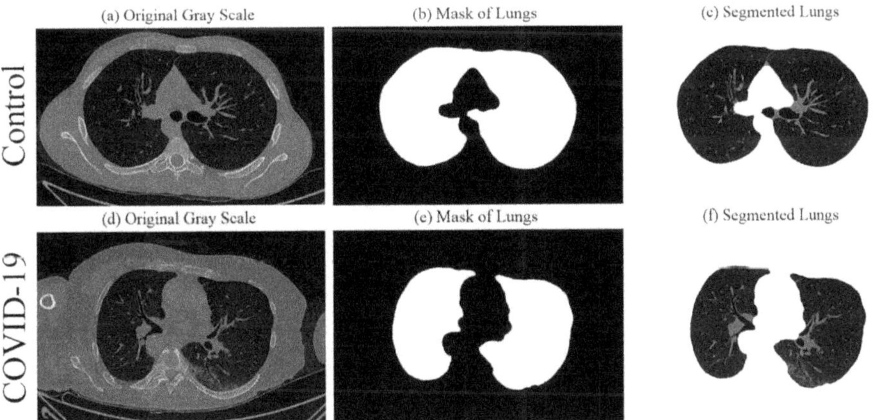


**Figure 12.3.** Segmentation of the lungs. (Reproduced with permission from [31]. Copyright 2021 Springer.)

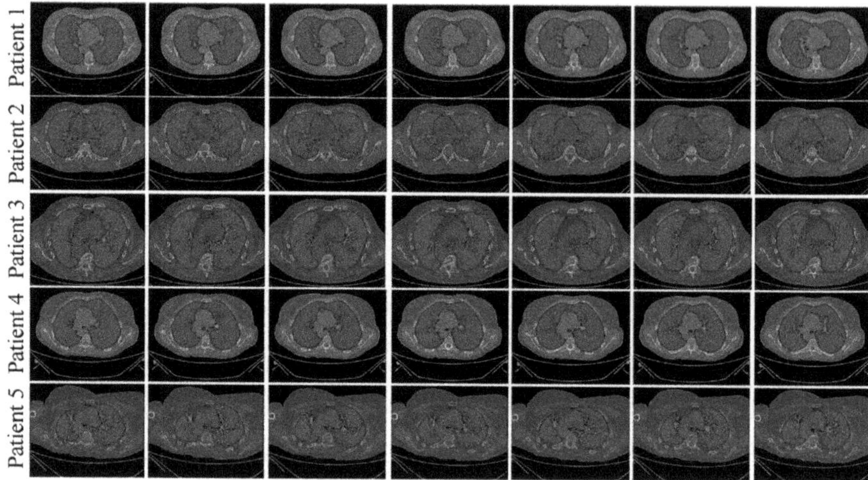

**Figure 12.4.** Segmentation of the lungs with color representation. (Reproduced with permission from [31]. Copyright 2021 Springer.)

## 12.2.5 Augmentation

The deep learning architectures require larger data samples for better training and performance, so we augmented the original cohorts into 2× to 6× folds using geometrical augmentation techniques such as flip random and rotate by 180 degrees. Table 12.2 presents the augmentation details of the cohort. We used these augmentation folds for 3D optimization. Figures 12.5 and 12.6 present the segmented lung samples of four patients.

## 12.2.6 Models

The global architecture of the proposed model is shown in figure 12.7. It contains four sections. The first one is acquisition. The second module is pre-processing. The third section is the main section showing the transfer learning paradigm adopted in this study. The fourth section describes the characterization and evaluation techniques used in the study.

*AlexNet*

We used the most popular model in deep learning, called AlexNet, for the classification of COVID-19. This architecture was introduced by Krizhevsky *et al* [32] in 2012 for solving the ImageNet challenge to classify high-resolution natural images into 1000 classes. AlexNet uses the standard pair of convolution and max-pooling techniques. It achieves an error rate of 15.3% in the top-five. Figure 12.8 presents the AlexNet architecture implemented in our study. We modified the end layer of AlexNet by adding a dropout layer (between FC6, FC7, and FC8) and one more dense layer. In the TL paradigm, we used ImageNet features in the top layers using the weight transfer method. We fed the target labels (COVID-19 labels) at FC6 and re-trained the model using K10 cross-validation (CV) with 100 epochs.

**Table 12.2.** Details of augmentation folds.

| Augmentation folds | Size of the cohort |
| --- | --- |
| Balanced (original) | Control: 2800<br>COVID-19: 2800<br>**Total: 5600** |
| Augmentation fold 2× (Aug2×) | Control: 5600<br>COVID-19: 5600<br>**Total: 11 200** |
| Augmentation fold 3× (Aug3×) | Control: 8400<br>COVID-19: 8400<br>**Total: 16 800** |
| Augmentation fold 4× (Aug4×) | Control: 11 200<br>COVID-19: 11 200<br>**Total : 22 400** |
| Augmentation fold 5× (Aug5×) | Control: 14 000<br>COVID-19: 14 000<br>**Total: 28 000** |
| Augmentation fold 6× (Aug6×) | Control: 16 800<br>COVID-19: 16 800<br>**Total: 33 600** |

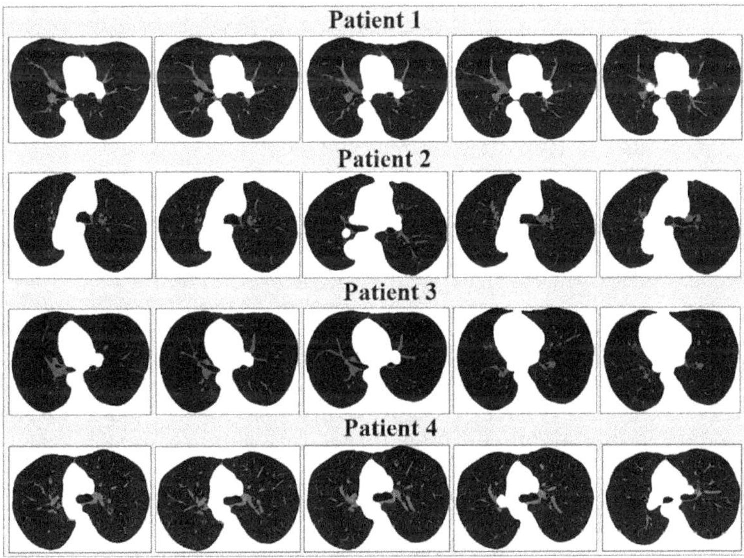


**Figure 12.5.** Sample of four patients in the control class.

### ResNet50

The residual network (ResNet) is a famous architecture from the 2016 ImageNet challenge, and it achieved a 3.57% error rate in top-five accuracy. He *et al* [33]

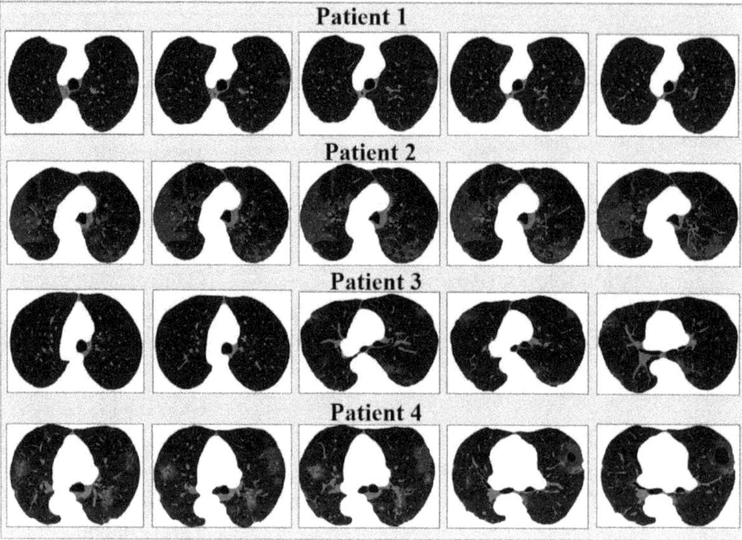

**Figure 12.6.** Sample of four patients in the COVID-19 class.

proposed a wide range of models with varying layers from 20 to 1202 layers. In the current study, we chose ResNet50, which contains 50 layers with 25 residual blocks. Residual blocks are the novel techniques introduced in this architecture. In addition, it contains skip net connections to reduce the memory requirement for the training.

In the TL paradigm, we used the ImageNet weights of pre-trained ResNet50. Then we changed the last layers and fed the target labels. First, we added a dropout layer and dense layer in the original ResNet50. Then we re-trained the model using target labels with K10 CV using the same number of epochs used in AlexNet. Figure 12.9 presents the ResNet50 architecture.

*DenseNet*

We considered various flavors of DL architectures, one of which was densely connected convolutional neural networks (DenseNet). Huang *et al* [34] proposed this model for the ImageNet Challenge of 2017 to solve the vanishing gradient problem, and it has better feature propagation with fewer training parameters. This architecture is made up of dense blocks that combine the features of the previous and current blocks. Unlike ResNet, DenseNet fuses all of the features rather than summarizing the previous blocks. The proposed architectures have feature maps ranging from 32 to 48 and depths ranging from 121 to 161. In ImageNet 2017, this model achieves a 6.15% error rate in the top 5%.

Our current study, similarly to previous models, used the TL paradigm in DenseNet121 and DenseNet169 for initial pre-trained weight transfer from ImageNet. First, we loaded all these weights to the initial layers. Then we added dropout and dense layers for re-training the model with target labels. As a result, our DenseNet model contained 129 and 161 layers; after adding the new layers, the number of layers became 131 and 163. Figure 12.10 shows the proposed DenseNet model.

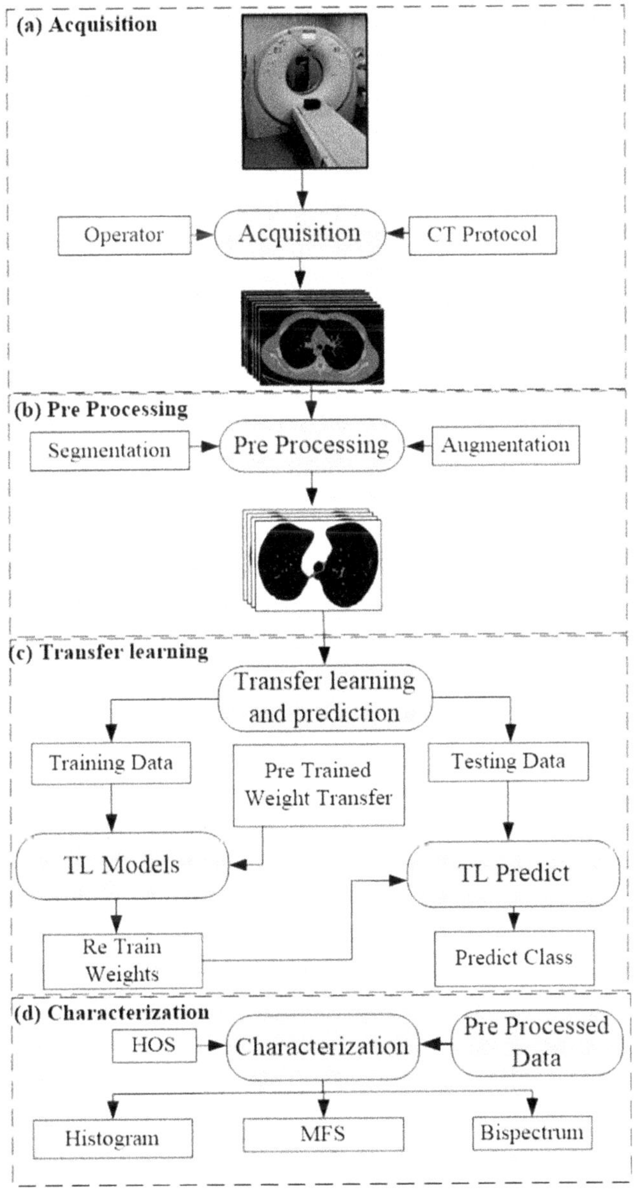



**Figure 12.7.** The global architecture of the proposed model.

*XceptionNet*
Chollet *et al* [35] from Google proposed the popular model, XceptionNet, based on InceptionNet, using a depth-wise separable convolution technique for solving the ImageNet Challenge in 2017. It achieved a top-one accuracy of 0.945 in top-five accuracy. XceptionNet contains 36 layers; we added three more layers (dense layer, dropout layer, dense layer) at the end for re-training the model under the TL

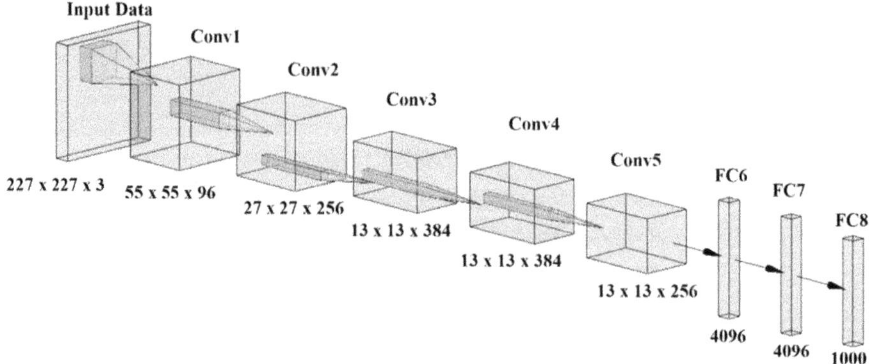

**Figure 12.8.** AlexNet architecture.

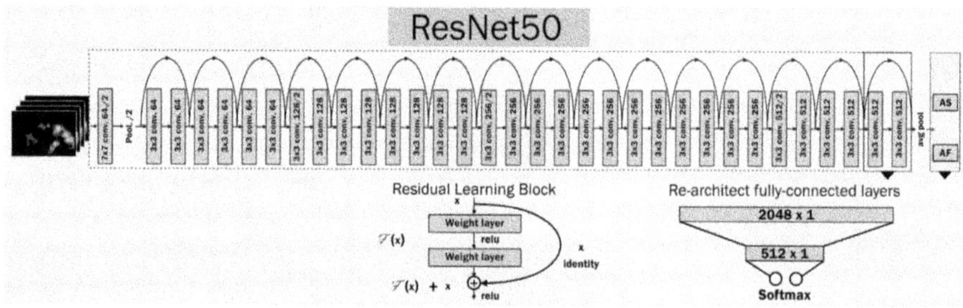

**Figure 12.9.** ResNet50 architecture.

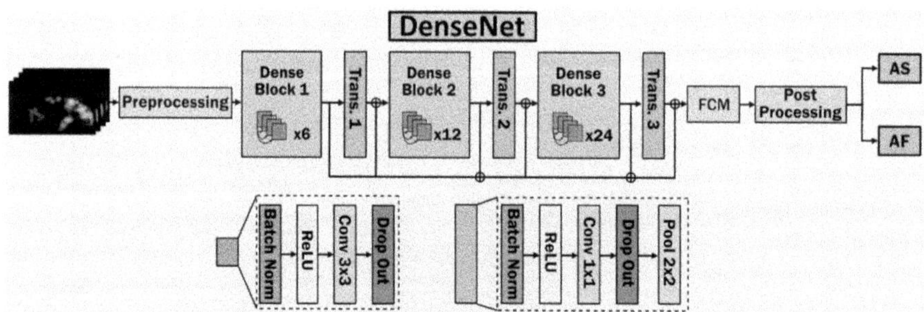

**Figure 12.10.** DenseNet121 and DenseNet161 architectures.

paradigm. We used K10 CV for re-training the XceptionNet with initial weights from ImageNet. Figure 12.11 presents the proposed XceptionNet.

*InceptionV3*

The InceptionV3 was created by Szegedy *et al* [36] to solve computer vision problems such as label smoothing and propagating label information down the neural network. Under the class of inception models, the author proposed several architectures. InceptionV3 (IV3) has the lowest error rate, with a rate of 4.2% among the top 1% accuracy. Under the TL paradigm, we considered IV3 with initial weights trained on ImageNet. As in previous models, we added the same number of layers for re-training the

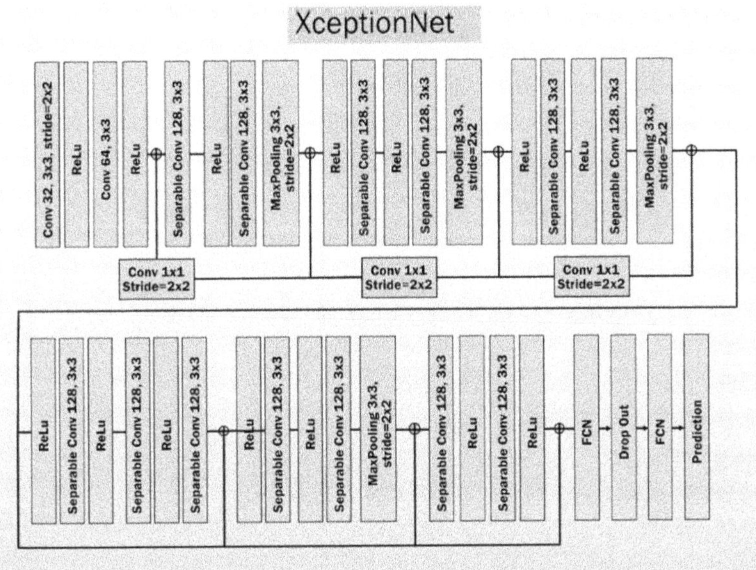



Figure 12.11. The proposed architecture of XceptionNet.

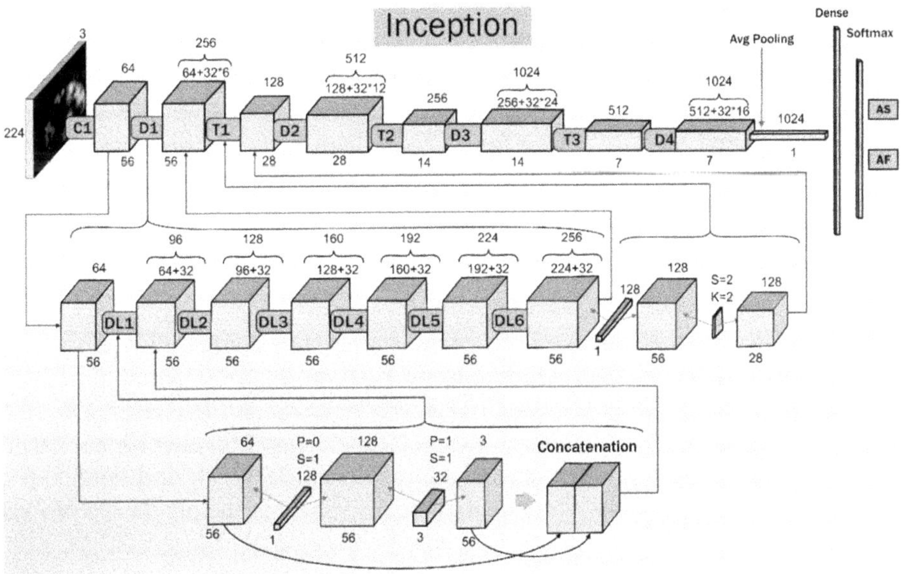



Figure 12.12. The modified architecture of InceptionV3.

IV3 for the COVID-19 cohort. In addition, we considered a dropout rate of 50% and K10 CV for re-training. Figure 12.12 presents the proposed IV3 model.

*VGG16*

Finally, we considered the most popular network, VGG16, for the classification of COVID-19 under the TL paradigm. It was first proposed by Simonyan *et al* [37]

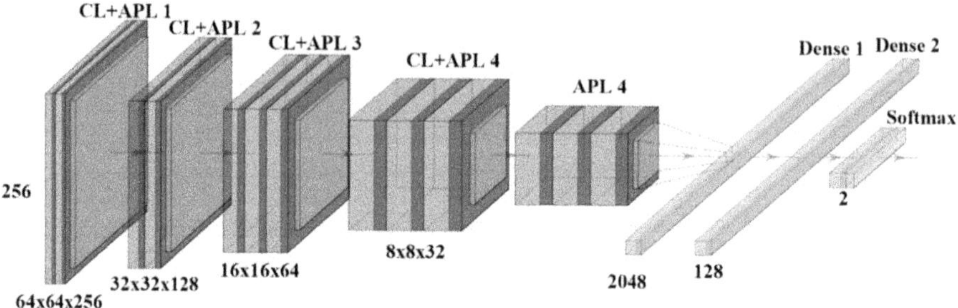

Figure 12.13. The modified architecture of VGG16.

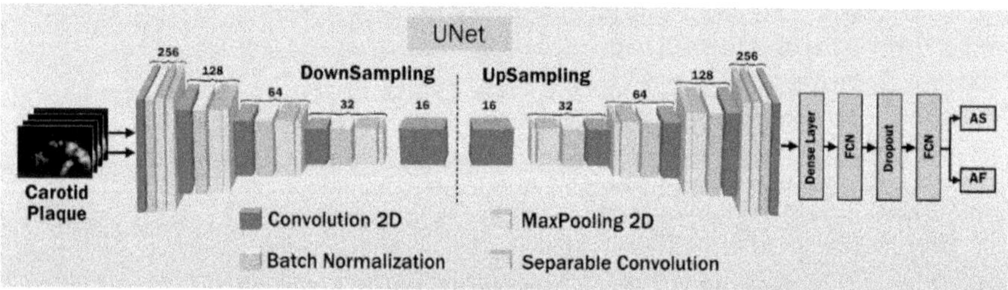


Figure 12.14. SuriNet architecture.

from the visual geometry group (VGG) for solving the ImageNet Challenge of 2014. The authors introduced $3 \times 3$ filters in a CNN for increasing the depth of the model. As a result, this model achieved an error rate of 7.0 in the top-five error rate. In our study we considered VGG16 under the TL paradigm with initialized weights from ImageNet. Similar to our earlier work on all the previous proposed models, we added two dense layers and one dropout layer for re-training the model with target labels (COVID-19) using K10 CV. Figure 12.13 represents the proposed VGG16 model.

*SuriNet*

The performance of the suggested TL architectures was compared against a new SuriNet DL design. SuriNet uses separable and batch normalization neural networks. We employed a max-pooling layer after each combination of these layer sets. In the beginning, we utilized a few filters, i.e. 32, and later increased the number to 256. The shape of the architecture is like a funnel, shown in figure 12.14, and details of the training parameters of SuriNet are shown in table 12.3. As in the previously proposed TL models, SuriNet was also trained and tested on target labels using K10 CV.

## 12.3 Results

We ran all the proposed models on all the augmentation folds, i.e. from Aug1× to Aug6× using K10 CV with 100 epochs. In total we used 7000 ($7 \times 10 \times 100$) runs for finding the optimization of every proposed model. Figure 12.15 presents the optimization of every proposed model. The black arrow indicates the optimization point. Table 12.4 presents the corresponding values. Among all the models,

**Table 12.3.** SuriNet architecture parameters.

| Layer type | Shape | # of parameters |
| --- | --- | --- |
| Convolution 2D | 128 × 128 × 32 | 896 |
| Batch normalization | 128 × 128 × 32 | 128 |
| Separable convolution 2D | 128 × 128 × 64 | 2400 |
| Batch normalization | 128 × 128 × 64 | 256 |
| Max-pooling 2D | 64 × 64 × 64 | 0 |
| Separable convolution 2D | 64 × 64 × 128 | 8896 |
| Batch normalization | 64 × 64 × 128 | 512 |
| Max-pooling 2D | 32 × 32 × 128 | 0 |
| Separable convolution 2D | 32 × 32 × 256 | 34 176 |
| Batch normalization | 32 × 32 × 256 | 1024 |
| Max-pooling 2D | 16 × 16 × 256 | 0 |
| Separable convolution 2D | 16 × 16 × 64 | 18 752 |
| Batch normalization | 16 × 16 × 64 | 256 |
| Max-pooling 2D | 8 × 8 × 64 | 0 |
| Separable convolution 2D | 8 × 8 × 128 | 8896 |
| Batch normalization | 8 × 8 × 128 | 512 |
| Max-pooling 2D | 4 × 4 × 128 | 0 |
| Separable convolution 2D | 4 × 4 × 256 | 34 176 |
| Batch normalization | 4 × 4 × 256 | 1024 |
| Max-pooling 2D | 2 × 2 × 256 | 0 |
| Flatten | 1024 | 0 |
| Dense | 1024 | 1049 600 |
| Dropout | 0.5 | 0 |
| Dense | 512 | 524 800 |
| Dropout | 0.5 | 0 |
| Dense (softmax) | 2 | 1026 |
| **Total trainable parameters** | | **1 687 330** |

DenseNet121 shows the best performance of 89.75% ± 5.85% accuracy. Similarly IV3 also exhibits 89.54% ± 4.56% accuracy. In contrast, the DL based SuriNet shows 85.75% ± 9.56% accuracy, although in SuriNet a few combinations exhibit the accuracy of 98.75%. We computed the AUC for all the models shown in figure 12.16 by plotting the receiver operating characteristic (ROC). We achieved the best AUC from DenseNet121 with 0.902 ($p < 0.0001$), IV3 0.895 ($p < 0.0001$), and SuriNet 0.857 ($p < 0.0001$).

*Power analysis*
We validated our study using power analysis do determine the number of samples required for a study. We considered the following equation for the power analysis to find the sample size:

$$\text{sample size} = \frac{2 \times \left(Z_\alpha + Z_{1-\beta}\right)^2 \times \sigma^2}{\Delta^2}. \tag{12.1}$$

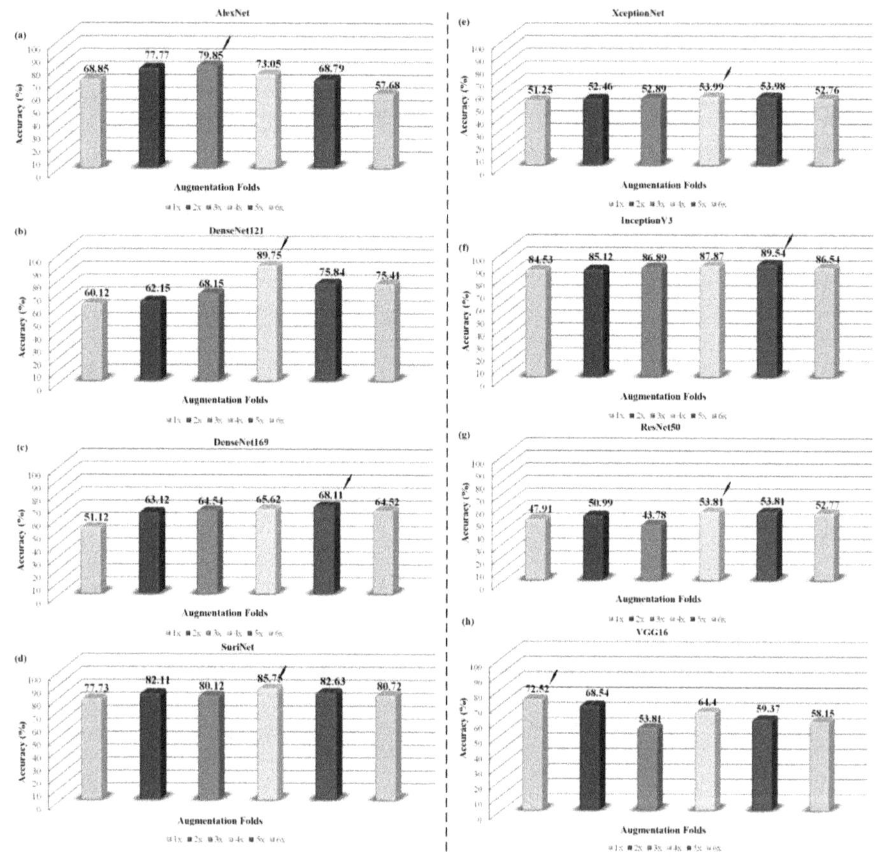

**Figure 12.15.** 3D-bar graph of proposed AI model optimization: (a) AlexNet, (b) DenseNet121, (c) DenseNet169, (d) XceptionNe, (e) InceptionV3, (f) ResNet50, and (g) SuriNet.

**Table 12.4.** 3D optimization of COVID-19 classification.

| AI model | Balanced | Aug2× | Aug3× | Aug4× | Aug5× | Aug6× |
|---|---|---|---|---|---|---|
| AlexNet | 68.85 | 77.77 | 79.85 | 73.05 | 68.79 | 57.68 |
| DenseNet121 | 60.12 | 62.15 | 68.15 | 89.75* | 75.84 | 75.41 |
| XceptionNet | 51.25 | 52.46 | 52.89 | 53.99 | 53.98 | 52.76 |
| InceptionV3 | 84.53 | 85.12 | 86.89 | 87.87 | 89.54* | 86.54 |
| ResNet50 | 47.91 | 50.99 | 43.78 | 53.81 | 53.81 | 52.77 |
| SuriNet | 77.73 | 82.11 | 80.12 | 85.75* | 82.63 | 80.72 |
| VGG16 | 72.52 | 68.54 | 53.81 | 64.4 | 59.37 | 58.15 |
| DenseNet169 | 51.12 | 63.12 | 64.54 | 65.62 | 68.11 | 64.52 |

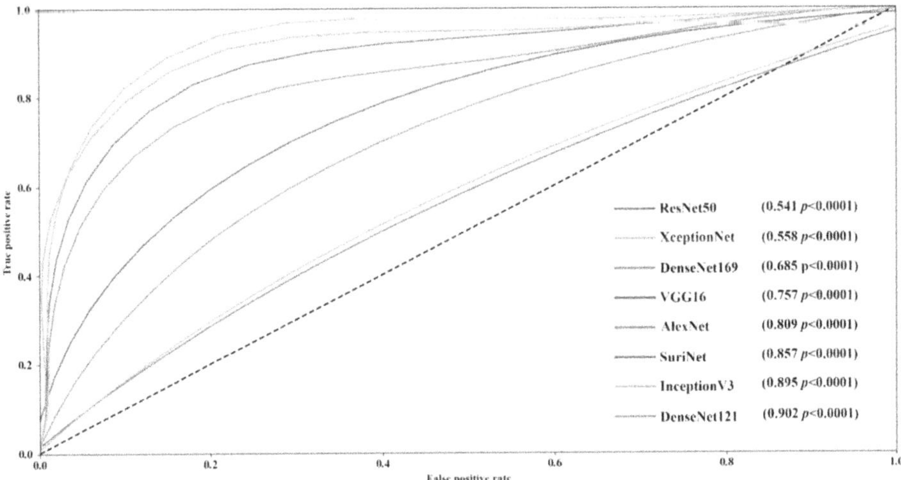


**Figure 12.16.** ROC curve analysis.

**Table 12.5.** Cohen's kappa analysis of proposed AI models.

| AI model | Kappa ($K_a$) |
| --- | --- |
| DenseNet121 | 0.784 |
| SuriNet | 0.733 |
| InceptionV3 | 0.731 |
| DenseNet169 | 0.562 |
| AlexNet | 0.559 |
| XceptionNet | 0.096 |
| ResNet50 | −0.545 |

Here we considered $Z_\alpha = 3.2905$ for type 1 error having a value of 1%, and $Z_{1-\beta} = 1.6449$ for type II error having a value of 1%, and we considered the standard deviation $\sigma = 5.65$ and mean difference $\Delta = 0.727$. By substituting these values in equation (12.1), we obtain the required number of samples as 2942.4. Thus the balanced (original) cohort contains 5600 samples, which is 47.45% higher than the required number of samples.

*Cohen's kappa analysis*
In our study, we used eight AI models and it is necessary to study the inter-rater agreement between them, so we used Cohen's kappa analysis [38]. It provides the inter-rater reliability. The value of kappa represents an agreement between the raters. If it is ⩽ 0 this indicates no agreement and 0.01–0.20 is slight agreement, 0.21–0.40 is fair agreement, 0.41–0.60 is moderate agreement, 0.61–0.80 is substantial agreement, and 0.81–1.00 is perfect agreement between raters. We calculate the average kappa ($K_a$) value of the model at their optimization point. Table 12.5 represents the Cohen's kappa analysis of our study in increasing order.

**Table 12.6.** DOR analysis of the proposed AI models.

| AI model | DOR |
| --- | --- |
| DenseNet121 | 254.45 |
| SuriNet | 135.78 |
| InceptionV3 | 128.79 |
| DenseNet169 | 63.63 |
| AlexNet | 6.16 |
| XceptionNet | 3.12 |
| ResNet50 | 0.77 |

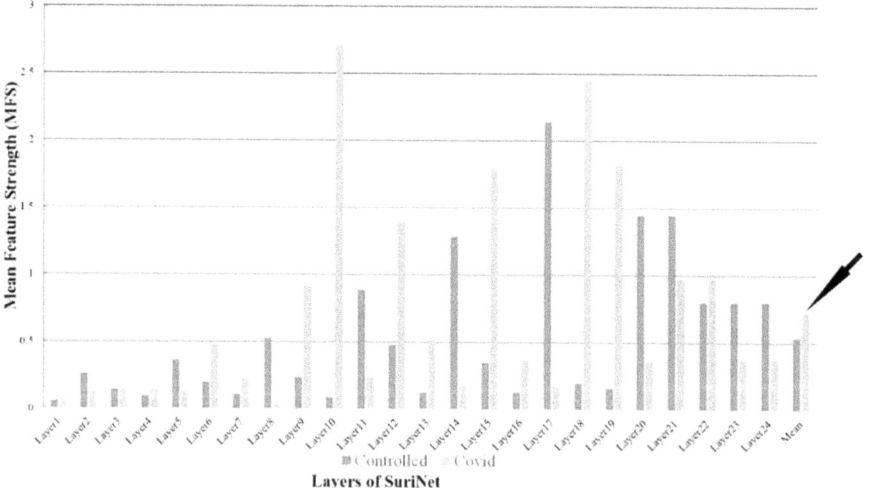


**Figure 12.17.** Mean feature strength.

*Diagnostic odds ratio*

The diagnostics odds ratio (DOR) is a popular evaluation metric in medical image processing. It gives the effectiveness of the diagnostic test. We computed the DOR for all the AI models in this study using the equation

$$DOR = (TP/FN) / (FP/TN). \tag{12.2}$$

We computed the average true positive (TP), false negative (FN), false positive (FP), and true negative (TN) and substituted in equation 12.2. Table 12.6 presents the DOR values of the AI model at their optimization point in increasing order.

## 12.4 Characterization

This section discussed the characterization of COVID-19 tissue using an AI based technique called mean feature strength (MFS), and it was validated using a signal processing based higher-order spectrum called bispectrum.

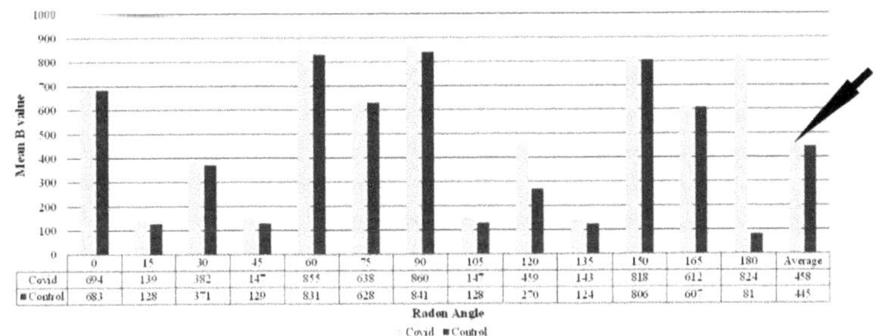

**Figure 12.18.** Bispectrum analysis.

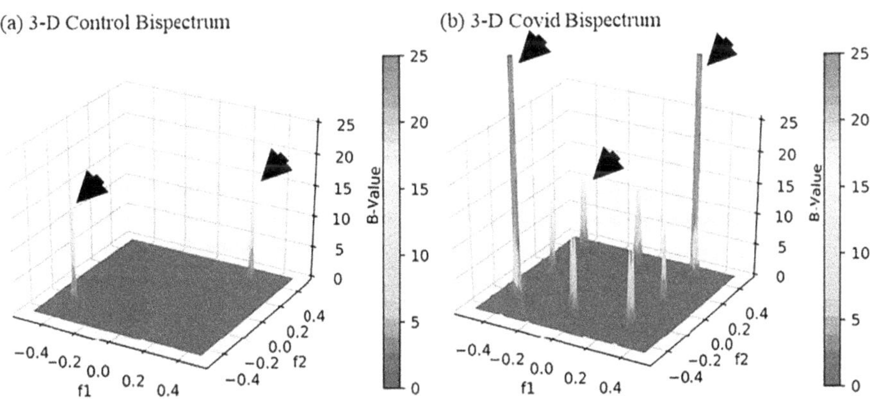


**Figure 12.19.** Higher-order spectrum of (a) control and (b) COVID-19.

*Mean feature strength (MFS)*

Among all the proposed AI models, TL models are not considered for MFS calculation due to their mixed features, because the initial features are from ImageNet and target labels. So for calculating MFS we choose the DL model SuriNet. We computed the feature strength for every model using one combination training sample in K10 cross-validation. Figure 12.17 exhibits the mean MFS at every layer. It shows that the MFS of COVID-19 is higher than for the control class by 25.09%.

*Higher-order spectrum*

We validated AI based characterization using signal-based well know method higher-order spectrum so-called bispectrum. We computed the mean B value of the one combination training samples of K10 cross-validation. We varied the radon angle from 0 to 180 degrees. The mean B value was shown in figures 12.18 and 12.19. It is observed that the mean B value of the COVID-19 is higher than the control by 2.83 percentage.

*Visualization of the DL features*

We considered the SuriNet architecture for visualizing the learned features. Figures 12.20 and 12.21 represent the extracted features from the control and COVID-19 images. Magenta represents the filter and yellow represents the learned

**11$^{\text{th}}$ Layer**

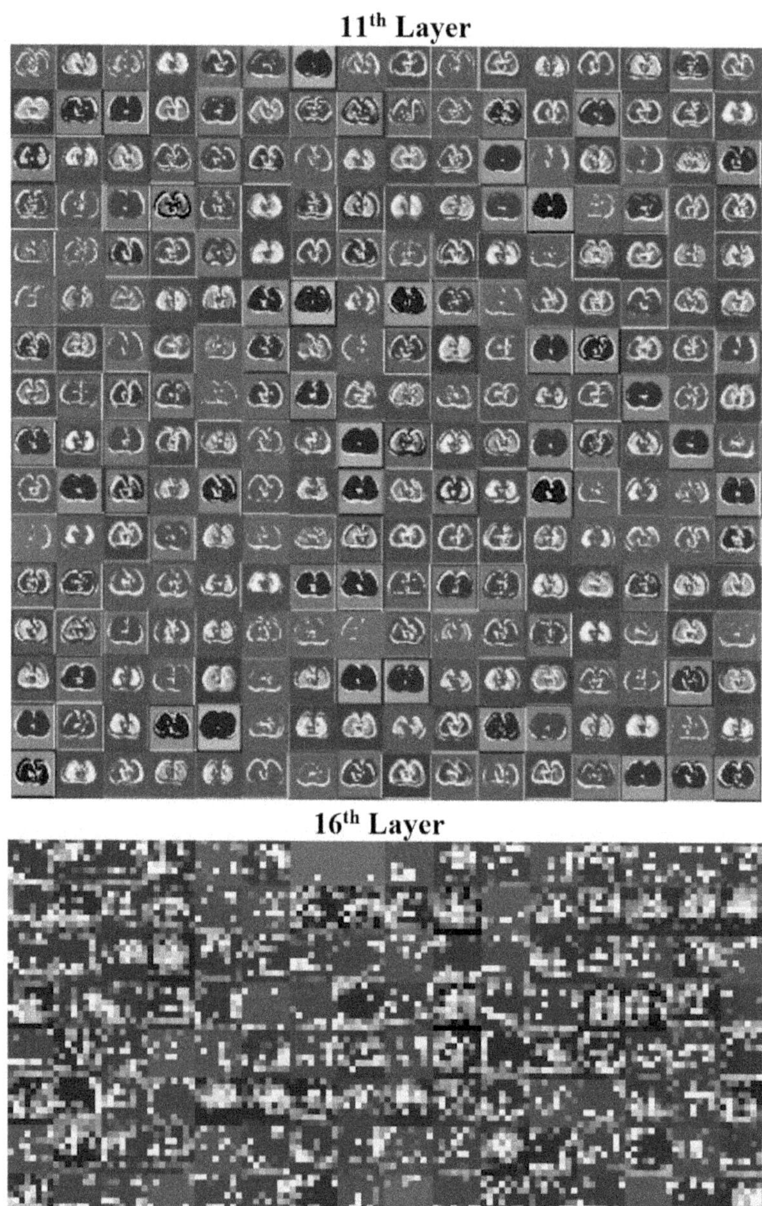

**16$^{\text{th}}$ Layer**

**Figure 12.20.** Sample of control class visualization of SuriNet.

feature. Figures 12.20 and 12.21 show that 256 filters output in a 16 × 16 grid at the eleventh layer and 128 filters in an 8 × 16 grid at the sixteenth layer (before the vectorization) in SuriNet. The grid of the filter output contains mixing of yellow and magenta. Some filter output shows that they are considering the background information of the images. For example, in figure 12.20, the seventh output in the first row shows that the learned region (yellow) is background, similar to the output;

**11ᵗʰ Layer**

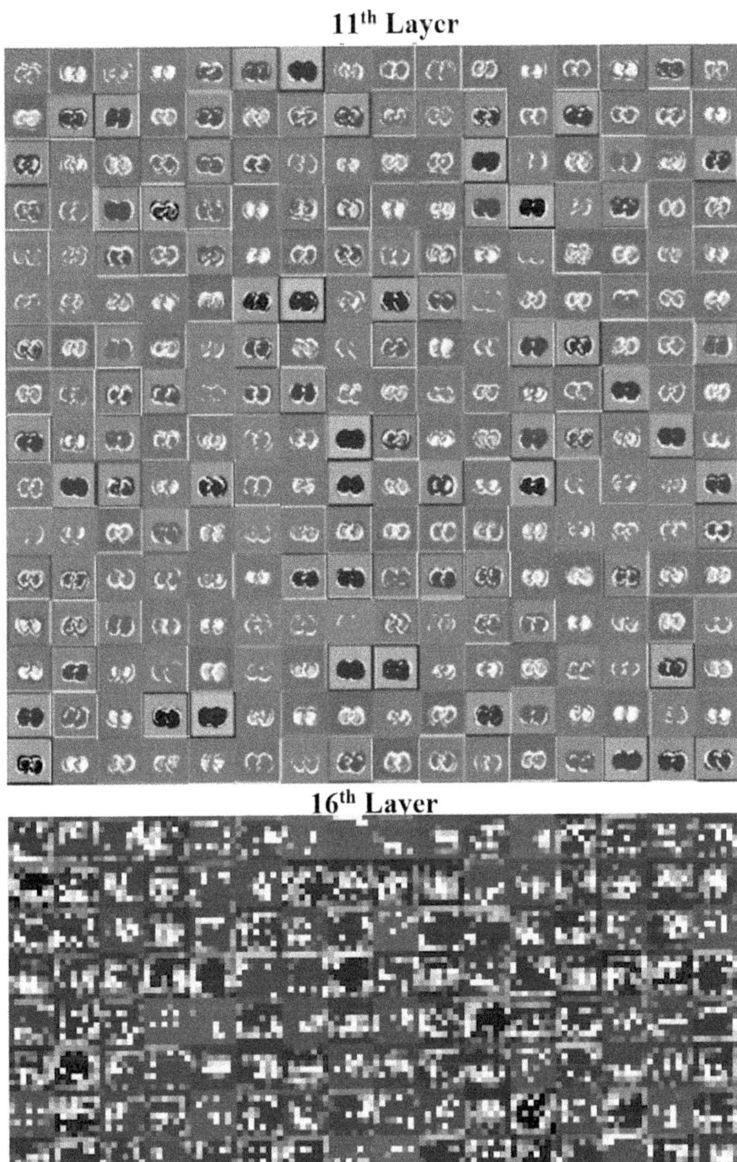

**16ᵗʰ Layer**

Figure 12.21. Sample of COVID-19 class visualization of SuriNet.

we can see a few places like this in figures 12.20 and 12.21. Similarly, the essential features are extracted from the lung CT scans, which will be seen in the dark yellow color inside the lung region, i.e. in figure 12.20 the fourth filter output in row one. These kinds of outputs from the filters are more common in the COVID-19 class than the control class. Visually it is represents that the features extracted for COVID-19 are greater than for the control. The quantification of these features is represented by the MFS.

## 12.5 Discussion

*Benchmarking*

We compared our proposed AI models to existing works. Table 12.7 presents the existing work on COVID-19 using AI models. Column 3 represents the dataset used in the study, and column 4 presents the model used. Columns 5 and 6 provide the accuracy achieved and the performance metrics. Polsinelli *et al* [39] proposed the SqueezeNet architecture to classify 360 CT scans in pneumonia and healthy and achieved an accuracy of 0.83 and sensitivity of 0.85. Hasan *et al* [40] used a histogram threshold to isolate the background of the CT scans and then extracted features using a Q-deformed entropy algorithm. The extracted features are fed to the long short-term memory (LSTM) neural networks for classifying pneumonia versus health. The authors achieved an accuracy of 99.68%. Ahuja *et al* [21] proposed a TL-based ResNet18, 50, 101, and SqueezeNet classifier for classifying the 349 CT scans

**Table 12.7.** Benchmarking table.

| Authors | Dataset | Model | Accuracy | Performance |
|---|---|---|---|---|
| Polsinelli *et al* (2020) [39] | 757 CT scans (COVID-19: 360, pneumonia: 397 | SqueezeNet | 0.83 | F1 score: 0.8333 |
| Hasan *et al* (2020) [40] | 321 chest CT scans (COVID-19: 118, pneumonia: 96, healthy: 107) | LSTM | 1.00 | — |
| Ahuja *et al* (2020) [21] | 728 lung CT scans (COVID-19: 349, non-COVID-19: 397) | ResNet18 | 0.99 | — |
| Singh *et al* (2020) [41] | 702 lung CT scans (COVID-19: 344, non-COVID-19: 358) | Bagging ensemble with SVM | 0.957 | AUC: 0.958, F1 score: 95.3% |
| Purohit *et al* (2020) [42] | 690 lung CT scans (COVID-19: 345, non-COVID-19: 345) | Proposed CNN | 0.95 | Sensitivity of 94.78% and specificity of 95.98% |
| Agarwal *et al* (2021) [31] | 60 Italian patients' lung CT scans (control: 30, COVID-19: 30) | CNN, RF | 0.994, 0.994 | 0.991, 0.998 ($p < 0.0001$) |
| Saba *et al* (2021) [30] | 60 Italian patients' lung CT scans (non-COVID-19: 30, COVID-19:100) | iCNN | 0.996 | 0.99 ($p < 0.0001$) |
| **Proposed study** | **5600 cohort (2800 control, 2800 COVID-19)** | **DenseNet121** | **89.75** | **0.902 AUC ($p < 0.0001$)** |

decomposed into three-level stationary wavelet transformation. Among all the classifiers, ResNet18 achieved an accuracy of 0.99. Singh *et al* [41] studied the effect of COVID-19 on the comorbid disease diabetes. The authors used 344 COVID-19 affected CT scans and 358 non-COVID-19 affected CT scans to extract features and fed these to the bagging ensemble for classifying COVID-19 versus non-COVID-19 scans, achieving an accuracy of 0.957 with an AUC of 0.958. Purohit *et al* [42] proposed a CNN using the multi-image augmentation technique and classifying COVID-19 versus non-COVID-19 CT scans. The authors achieved an accuracy of 0.95 and a sensitivity of 94.78%. Agarwal *et al* [31] proposed nine kinds of classifiers (ANN, DT, RF, CNN, VGG16, Denesenet121, DenseNet169, DenseNet201, and MobileNet) to classify the CT scans into control versus COVID-19 using a novel block imaging technique. The authors achieved an accuracy of 0.994 from both the DCNN and RF. Saba *et al* [30] proposed six AI models for classifying the COVID-19 versus control class using 130 CT lung scans. The authors characterized the COVID-19 lungs using a bispectrum. They achieved an accuracy of 0.996 using an improved CNN (modified sigmoid function) classifier.

*Strengths, weaknesses, and extensions*
We characterized the COVID-19 tissue using AI based MFS and signal processing based bispectrum. We achieved the best accuracy of 89.75% ± 5.85% using DenseNet121. We compared the efficiency of the TL model with the novel DL architecture SuriNet. The main weakness of this study is that the number of samples required for efficient DL classifiers should be greater. We have moderate samples in our cohort, so we augmented it. The extensive dataset on COVID-19 helps the model to understand more complex patterns and efficiently classify the COVID-19 CT scans. The hybrid DL models are efficient for understanding more complex patterns in the cohort. Big Data with the hybrid model will bring more insight into the detection of COVID-19 from CT scans.

*Conclusion*
We classify and characterize the control versus COVID-19 CT scans successfully using TL and DL classifiers. We have the highest accuracy for DenseNet121 of 89.75% ± 5.85% with an AUC of 0.902 ($p < 0.0001$). Compared to this model, IV3 exhibits 89.54% ± 4.56% accuracy with an AUC of 0.895 *($p < 0.0001$)*. The DL based SuriNet exhibits a best accuracy of 85.75% ± 9.56%. Among all the combinations of K10, SuriNet shows 98.75%, the best accuracy for a few cases. The bispectrum shows that the COVID-19 mean B value is higher than the control by 2.85%, and using AI based MFS shows 25.09%.

# References


[1] WHO 2020 *Coronavirus Disease (COVID-19) Situation Reports* https://who.int/emergencies/diseases/novel-coronavirus-2019/situation-reports

[2] Gabutti G *et al* 2020 Coronavirus: update related to the current outbreak of COVID-19 *Infect. Dis. Ther.* **9** 241–53

[3] Wang H-Y *et al* 2020 Potential neurological symptoms of COVID-19 *Therap. Adv. Neurol. Disorders* **13** 10.1177/1756286420917830

[4] Cascella M *et al* 2022 Features, evaluation, and treatment of coronavirus (COVID-19) *StatPearls*

[5] Pan C *et al* 2020 Lung recruitability in COVID-19-associated acute respiratory distress syndrome: a single-center observational study *Am. J. Respir. Crit. Care Med.* **201** 1294–7

[6] Meng L *et al* 2020 Intubation and ventilation amid the COVID-19 outbreak: Wuhan's experience *Anesthesiology* **132** 1317–32

[7] Huang C *et al* 2020 Clinical features of patients infected with 2019 novel coronavirus in Wuhan, China *Lancet* **395** 497–506

[8] Devaux C A, Rolain J-M and Raoult D 2020 ACE2 receptor polymorphism: susceptibility to SARS-CoV-2, hypertension, multi-organ failure, and COVID19-19 disease outcome *J. Microbiol., Immunol. Infect.* **53** 425–35

[9] Zaim S *et al* 2020 COVID-19 and multi-organ response *Curr. Prob. Cardiol.* **45** 100618

[10] Saba L and Suri J S 2013 *Multi-Detector CT Imaging: Principles, Head, Neck, and Vascular Systems* vol 1 (Boca Raton, FL: CRC Press)

[11] El-Baz A and Suri J 2019 *Lung Imaging and CADx* (Boca Raton, FL: CRC Press)

[12] El-Baz A and Suri J S 2011 *Lung Imaging and Computer Aided Diagnosis* (Boca Raton, FL: CRC Press)

[13] World Health Organization 2020 *Use of Chest Imaging in COVID-19: A Rapid Advice Guide* (Portland, OR: Pacific Northwest Evidence-Based Practice Center, Oregon Health and Science University)

[14] Hatamabadi H *et al* 2020 Lung ultrasound findings compared to chest CT scan in patients with COVID-19 associated pneumonia: a pilot study *Adv. J. Emerg. Med.* **5** e19

[15] Revel M-P *et al* 2020 COVID-19 patients and the radiology department—advice from the European Society of Radiology (ESR) and the European Society of Thoracic Imaging (ESTI) *Eur. Radiol.* **30** 4903–9

[16] Yan C *et al* 2020 Lung ultrasound vs chest x-ray in children with suspected pneumonia confirmed by chest computed tomography: a retrospective cohort study *Exp. Therap. Med.* **19** 1363–9

[17] Hu S *et al* 2020 Weakly supervised deep learning for COVID-19 infection detection and classification from CT images *IEEE Access* **8** 118869–83

[18] Li Z *et al* 2020 From community acquired pneumonia to COVID-19: a deep learning based method for quantitative analysis of COVID-19 on thick-section CT scans *Eur. Radiol.* **30** 6828–37

[19] Chen J *et al* 2020 Deep learning-based model for detecting 2019 novel coronavirus pneumonia on high-resolution computed tomography *Sci. Rep.* **10** 1–11

[20] Chaganti S *et al* 2020 Quantification of tomographic patterns associated with COVID-19 from chest CT, arXiv:2004.01279

[21] Ahuja S *et al* 2021 Deep transfer learning-based automated detection of COVID-19 from lung CT scan slices *Appl. Intell.* **51** 571–85

[22] Maghdid H S *et al* 2020 Diagnosing COVID-19 pneumonia from x-ray and CT images using deep learning and transfer learning algorithms, arXiv:2004.00038

[23] Jaiswal A *et al* 2020 Classification of the COVID-19 infected patients using DenseNet201 based deep transfer learning *J. Biomol. Struct. Dyn.* **39** 5682–9

[24] Caldwell M and Griffin L D 2019 Limits on transfer learning from photographic image data to x-ray threat detection *J. X-Ray Sci. Technol.* **27** 1007–20

[25] Saba L, Sangala S S, Gupta S K, Koppula V K, Johri A M, Sharma A M, Kolluri R, Bhatt D L, Nicolaides A and Suri J S 2021 Ultrasound-based internal carotid artery plaque characterization using deep learning paradigm on a supercomputer: a cardiovascular disease/stroke risk assessment system *Int. J. Cardiovasc. Imaging.* **37** 1511–28

[26] Saba L *et al* 2021 A multicenter study on carotid ultrasound plaque tissue characterization and classification using six deep artificial intelligence models: a stroke application *IEEE Trans. Instrum. Meas.* **70** 1–12

[27] Sanagala S S *et al* 2019 A fast and light weight deep convolution neural network model for cancer disease identification in human lung(s) *18th IEEE Int. Conf. On Machine Learning And Applications* (Piscataway, NJ: IEEE)

[28] Skandha S S *et al* 2020 3-D optimized classification and characterization artificial intelligence paradigm for cardiovascular/stroke risk stratification using carotid ultrasound-based delineated plaque: Atheromatic™ 2.0 *Comput. Biol. Med.* **125** 103958

[29] Nolf E X *et al* 2003 An open-source medical image conversion toolkit *Eur. J. Nucl. Med.* **30** S246

[30] Saba L *et al* 2021 Six artificial intelligence paradigms for tissue characterisation and classification of non-COVID-19 pneumonia against COVID-19 pneumonia in computed tomography lungs *Int.J. Comput. Assist. Radiol. Surg.* **16** 1–12

[31] Agarwal M *et al* 2021 A novel block imaging technique using nine artificial intelligence models for COVID-19 disease classification, characterization and severity measurement in lung computed tomography scans on an Italian cohort *J. Med. Syst.* **45** 1–30

[32] Krizhevsky A, Sutskever I and Hinton G E 2012 ImageNet classification with deep convolutional neural networks *Commun. ACM* **60** 84–90

[33] He K *et al* 2016 Deep residual learning for image recognition *Proc. of the IEEE Conf. on Computer Vision and Pattern Recognition* 770–8

[34] Huang G *et al* 2017 Densely connected convolutional networks *Proc. of the IEEE Conf. on Computer Vision and Pattern Recognition* 4700–8

[35] Chollet F 2017 Xception: Deep learning with depthwise separable convolutions *Proc. of the IEEE Conf. on Computer Vision and Pattern Recognition* 1251–8

[36] Szegedy C *et al* 2016 Rethinking the inception architecture for computer vision *Proc. of the IEEE Conf. on Computer Vision and Pattern Recognition* 2818–26

[37] Simonyan K and Zisserman A 2014 Very deep convolutional networks for large-scale image recognition, arXiv:1409.1556

[38] Anthony D 2010 Introductory statistics for health and nursing using SPSS *Nurse Res.* **17** 89

[39] Polsinelli M, Cinque L and Placidi G 2020 A light CNN for detecting COVID-19 from CT scans of the chest *Pattern Recognit. Lett.* **140** 95–100

[40] Hasan A M *et al* 2020 Classification of COVID-19 coronavirus, pneumonia and healthy lungs in CT scans using $q$-deformed entropy and deep learning features *Entropy* **22** 517

[41] Singh A K *et al* 2020 Diabetes in COVID-19: prevalence, pathophysiology, prognosis and practical considerations *Diab. Metab. Synd.: Clin. Res. Rev.* **14** 303–10

[42] Purohit K *et al* 2022 COVID-19 detection on chest x-ray and CT scan images using multi-image augmented deep learning model *Proceedings of the Seventh International Conference on Mathematics and Computing* (Cham: Springer) pp 395–413